Void Where Prohibited

Revisited

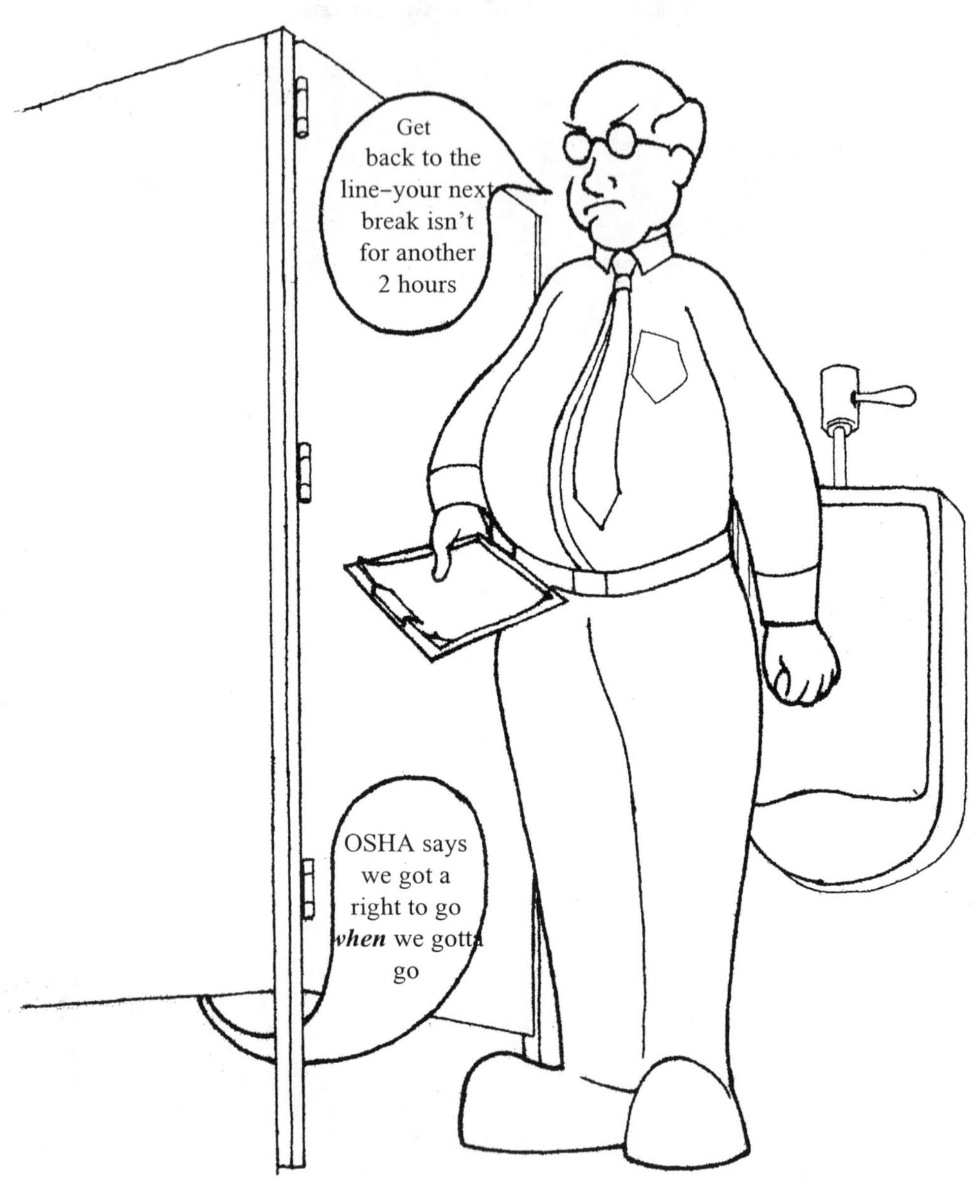
Get back to the line–your next break isn't for another 2 hours
OSHA says we got a right to go *when* we gotta go

Void Where Prohibited Revisited

The Trickle-Down Effect of OSHA's At-Will Bathroom-Break Regulation

Marc Linder

Fănpìhuà Press
Iowa City
2003

Printed in the United States of America

Cover illustration: Schuyler Rahe-Dingbaum

The cartoon by Walter Bartlett on page 263 originally appeared in *The Macon Telegraph* on Mar. 19, 1979, at page 4D, cols. 3-5, and is reprinted with the permission of *The Macon Telegraph*.

The cartoon by Hekate, "Why Don't They Just Call In?" is reprinted on page 285 with the permission of Hekate.

Suggested Library of Congress Cataloging
Linder, Marc, 1946—
Void where prohibited revisited:
The Trickle-down effect of OSHA's at-will
bathroom-break regulation/by Marc Linder.
xii, 382 p.; 23 cm.
Includes bibliographical references and index.
ISBN 0-9719594-0-4
1. Rest periods. 2. United States. Occupational Safety and Health Administration. 3. Employee rights.
HD5112.L561 2003
331.25'76—dc21
Library of Congress Control Number: 2003090222

Over the past several months we have noticed an increase of time spent off the floor and away from work stations.... This can no longer be tolerated. ... You will be excused to go to the restroom in an emergency situation only—daily is not an emergency, but a habit which you will need to break.[1]

Human beings have a longstanding practice of excessive breaks.[2]

How does a practice end with a signed piece of paper? No it didn't: it did not end at that time.[3]

In the Soviet enterprises we should carry on a resolute and uncompromising struggle against violations of labour discipline, breaking factory rules, stealing of public property, waste of materials, slowdowns and unauthorized breaks. We should have the names of the worst offenders written on a blackboard, or even have them dismissed or imprisoned, because they purposely destroy public property which should be sacred. These acts make them enemies of the people.[4]

---

[1]Written policy at Excel Corporation slaughterhouse in Friona, Texas (which was obtained through discovery during litigation and furnished by attorney Philip Russ on Dec. 10, 1999): "It was prepared by the supervisors and posted on the bulletin board for a time and I retrieved several copies from individual personnel files. It seems when someone was disciplined for wanting to relieve themselves (apparently when it wasn't an emergency) this memo would be placed in their file. I do not have a definite date but there was some discussion of it in the deposition of either the HR manager or the plant manager at Friona.... The pages were not numbered just a two page memorandum." Email from Philip Russ to Marc Linder (Sept. 12, 2002).

[2]Carlos Tejada, "Work Week," *Wall Street Journal*, Aug. 28, 2002, at B3, col. 1 (quoting Jack Allen, human resources director, Jim Beam Clermont, KY plant).

[3]Commonwealth of Kentucky, Occupational Safety and Health Review Commission, Administrative Action No. 02-KOSH-0015, Secretary of Labor, Commonwealth of Kentucky v. Jim Beam Brands Co. , KOSH 3681-01, "Transcript of Hearing" at 202 (Aug. 29, 2002) (Jeff Conder, plant manager, Jim Beam Brands Co., Clermont, KY plant).

[4]Liu Shaoqi, "A New Form of Labour Demands a New Attitude," in *Selected Works of Liu Shaoqi* 1:30-33 at 32 (1981 [Mar. 20, 1934]).

Contents

Tables

Preface

The Inaccessible Abode of Elimination Within the No Longer Hidden Abode of Production

> [I]n today's workplace, as Linder and Nygaard show, the spirit...of the old regime persists.[1]

The ruckus caused by the appearance of *Void Where Prohibited* at the end of 1997 prompted the Occupational Safety and Health Administration—which, incredible as it may seem, during the first quarter-century of its existence had not recognized workers' right to use the toilets that companies were obligated to provide—to issue a Memorandum on April 6, 1998, announcing that in the future employers would be required to make those toilets available so that workers can use them "when they need to do so."[2] *Void Where Prohibited Revisited* is an attempt, five years later, to assess the real-world impact of OSHA's issuance of that new interpretation of its sanitation standard.

The market ("the sphere of circulation") was regarded by Karl Marx as a noisy realm dwelling on the surface of events and accessible to everyone, but one where the seeming freedom and equality of sellers and buyers (especially of labor power) inverted social reality. In contrast, "the hidden abode of production," on whose threshold hung the sign, "No admittance except on business," was the place where the secret of surplus value production would finally have to be revealed.[3] One secret of the hidden abode of production that Marx did not study enshrouds the inaccessible abode of elimination on whose threshold today hangs the sign: No admittance except on scheduled breaks. Workplace toilets have not yet been turned into profit centers, but some employers presumably believe that preventing workers from stopping work to go there does increase profits.

Exposing the hidden abode of production together with its secret of surplus

---

[1]Corey Robin, "Lavatory and Liberty: The Secret History of the Bathroom Break," *Boston Globe*, Sept. 29, 2002, at D1.

[2]See below Appendix II.

[3]Karl Marx, *Das Kapital: Kritik der politischen Oekonomie*, Vol. 1: Book I: *Der Produktionsprocess des Kapitals* 140 (1867).

value has not toppled capitalism. The taboo that Marx broke may have been more profound and central to the inner workings of society, but, as the French daily *Le Monde* recently observed, each era has the seminal conflict it deserves, and today's may be the struggle for the right of free access to workplace toilets.[4] Emblematically, the authors of a recent book on the living-wage movement seemed to be referring to a select honor roll when they praised a high-wage poultry-slaughter firm with a progressive management "ethos" "whose workers may use the rest room at any time they please," in contrast to a low-wage animal-slaughter company whose employees are "generally bothered by supervisors for their occasional need to use the rest room."[5] And a labor arbitrator (who was also the director of personnel at Harvard University), as if applying Marx's analysis of the contradiction between the right of the capitalist as buyer of labor power to make the workday as long as possible and the right of the worker as seller to limit the day to a normal length—between which only force or government regulation decides[6]—observed of a dispute over bathroom access at a Massachusetts manufacturing plant:

> We might, perhaps, learn from our French cousins who refer to the personal requirements of people as "les petits besoins," or the *little* needs of the employees, which adjective carries with it the implication that there is no real conflict, or should be none, between these little needs and the major purposes of a company or enterprise. In this case, however, we do have a conflict...between the management's running of its business and the employee's management of himself. Each basically would appear to have a clear right...and these rights should not be in conflict, but in fact, are.[7]

Workers who have themselves been painfully initiated into the secret history of the inaccessible abode of elimination have often felt constrained by embarrassment to help keep the secret, but *Void Where Prohibited* revealed that many employers restrict their employees' access to the toilet. Now that the secret is out and OSHA has had five years to enforce compliance with its new-found understanding of one of the foundations of workplace sanitation, hygiene, and health, the time has come to determine the extent to which the Memorandum has empowered workers to gain unilateral control over their bladders and colons.

---

[4]Maurus Veronique, "La course à la productivité au coeur du conflit des Bigard," *Le Monde*, Aug. 14, 1995 (Lexis).

[5]Robert Pollin and Stephanie Luce, *The Living Wage: Building a Fair Economy* 151, 155, 156, 157-58 (rev. ed. 2000 [1998]).

[6]Marx, *Das Kapital* at 202, 280-81.

[7]Gremar Mfg. Co., 46 Labor Arbitration Reports (BNA) 215, 218 (1965). The arbitrator, John W. Teele, was director of personnel at Harvard University from 1949 to 1969. Telephone interview with his successor, John Butler, Lexington, MA (Jan 27, 2003).

The book is largely structured chronologically, punctuated by several historical and comparative excursions. Part I is devoted to the pressure that was mobilized in 1997-98 to force OSHA to abandon its preposterous position that its toilet standard did not require employers to let workers use the toilets. Following general and specific remarks in Chapter 1 on how the words in a law get translated into real-world effects, Chapter 2 examines the impact of *Void Where Prohibited* on Federal OSHA's decision to reverse itself. The prelude to that reversal that took place in Iowa, also under the influence of *Void Where Prohibited*, is related in Chapter 3. Part II interrupts the narrative to interpolate two crucial background facts. Chapter 4 tells the ironic tale of how during the Carter administration, OSHA, under its most prolabor Administrator, tried and almost managed to delete its toilet standard altogether; had that deletion project succeeded, no norm would have been available two decades later from which to derive a right to go to the bathroom. Although the reach of OSHA's toilet standard is very broad, Chapter 5 explains which sectors of the economy are not covered by the agency's right-to-void standard. Part III is devoted to the actual promulgation of the Memorandum. Following analysis of the text of the Memorandum in Chapter 6, its status and validity from the perspective of possible challenges under administrative law are discussed in Chapter 7. Initial reactions from various groups to the Memorandum form the subject of Chapter 8. Part IV then interrupts the chronology again, this time to insert two international comparative studies—of a surprisingly backward fee-to-pee regime in Canada (Chapter 9) and the radical fundamental human right to void at work in France (Chapter 10). Part V shifts to OSHA's enforcement of its new right-to-void standard. Based on a unique collection of unpublished enforcement documents, Chapter 11 provides a detailed statistical and descriptive account of all the relevant citations issued by Federal OSHA and state OSHA agencies. The next four chapters deal with four state OSHA programs whose enforcement efforts merit special attention: Iowa's citations against animal slaughter plants (Chapter 12); the high-profile Jim Beam dispute in Kentucky (Chapter 13); Washington's unique adoption of the right to void with regard to bus drivers before 1998 (Chapter 14); and California's unique refusal to enforce Federal OSHA's new standard (Chapter 15). Finally, Part VI tries to assess the progress that has been achieved. Chapter 16, in an attempt to develop a broad qualitative picture of the extent to which OSHA and labor unions have been able to vindicate the right to void, offers the results of a large number of interviews with officials in various organizations, while Chapter 17 analyzes the state of the law, devoting special attention to the possible legal ramifications of employers' efforts to discipline workers for "abusing" bathroom breaks.

For readers' convenience, the full texts of the relevant OSHA standards have been assembled in the Appendices.

Acknowledgments

A large number of workers in scores of Federal and State OSHA offices made this study possible by furnishing copies of the hundreds of public-record citations and redacted inspector worksheets and narratives that form the basis of the book's analysis of OSHA's enforcement of its Memorandum of April 6, 1998. Several of them deserve special gratitude for expediting the processing of Freedom of Information Act requests and/or persistence in locating hard-to-find files: Elizabeth Slatten (Austin), Mary Bryant (Iowa), Margaret Miles (Kentucky), Laurie Lorish (Michigan), Tamra Larue (Mobile), Grace Kropp (North Carolina), Cheryl Gray and Bonita Winningham (Omaha), Chris Ottoson (Oregon), Ronnie Bilczewski (Parsippany), Luis Mireles (San Diego), Lisa Tilley (Savannah), Mike Maenza (Tennessee), Jule Jones (Toledo), Susana Freund (Ventura), Jay Withrow (Virginia), and Barbara Harris-William (Washington). Andrea Howard of the California Occupational Safety and Health Standards Board performed almost instantaneous regulatory history searches and faxed the material almost faster than the recipient could run to the fax machine.

A belated expression of gratitude is owed Elaine Bynum (now retired) of the Docket Office in OSHA's national office in Washington, D.C., who in 1997 made available copies of the transcript of the public hearings on toilets held on November 8-10, 1972 and of the comments submitted, analysis of which significantly enriched *Void Where Prohibited*, and other parts of which have been used here. One of her successors, Vanessa Reeves, undertook the heroic (though ultimately fruitless) task of trekking to the National Archives in Suitland, Maryland, to look through 67 boxes of totally disorganized material from Docket S-250 on OSHA's 1978 revocation project.

Elizabeth Slatten, the Assistant Area Director of Federal OSHA's Austin Area Office tirelessly and cheerfully responded to countless requests for information about various aspects of OSHA's operations, policies, and procedures.

The incredible speed with which Denise Partee, Deputy Clerk, U.S. District Court for the Middle District of Georgia, Macon Division, retrieved from the regional archive, copied, and sent off pleadings in *Bagley v. Cagle's* made it possible to discuss and analyze that important case.

In order to gain a better understanding of the conflicts that are often abstractly encapsulated in OSHA citations, as many persons as possible with direct knowledge of the events were interviewed. If more union officials and workers are quoted than managers, the reason is that, unsurprisingly, whereas virtually no unionist declined to talk, many company officials refused to provide their version. Several labor union officials who were particularly helpful deserve special mention: Jo Anne Kelley, president of UFCW Local 111-D at the Jim Beam plant in Clermont, Kentucky; Jackie Nowell, director, and Robyn Robbins, assistant director, of the occupational safety and health office, and Peter Ford, assistant general counsel, of the UFCW; Jim Fredericks of the occupational safety and health department of the United Steelworkers; Dave LeGrande of the occupational safety and health department of the Communications Workers; Susan Stoner, general counsel of Amalgamated Transit Union, Division 757, in Portland, Oregon; Claude Grey, vice president, and Donny Brown, business agent, Teamsters Local 391, in Goldsboro and Winston-Salem, North Carolina; and Edgar Fields, international representative, Retail, Wholesale & Department Store Union, Atlanta.

At his own expense, attorney Jean Claude Richard of Marseille generously furnished a complete stranger a copy of the French labor tribunal decision discussed in chapter 10, while simultaneous interpreter Corinne Laloux of Jouy-en-Josas corrected some of the translations in the same chapter.

Dr. Ingrid Nygaard answered numerous questions about urination, while Larry Norton discussed several tricky legal issues.

Matt Stilwell expertly scanned in and adjusted the cover cartoon, which the noted Midwestern speed artist Schuyler Rahe-Dingbaum had sketched in a not-too-bad 11.5 minutes.

Note on Nomenclature and Sources

Throughout the text "OSHA" is used to refer both to the Occupational Safety and Health Administration and to the Occupational Safety and Health Act. The latter is intended very infrequently and the context always makes the reference clear. Similarly, for the sake of simplicity and uniformity, all federally approved state occupational safety and health agencies are referred to as "Kentucky OSHA," "Washington OSHA," and so on, even though they may have other formal designations (which are pointed out in the text or footnotes).

The vast bulk of the sources cited in this study of the development of workplace voiding rights during the past five years fall into the following categories: OSHA's unpublished and/or archival administrative and enforcement records; telephone interviews and email correspondence; administrative and judicial hearing transcripts; newspapers, magazines, and websites; and state and federal statutes, regulations, and judicial, administrative, and arbitral decisions. In contrast, relatively little use has been made of published sources such as books and journal articles. Although a bibliography, which would be of little practical value to readers, has therefore been dispensed with, the full bibliographical data of each source are given with its first citation in each chapter. In addition, Appendix V describes in detail the OSHA inspection reports that form the basis of the analysis of the agency's enforcement efforts.

Part I

Pressure

"Void Where Prohibited"...opened the floodgates of workers' wrath about a sensitive issue.[1]

---

[1]Susan Yerkes, "Whazzup," Mar. 26, 1998, on http://www.expressnews.com/whazzup/whazzup1.shtml (visited June 11, 1998).

1

Books Have Their Fates—And So Do Laws

Just think...if...millions of Americans now actually can go to the bathroom when they want to, you will die having accomplished something really good—not something most people can say.[1]

The question this book examines is summed up by that little conditional word "if": Can in fact millions of workers in the United States stop work when they need to void now that the Occupational Safety and Health Administration has imposed on employers an obligation to let them go?

Even before it appeared at the end of 1997, *Void Where Prohibited: Rest Breaks and the Right to Urinate on Company Time* had already begun mobilizing public opinion to pressure OSHA to abandon its preposterous and outrageous position that its industrial sanitation standard, which required employers to provide toilets,[2] did not obligate companies to let workers use those toilets. On April 6, 1998, OSHA, finally listening to reason, issued a Memorandum declaring that the "standard requires employers to make toilet facilities available so that employees can use them when they need to do so."[3] Thus with a few keystrokes on a computer, a governmental agency was able, literally from one day to the next, to create a right for tens of millions of workers in the United States to stop work when they need to void.

Or was it? Was this instant establishment of at-will bathroom breaks worth the paper (or cyberspace) it was written on? How do labor-protective regulations get enforced in the real world consisting of: aggressive and powerful employers opposed to governmental interference with their managerial prerogative to control their employees' time unilaterally; a private-sector workforce—90 percent non-unionized—largely afraid to assert their rights or even to file a complaint with the agency; and an understaffed OSHA, which, preoccupied with preventing what it deems other far more urgent safety and health hazards, even when it does receive

[1]Email from Dr. Ingrid Nygaard to Marc Linder (Apr. 13, 1998).

[2]29 CFR sect. 1910.141(c)(1)(i) (2002); see below Appendix I.

[3]See below Appendix II.

complaints, fails to pursue them with all possible rigor and vigor (or, in the unique case of Cal/OSHA, outright refuses to fulfill its obligation to "ensure that State standards and their interpretations remain 'at least as effective' as the Federal standard")?[4]

Void Where Prohibited was characterized by *The New York Times* as a "grim" tour d'urinal,[5] by Litigation Management Inc. (which provides medical information to corporations) as "highly entertaining,"[6] by the *National Law Journal* as a "not-at-all frivolous book...cover[ing]...bathroom breaks from A to P,"[7] and by *The Non-violent Activist* as an "impassioned plea."[8] The book has enjoyed an interesting and unusual fate for a scholarly work—which *Labor Studies Journal* viewed as "combin[ing] the muckraking tradition with a theoretical sharpness worthy of Karl Marx"[9]—having impelled the U.S. national regulatory bureaucracy to overrule itself and to vindicate the right of tens of millions of workers to void "when they need to do so." A nursing journal credited the book with having focused research on the impact of urinary incontinence on working men and women.[10] In an article about the lawlessness of the workplace in the United States, Germany's leading weekly newspaper, *Die Zeit*, cited the book findings as the most extreme of its "unbelievable histories from the world of work."[11] In the words of the *University of Pennsylvania Journal of Labor and Employment Law*: "It is a rare scholarly work that can arouse a slumbering agency to act, but Void Where Prohibited...did just that, on both the state and federal level."[12] Or

---

[4]See below Appendix II.

[5]Mary Walsh, "Blue-Collar Urgency: Bathroom Rights," *N.Y. Times*, Nov. 22, 2000, at G1.

[6]Elizabeth Juliano and James Fell, Review of *Void Where Prohibited*, in *M.I.M. Reporter: Quarterly Review of Medical Information Management for Litigation*, 1(2):7 (June 1998), on http://www.litigation-mgmt.com/lmimain/AcrobatMIM/revised/secure/MIMV1N2.pdf.

[7]"There's No Escaping the Law, So Read All About It," *National Law Journal*, Aug. 3, 1998, at A25 at A26 (book review).

[8]Phyllis Eckhaus, "Piss Poor" (Book Review), *Nonviolent Activist*, Nov.-Dec. 1998, at 19.

[9]Bruce Nissen, Book Review, *Labor Studies Journal* 26(1):81-82 at 82 (2001).

[10]Sheila Fitzgerald, Mary Palmer, Susan Berry, and Kristin Hart, "Urinary Incontinence: Impact on Working Women," *AAOHN Journal* 48(3):112-18 at 112 (Mar. 2000).

[11]Christian Tenbrock, "Vorsicht, Kamera! Unglaubliche Geschichten aus der Arbeitswelt in den Vereinigten Staaten," *Die Zeit*, June 8, 2000, on http://www.fortunecity.de/parkalleen/mozartweg/682/presse/zeit0024.htm.

[12]Rebecca Schleifer, Review of *Void Where Prohibited*, in *University of Pennsylvania Journal of Labor and Employment Law* 2:603-608 at 604 (2000). See also Anne Scott, "The Continuing Saga of the Potty Wars," *Business Record* (Des Moines), Jan. 26, 1998,

as the British occupational safety and health magazine, *Hazards*, put it:

It is a rare book on working conditions that can be said to make a real and lasting difference. *Void where prohibited*....did though. [The] book caused a major stink in the US, leading OSHA...to issue new guidance on a worker's right to pee breaks. But what Void where prohibited graphically illustrates...is how robbing workers of their dignity is a very common and deliberate management technique. It comes to something when your boss has control of your bladder too. This book is about how humiliation is a big part of today's management armoury.[13]

The much-quoted saying, "books have their fates" ("habent sua fata libelli") of Terentianus Maurus, a second-century A.D. Roman grammarian, begins with a phrase that moderns often omit: "pro captu lectoris." The adage can thus be taken to mean that a book's fate hinges in part on its readers' powers of comprehension.[14] To be sure, the fate of a (labor) law is hardly a matter of the relative strengths of its disputing readers' powers of comprehension. 'Pure reason' is subordinated to the diametrically opposed class interests of employers and employees, whose strategies for determining the outcome of disputes over the

---

at 5 (Lexis) ("Nygaard and Linder are driving some changes"); Maggie Jackson, "Bathroom Breaks Legal, Thanks to UI Professor," *Daily Iowan*, Mar. 13, 1998, at 4A; Valica Boudry, "Professor Helps Secure Workers' Basic Rights," *fyi,* Jan. 30, 1998, at 4 ("OSHA's decision to cite a Hudson Foods Inc. poultry processing plant...was due in part to research done by...Linder"); "OSHA Clarifies Bathroom Policy," Newsline (Apr. 3, 1998), on http://www.penton.com/oh/member/newsline/4-3.html; "UI Push Forces Break Changes," *Iowa City Press Citizen*, Apr. 9, 1998, at 1A, col. 1 ("Linder and...Nygaard initiated OSHA's reexamination of its position"); Stephanie Ball, "Taking a Leak Just Got Easier," *Daily Iowan*, Apr. 10, 1998, at 1A, col. 5 ("If you gotta go, you gotta go. And employees across the nation may have an easier time taking bathroom breaks thanks to a book by...Marc Linder and Ingrid Nygaard"); "Book Led to New OSHA Rules," *Cedar Rapids Gazette*, Aug. 16, 1998 ("the outrage engendered by Linder and Nygaard's book prompted needed reform"); Gregory Weaver, "OSHA Deems If Employers' Restroom Policies Are Too Restrictive," *Indianapolis Star*, May 8, 2002, at 6B (Lexis) ("The guidelines were issued after the agency apparently was embarrassed by a book that took it to task for allowing workers to be denied restroom breaks for entire eight-hour and 12-hour shifts--even though it had rules requiring employers to 'provide' a restroom"); Al Karr, "OSHA to Clarify Rest Room Access Rule," *Safety + Health* 157(3):14 (Mar. 1998); Lisa Finnegan, "OSHA Directive Clarifies Bathroom Use Policy," *Occupational Hazards* 60(5):13 (May 1998) (Westlaw)."Bathroom Use," *Safety and Health Practitioner* 18(3):9 (Mar. 1, 2000) (Westlaw).

[13]Review of *Void Where Prohibited*, *Hazards*, No. 65, at 20 (1999).

[14]For a full discussion, see Wolfgang Milde, *Habent sua fata libelli: Zur Geschichte eines Zitats* (1988).

meaning of laws and regulation are a classic example of the merely instrumental use of arguments.

But initially a law—and even more so an administrative agency's interpretation of its own regulation—is, like a book, merely words printed on a piece of paper, without the ability to compel a universally accepted meaning, let alone to command or coerce compliance (which printed pages in non-legislative books do not even purport to do). Nevertheless, many people believe that the mere act of publication by OSHA of a memorandum interpreting its own sanitation standard for general industry, which requires employers to provide a certain number of toilets, as also "requir[ing] employers to make toilet facilities available so that employees can use them when they need to do so"[15] automatically means that from then on workers would actually be able to void "when they need to do so."

However, if compliance were instantaneous, it would not, for example, have been necessary in 1999 for Hispanic workers in animal slaughter plants in Nebraska to complain that they "end up urinating in their pants while working on the line" because "they're given inadequate bathroom breaks."[16] Nor would it have been necessary for the state's Republican governor, who had campaigned to reduce the size of government,[17] to issue a non-enforcible Nebraska Meatpacking Industry Workers Bill of Rights, under which an "employer agrees to provide to employees...[a]dequate time for necessary restroom breaks."[18] But in the absence of a frictionless world of self-enforcing universal compliance and/or ubiquitously and perpetually patrolling peripatetic police, as a leading comparative labor law scholar observed, "in labour relations legal norms cannot often be effective unless they are backed by social sanctions..., that is by the countervailing power of trade

---

[15]See below Appendix II.

[16]Ted Kirk, "Critics: Hispanics Exploited in Omaha Meatpacking Jobs," *Lincoln Journal Star*, Sept. 5, 1999 (a copy of the web version, which is no longer on-line, was faxed to the author by Jose Santos, the Nebraska Department of Labor Meatpacking Industry Worker Rights Coordinator (Jan. 3, 2003)). See also Mike Sherry, Cindy Gonzalez, and Leslie Reed, "Meatpacking Inquiry Opened," *Omaha World-Herald*, Sept. 11, 1999, at 59 (Lexis); Nancy Hicks, "Meatpackers' Job Conditions to Be Studied," *Omaha World-Herald*, Sept. 10, 1999, at 17 (Lexis).

[17]http://www.mikejohanns.com.

[18]Nebraska Workforce Development, "Nebraska Meatpacking Industry Worker Bill of Rights" sect. 3 (June 28, 2000). The spirit of this project was captured by a statement by the lieutenant governor, who reported on conditions in meatpacking plants: "I discovered in visits with workers concerns that were genuine because the workers believed they were genuine. I recognize that like many complex societal challenges, there is no magic bullet to solving the issues at hand. There are no easy solutions." From the Office of Mike Johanns, News Release (Jan. 24, 2000), on http://gov.nol.org/Johanns/News/jan00/recommendwbor.htm.

unions and of the organised workers to withhold their labour."[19] Consequently, with weak unions and a small and shrinking proportion of unionized workers supporting a "half-heartedly enforced standard,"[20] the enunciation of a new policy alone is unlikely to modify the behavior of firms that deprive workers of toilet breaks since they tend to be labor scofflaws in general.[21]

To coincide with the first anniversary of OSHA's toilet access interpretation a newspaper reporter in Tucson followed up on what the author, emphasizing to her that the OSHA Memorandum "'is just a piece of paper until workers try to assert that right,'" termed "the most important question...in all of this"—namely, whether "OSHA's interpretation is worth the paper it's written on." RuthAnn Hogue soon discovered that whereas Federal OSHA took the position that the standard "appears to be effective" because "'there was enough information out there that employers know they have to give them access or workers aren't complaining,'" "some Tucsonans say the law is little more than an ineffective piece of paper. ... Conditions at some companies in Tucson today mirror those that sparked OSHA's move to increase restroom access."[22] Specifically she found:

> Schoolteachers report having accidents in the classroom, and suffering kidney and bladder infections because they often can't use the restroom more than once a day.
>
> Pharmacists who work 12-hour shifts alone must often wait until closing time to relieve themselves. State law doesn't allow them to leave the pharmacy unattended, and pressure from employers often prevents them from locking up briefly.
>
> Call center employees say pay incentives often penalize them for logging off the phone system between scheduled breaks.[23]

For example, at one Tucson call center—in Britain, too, half of call-center workers surveyed "said they deferred taking toilet breaks because of management"[24]—pressure for increased output cascading down from higher levels of management led to the firing for "'unprofessional behavior'" of an employee for questioning the imposition of a rule permitting only three restroom breaks daily

---

[19]Otto Kahn-Freund, *Labour and the Law* 11 (1972).

[20]Barbara Ehrenreich, "Warning: This Is a Rights-Free Workplace," *The New York Times Magazine*, Mar. 5, 2000, at 88-92 at 88 (citing Marc Linder).

[21]Simon Nadel, "Restricting Regular Bathroom Access Can Spur Negative Workplace Repercussions," *U.S. Law Week* 66(37):2579-80 at 2580 (Mar. 31, 1998) (quoting Deborah Berkowitz, UFCW director of health and safety).

[22]RuthAnn Hogue, "No Relief in Sight," *The Arizona Daily Star*, Apr. 11, 1999, at 1D, col. 2

[23]Hogue, "No Relief in Sight," at 1D, col. 3.

[24]Christine Norton, "Loo Breaks at Work—Is There a Problem?" *European Health and Safety Magazine* 2(7):26-27 (Nov. 2002).

(including lunch) and restriction of fluid intake to minimize the need to void.[25] (In all seriousness a plaintiff-side attorney specializing in the Americans with Disabilities Act soon suggested that before claiming that it would be an undue hardship, employers accommodate telemarketing employees who had to use the toilet frequently by enabling them to "take a portable phone to the bathroom....")[26]

Within two weeks the mere publication of Hogue's article triggered at least a dozen telephone complaints of denial of toilet access to OSHA in Tucson compared to a previous average of three every six months. All the complainants worked in the teleservice industry, which was not adverse to trying to absolve itself of blame by alleging that workers were in fact "reluctant to use the restroom because they are wary of upsetting co-workers trying to reach a common production goal." Workers, however, explained to Hogue that "while their employer posts memos stating restroom-friendly policies, taking time off the phone to use the restroom counts against them and can affect promotions or pay." OSHA rejected such policies as inconsistent with the toilet access standard: "'The employer should not deduct pay from them for using the restroom. That would discourage a lot of employees from using the restroom when they need to.'"[27]

The purpose of this follow-up study to *Void Where Prohibited* is to shed light on the process by which labor rights on paper become a firm part of the physical, political, and socio-economic reality of working life—or, in the words of one reviewer, of whether there was a "happy ending for workers."[28] The inquiry will also examine whether the author's skepticism of OSHA's efforts, voiced days before the agency issued its new interpretation, and especially the prediction that it "will back down under pressure from businesses," have been confirmed by events.[29] Even after OSHA published its interpretive Memorandum, the author remained "'quite pessimistic about enforcement'" against the background of its tradition of foot-dragging on the issue.[30] Alternatively, this study will examine

---

[25]Hogue, "No Relief in Sight."

[26]"Accommodating Bathroom Use Challenges Employers," *Successful Job Accommodation Strategies* 4(4) (Aug. 1998) (Lexis).

[27]RuthAnn Hogue, "OSHA Is Flush with Restroom Complaints," *Arizona Daily Star*, Apr. 25, 1999, at 1D, at col. 1, 3D (quoting Art Morelos, compliance officer, OSHA, Tucson).

[28]Review of *Void Where Prohibited*, *Update*, Summer 1998, at 14 (National Employment Law Project).

[29]Brian Tumulty, "Bathroom Breaks on OSHA's Agenda," *Journal and Courier* (Lafayette, Indiana), Apr. 4, 1998. This article was also sent out as Brian Tumulty, "Teachers Wrestle With Issue of Bathroom Breaks: OSHA Might Come to Rescue," Gannett News Service, Apr. 1, 1998 (Westlaw).

[30]Janet Gemignani, "Can Potty Laws Unlock the Bathroom Door"? *Business & Health*, 16(5):12 (May 1, 1998) (Westlaw).

whether, as a former Pittsburgh area director of the Wage and Hour Division of the U.S. Department of Labor turned labor standards consultant put it in early 1999, "companies started taking the OSHA regulations seriously when the agency began citing firms for not allowing employees enough time to go to the bathroom."[31] This review of the actual formulation of OSHA's policy and enforcement also analyzes the extent to which enforcement has been constrained by the concern with "excessive government intrusion into the workplace and the potential disruption of business operations by employees who abuse breaks" expressed by employers even before OSHA acted.[32]

The dynamics of the enforcement process, however, are fundamentally misconceived when they are viewed in the manner of a sympathetic reviewer, who asserted: "In the face of public apathy and, except for Linder and Nygaard, with few from the academy calling on the public's attention, we can expect a number of employers to continue to treat their employees little better than livestock."[33] In fact, neither the public nor academia is crucial at this point. It is workers themselves and especially unions and their members that must impose norms of worker autonomy and co-determination on management both through day-to-day struggles at the workplace and by pushing OSHA to inspect, cite, and impose significant monetary penalties on employers that violate safety and health standards.

---

[31]Jane-Ellen Robinet, "PNC Park Seeks Potty Parity for Different Needs of Men, Women," *Pittsburgh Business Times & Journal* 18(31):45 (Feb. 19, 1999) (Westlaw).

[32]Kirstin Grimsley, "A Tough Break When Nature Calls," *Washington Post*, Nov. 26, 1997, at C9, col. 1.

[33]Matthew Finkin, Review of *Void Where Prohibited*, *Industrial and Labor Relations Review* 53(2):338-39 at 339 (Jan. 2000).

2

Void Where Prohibited and Its Impact on OSHA's Interpretive About-Face

> This book reads like a thriller. It's hard to put down. It could easily be a sequel to George Orwell's 1984. I can imagine a creative writing teacher telling his or her students that it doesn't seem believable. The only trouble is that the book is non-fiction. Every chilling word is true, incredible though it may be.[1]

> Paving the path to provide presidential poop patrol protection, we have a heroic prophetic prince promoting propaganda to prevent persecution of plentiful poopers. Professor Marc Linder, of the University of Iowa College of Law recently published a book entitled Void Where Prohibited, portraying the plight of the oppressed people plundered of plentiful poop time in the potty.[2]

The origins of *Void Where Prohibited* go back to 1995 when the author received a telephone call from a migrant farmworker legal services lawyer in New Mexico concerning the complaints of workers at a dairy, whose employer permitted them no breaks of any kind during their 12-hour shifts—not to rest, not to eat, and not to void: "When they have to relieve themselves, the boss tells them to try not to take a break."[3] Initial research uncovered the surprising fact that no federal or state labor standards law mandated any such breaks for run-of-the-mill workers.

---

[1]Susan Lippman, Book Review of *Void Where Prohibited*, *NY Hard Hat News*, Apr. 2001, at 7.

[2]Joe Murray, "The Presidential Potty Patrol," on http;//www.sodbuster.com/Opinions/joe/OPI031898JM.htm.

[3]Letter from Kay Drought, Texas Rural Legal Aid, Plainview, TX, to Olga Pedroza, Centro Legal Campesino, Las Cruces, NM (Sept. 19, 1995) (faxed to author Sept. 22, 1995). See also Stephen Franklin, "Gimme a Break: A Minute or Two to Go to the Bathroom Isn't Necessarily a Worker's Right," *Chicago Tribune*, Feb. 8, 1998, sect. 6, at 5. Contrary to Heather Newman, "Workers Have a Right to Visit Bathroom," Detroit *Free Press*, Apr. 18, 1998, on http://www.freep.com/business/qpotty18.htm, the author did not work at the dairy.

According to a 1976 study of time-use at work, the dairy workers in New Mexico were hardly unique. That survey revealed that no coffee or other scheduled breaks were reported by 26 percent of clerical, 29 percent of craftsman, 8 percent of operative, and 40 percent of unskilled respondents; among those who reported such breaks, they averaged 17, 20, 24, and 14 minutes per eight-hour day, respectively. Among the same groups no unscheduled breaks (including informal breaks, socializing, personal business, and lunch hours beyond 60 minutes) were reported by 32, 29, 34, and 42 percent, respectively; among those who reported such breaks, they averaged 28, 39, 29, and 19 minutes daily, respectively. Thus overall these four groups reported total daily break time of 46, 58, 53, and 33 minutes, amounting to 10.0, 12.2, 10.9, and 8.2 percent of their average daily time at work (excluding lunch) of 466, 491, 486, and 438 minutes, respectively.[4]

If hours and rest period laws offered no relief, a more promising approach was opened by an OSHA regulation (or "standard"), 29 Code of Federal Regulations section 1910.141(c)(1)(i), under which "toilet facilities, in toilet rooms separate for each sex shall be provided in all places of employment in accordance with Table J-1 of this section," which requires employers to provide 1 toilet in workplaces with 1 to 15 employees, 2 toilets in workplaces with 16 to 35 employees, and so on.[5]

However, when contacted, officials at OSHA took the position that this regulation did not require employers actually to let workers use those toilets that had to be "provided" or confer a right on workers to use them.[6] Nor is it clear that

---

[4]Frank Stafford and Greg Duncan, "The Use of Time and Technology by Households in the United States," in *Time, Goods, and Well Being* 245-88, tab. 10.3 at 256 (F. Juster and Frank Stafford eds. 1985 [1980]). The data showing that 84 percent of union members reported scheduled breaks compared with 60 percent of nonunionists were unfortunately not disaggregated by occupation.

[5]29 CFR sect. 1910.141(c)(1)(i) (2002).

[6]Marc Linder and Ingrid Nygaard, *Void Where Prohibited: Rest Breaks and the Right to Urinate on Company Time* 55-56 (1998). Seven years after the author began questioning OSHA officials about its interpretation of sect. 1910.141(c)(1)(i), OSHA's Assistant Regional Administrator for Compliance Programs in the Atlanta Region stated that in the period before OSHA changed its interpretation, he had considered citing a Tyson poultry slaughter plant for its refusal to permit a worker to go to the bathroom (until the supervisor furnished relief) who then defecated on himself on the grounds that it violated OSHA's general housekeeping standard: "All places of employment, passageways, storerooms, and service rooms shall be kept clean and orderly and in a sanitary condition." 29 CFR sect. 1910.22(a)(1) (2002). However, when it could not verify that the worker had actually defecated at the workstation, OSHA dropped the matter. Since this subpart is titled, "Walking-Working Surfaces," unless fecal matter or urine dripped out of the

unions during the preceding quarter-century of OSHA's existence had urged OSHA to take a different position. A number of arbitration decisions suggest that union themselves had not considered the issue as subject to government regulation. Thus in a 1974 case involving assembly-line workers, the United Steelworkers filed a grievance alleging that the company had failed to provide adequate relief for workers "tied to the line," who "frequently...had to wait as long as 30 and 45 minutes...." Although the arbitrator upheld the grievance on another basis, one contract provision on which the union unsuccessfully sought to ground its grievance read: "The Company will accept and consider safety and health recommendations made by the Union Safety Committee in order that all legal safety and health requirements established by the law will be followed." The arbitrator rejected appeal to this provision because: "There was no evidence as to what such legal requirements are nor any allegation that the Company had violated them. If the Union believes that the Company's relief practice or its bathroom facilities do not meet legal requirements, the Union is free to complain to the U.S. Department of Labor or the Tennessee Department of Labor."[7]

Similarly, in 1980, a cabinet-manufacturing company, which was "merely attacking the problem of too much time lost in loitering in the washrooms," posted a rule requiring workers to punch out and back in when going to and from the bathroom outside of scheduled breaks and warning that "[e]xcessive trips or too long trips may cause disciplinary action to be taken." The United Brotherhood of Carpenters immediately filed a grievance, alleging, inter alia, that the terms "excessive" and "too long" were too vague to be fairly enforcible. In addition, women workers complained that they "wasted" time in the bathroom waiting for the only one of three stalls that "insured privacy," one being furnished with a curtain only and the other with nothing at all to shut it off. Because the employer claimed that its toilet facilities had been approved by Indiana OSHA, the arbitrator himself contacted the agency and discovered that workers were entitled to stalls with doors and walls to insure privacy.[8] But neither the arbitrator nor the union apparently thought to ask or file a complaint with OSHA about the lawfulness of the employer's rule limiting the frequency and length of bathroom breaks.

Recognizing that neither mere professorial reason nor even common sense

---

worker's pants onto the floor, such an interpretation would, as the official himself conceded, probably not have been upheld by an Administrative Law Judge. Telephone interview with Benjamin Ross (Dec. 18, 2002); email from Benjamin Ross to Marc Linder (Dec. 20, 2002). In any event, the housekeeping standard would have been inapplicable to the vastly larger number of workers who were denied access but did not defecate or urinate on themselves.

[7]Mor-Flo Industries, Inc., 62 Labor Arbitration Reports (BNA) 398, 399-400 (1974).

[8]Schmidt Cabinet Co., 75 Labor Arbitration Reports (BNA) 397, 398-99 (1980).

would suffice to impel so benighted a bureaucracy to change its position, the author decided to try to involve the labor movement. Since food processing firms were among the worst oppressors of working-class bladders, the chief organizer of the industry's workers, the United Food and Commercial Workers, appeared to be the most credible leader of the struggle to clear the path to factory bathrooms. The author ultimately persuaded Deborah Berkowitz, the director of the Office of Occupational Safety and Health of the Field Services Department of the UFCW, to commit her considerable personal and organizational energy, resources, and political influence with a Democratic administration to an OSHA lobbying campaign for vindication of workplace voiding rights,[9] despite the fact that the union's legal department was skeptical of the strategy of seeking a clarification from OSHA lest the interpretation turn out to be more restrictive than the standard.[10] Then in connection with numerous conversations that Berkowitz and the author had with OSHA, on April 18, 1997, Berkowitz sent a letter to John Miles, Director of OSHA's Directorate of Compliance Programs,[11] requesting "a clarification of OSHA standard 1910.141, which requires employers to provide toilet facilities, regarding "employee rights to use those facilities as necessary." Berkowitz alerted Miles to the fact that:

> We have recently been made aware of situations in general industry where supervisors have denied employee requests to use the bathroom at work, except during scheduled breaks. In the food processing and packing industry, for example, scheduled breaks can be few and far between. As you know, there is no requirement in the Fair Labor Standards Act that employers provide regularly scheduled breaks, and there is no uniform practice under state law. Even where employers in the industry do afford breaks, it is not uncommon for employees to work over four hours—and in many cases much longer—without such a break.
>
> In response to a recent inquiry on this issue, Iowa OSHA responded that they believe they lack the power to require employers to allow workers to use the bathroom and therefore they could not cite any employer for refusing such employee requests. While we trust this is not the position of the National Office, we are requesting this clarification to make express that OSHA standard 1910.141 requires employers to allow workers to use the toilets when necessary.
>
> For your reference, I have enclosed a summary of industry comments taken from

---

[9]After spending a morning in November 1998 with Berkowitz's successor, Jackie Nowell, the editor of *Hazards* Magazine, the British occupational safety and health periodical, wrote the author: "I now know your work triggered all the activity in the US on this issue." Email from Rory O'Neill to Marc Linder (Dec. 9, 1998).

[10]Telephone interview with Jackie Nowell, director of safety and health, UFCW, Washington, D.C. (Oct. 18, 2002).

[11]Brian Tumulty, "Teachers Wrestle With Issue of Bathroom Breaks: OSHA Might Come to Rescue," Gannett News Service, Apr. 1, 1998 (Westlaw).

OSHA hearings held in 1972 on a related issue to the above question (prepared by Marc Linder, Professor of Law at the University of Iowa). Significantly, a number of employer representatives in making their comments on toilet distance requirements, not only believed that OSHA standard 1910.141 provided workers with the right to use a bathroom when needed, but also recommended that OSHA establish a system whereby employees on assembly lines could signal the need for relief.[12]

Although the UFCW never received an official reply to this letter from OSHA, Berkowitz, on the eve of OSHA's issuance of its new standard interpretation, correctly observed that the citation that the agency had issued on July 22, 1997 was "already an answer...."[13] On that date OSHA cited the Hudson Foods, Inc. chicken slaughter and processing plant in Noel, Missouri—where, despite having organized the workers and achieved a collective bargaining agreement, the UFCW, unable to resolve this or numerous other safety and health matters, had filed a complaint with OSHA—because "employees were denied necessary use of bathroom facilities. In isolated areas throughout the production plant, supervisors and/or leads do not allow workers relief from the production line in order to use the toilets, in effect, locking them out of or failing to provide bathroom facilities."[14] It is noteworthy that OSHA, whose national enforcement office, under UFCW's prodding, had been instrumental in deciding to cite this standard, tied the violation as closely as possible to the absence of the physical hardware called for by the standard; by analogizing an employer's oral-disciplinary denial of permission to stop working in order to go to the bathroom to placing a real lock on the door or failure to provide the proper number of toilets, OSHA sought to create as little discontinuity as possible with its previous policy, which grudgingly conceded that locking a toilet was tantamount to not providing it.[15] At this point OSHA apparently deemed it neither necessary nor prudent to elaborate on the health and safety basis of the standard or even to mention that workers had a right to stop work when they had to urinate or defecate.

The press erroneously reported that OSHA had fined Hudson Foods for the violation of section 1910.141(c)(1)(i),[16] but in fact it had imposed no monetary penalty for that particular ("other-than-serious") violation,[17] although OSHA did

---

[12]Letter from Deborah Berkowitz to John Miles (Apr. 18, 1997).

[13]Tumulty, "Teachers Wrestle With Issue of Bathroom Breaks."

[14]Hudson Foods, Inc., Inspection No. 300002250, Citation and Notification of Penalty (July 22, 1997), on http://155.103.6.10/cgi-bin/est/est1xp?i=300002250.

[15] Linder and Nygaard, *Void Where Prohibited* at 56.

[16]Frank Greve, "In an OSHA First, Agency Fines Plant for Inadequate Toilet Breaks," *Austin American-Statesman*, Aug. 23, 1997, at A1 (Lexis).

[17]According to OSHA, an other-than-serious violation "shall be cited in situations where the most serious injury or illness that would be likely to result from a hazardous

propose an initial penalty of $322,500 for many other violations that six months of inspection had unearthed; to be sure, after Tyson Foods announced a merger with Hudson on September 4, 1997, it achieved a reduction of this penalty by an administrative law judge to $57,000 in July 1999.[18] Nevertheless, OSHA itself, through its chief spokesman, characterized the action as "the first time the agency has punished anyone for denying workers toilet breaks." After having identified "the first alleged violator" in the 26 years since the toilet standard had gone into effect, OSHA inspectors were reported also to "have shown interest in toilet breaks at a Tyson Foods poultry processing plant in Jackson, Miss." And having quickly adopted a new-found appreciation of common sense, Chuck Adkins, OSHA's regional administrator in Kansas City explained to the press: "'If the standard requires toilets, it's implicit that you let people use them.'"[19] (Five years later, the Assistant Regional Administrator for Federal and State Operations in Adkins' office stated that although the Memorandum had not been issued until 1998, that interpretation had "always been there intuitively.")[20] Informally, however, OSHA inspectors appeared to apply a much looser subjective test. As one OSHA official in Missouri put it "on condition of anonymity," the case was "'surprising. It did engender a sense of outrage, or we probably wouldn't have addressed the issue.'"[21]

---

condition cannot reasonably be predicted to cause death or serious physical harm to exposed employees but does have a direct and immediate relationship to their safety and health." OSHA, Field Inspection Reference Manual, CPL 2.03, ch. III, sect. C.2.a., on http://www.osha.gov/Firm_osha_data/100007.html. The statute itself states that "a serious violation shall be deemed to exist in a place of employment if there is a substantial probability that death or serious physical harm could result from a condition which exists, or from one or more practices, means, methods, operations, or processes which have been adopted or are in use, in such place of employment unless the employer did not, and could not with the exercise of reasonable diligence, know of the presence of the violation." 29 USC sect. 666(k) (2000). The statute requires that an employer cited for a "serious violation" of a standard "shall be assessed a civil penalty of up to $7,000 for each such violation," whereas an employer cited for a violation "specifically determined not to be of a serious nature...may be assessed a civil penalty of up to $7,000 for each such violation." 29 USC sect. 666(b) and (c).

[18]Hudson Foods, Inc., Inspection No. 300002250. Tyson also succeeded in securing a reduction of another Hudson Noel penalty from $840,000 to $143,000 in 1997. Hudson Foods, Inc., Insp. No. 300002243 (Dec. 18, 1997). See also http://www.tysonfoodsinc.com/IR/newsinfo/viewNews.asp?article=238; http://www.sierraclub.org/factoryfarms/rapsheets/missouri/hudson_noel.asp.

[19]Greve, "In an OSHA First, Agency Fines Plant for Inadequate Toilet Breaks."

[20]Telephone interview with Steve Carmichael, Kansas City (Nov. 27, 2002).

[21]Philip Dine, "Missouri Plant Cited Over Access," *St. Louis Post-Dispatch*, Oct. 10, 1997, at 21C (Lexis).

Demonstrating that the vindication of voiding rights depends on the vigor with which unions fight for them, on August 19, 1997, just four weeks after OSHA had cited Hudson Foods, the UFCW announced that it was launching its own compliance program, including a survey of workers and filing of complaints with OSHA, in its organizing campaign among food processing workers in North Carolina. As Willie Baker, the union's southeastern regional director, put it: "The UFCW charge that brought the OSHA citation breaks new ground for worker rights. We will use the OSHA ruling to protect the rights of North Carolina Workers."[22]

By early December 1997, the well-informed Bureau of National Affairs *Daily Labor Report* was reporting that OSHA's communications director had informed it that the citation issued to Hudson Foods had necessitated a standard interpretation, which the agency was planning to issue "shortly" and "soon." The spokesman stated that the interpretation "'clarifies this situation'"; even though it would "not discuss the issue of restroom breaks," he noted that the "'basis for the interpretation and implicit [in the standard] is that employees must be able to use the restroom.'"[23]

The summer of 1997 brought another highly publicized case of bathroom denial to public attention. This time the employer was the City of New York, which was sued in state court by a class of public assistance recipients assigned to clean streets and remove debris as part of the Work Experience Program. The plaintiffs alleged that in violation of OSHA's toilet standard for mobile crews, the City failed to make "transportation immediately available to nearby toilet facilities."[24] Instead, the workers "must rely on the lucky event that there is a public toilet near the assigned route, perhaps get driven at a supervisor's whim to a bath-

---

[22]UFCW News, Press Release (Aug. 19, 1997); also available in part as "UFCW Wins OSHA Charge on Worker Bathroom Rights," U.S. Newswire, Aug. 19, 1997 (Lexis). The organizing campaign at Carolina Food Processors, the world's biggest hog slaughterhouse (and owned by Smithfield) was thwarted by numerous employer unfair labor practices. "UFCW Vows 'Smithfield Won't Get Away with Stealing' Hotly-Contested Election," U.S. Newswire, Aug. 22, 1997 (Lexis); Bob Williams, "Labor Struggle: Who Has the Muscle?" *News and Observer* (Raleigh), Aug. 31, 1997, at A23 (Lexis). Nevertheless, the UFCW is once again trying to organize Smithfield, where it is certain bathroom breaks are still being denied. Email from Jackie Nowell, UFCW, to Marc Linder (Jan. 13, 2003).

[23]"OSHA Interpretation of Standard on Bathroom Breaks Expected Shortly," *Daily Labor Report*, Dec. 9, 1997, at 236 D14.

[24]"The requirements of paragraph (c)(1)(i) of this section do not apply to mobile crews or to normally unattended work locations so long as employees working at these locations have transportation immediately available to nearby toilet facilities which meet the other requirements of this subparagraph." 29 CFR sect. 1910.141(c)(1)(ii).

room, go in the street, or 'hold it.'"[25] The named plaintiff filed an affidavit stating: "If we need to urinate or move our bowels, we have to squat behind a tree or bush or ask one of our co-workers to hold up a plastic bag to shield us from the passing cars. ... During my menstrual period, there is no place to go to change my pad. I have to wait until the end of our shift and by then my clothes are soaked with blood."[26] In August 1997 the state Supreme Court trial judge issued a preliminary injunction, ruling that the country's largest workfare program could not continue to employ recipients outdoors as long as the City failed to provide toilets.[27] However, the City appealed the decision, the injunction was held in abeyance during the appeal process, and by the following April—exactly one week after OSHA issued its long awaited ruling—*The New York Times* reported that some workers sometimes still had to wait until the end of their six-hour shift to use a bathroom. One supervisor told the newspaper "that senior managers had issued orders not to permit workers to leave to find toilets but that supervisors routinely ignored the orders."[28]

Half a year before the book was published, the right-wing *American Spectator* devoted almost a full column to reproducing the blurb from Cornell University Press's fall catalog, snidely calling the authors "prophets of the Golden Shower."[29] In October 1997, the magazine *American Health for Women* printed an article on incontinence, citing the as yet unpublished *Void Where Prohibited* for evidence that social conditions can exacerbate the condition and discussing it at some length in connection with workplace-related incontinence.[30]

The impact of *Void Where Prohibited* became international even before it appeared at the very end of 1997. As a result of a 13-page pre-publication condensation of the book in the November-December 1997 issue of the union-affiliated magazine *Working USA*,[31] the safety and health researcher for the Brit-

---

[25]Capers v. Giuliani at 8 (No. 97-402894, Sup. Ct. N.Y. County, 1997).

[26]Capers v. Giuliani, Affidavit of Tamika Capers at 3-4 (July 7, 1997).

[27]David Firestone, "Judge Orders Health and Safety Items for 5,000 in Workfare," *N.Y. Times*, Aug. 19, 1997, at B2, col. 1 (Lexis).

[28]Alan Finder, "City Slowly Improves Some Working Conditions," *N.Y. Times*, Apr. 13, 1998, B6, at 1 (Lexis). Although there was no formal adjudication, the City of New York acknowledged during litigation that it has an obligation to comply with the Public Employee Safety and Health Act with regard to these workers, and ultimately their working conditions improved. Email from National Employment Law Project to Marc Linder (Nov. 4-5, 2002).

[29]*The American Spectator* 30(7):85 (July 1997).

[30]Ronni Sandroff, "Urgent Matters: Incontinence Is Treatable, If Only Women Would Talk About It," *American Health for Women* 8(16):28 (Oct. 1997) (Lexis).

[31]Marc Linder and Ingrid Nygaard, "Void Where Prohibited," *Working USA*, Nov.-Dec. 1997, at 21-29, 68-71. Don Stillman, the editor, who did the condensing, was also

ish Transport and General Workers Union (TGWU), whose bus-driver and poultry-industry members had called the union about toilet access problems, was "prompted" by reading the article in the autumn to write to the Director General of the Health and Safety Executive, which is the British government agency charged with responsibility for occupational safety and health.[32] The controlling law in Britain merely states:

(1) Suitable and sufficient sanitary conveniences shall be provided at readily accessible places.
(2) [S]anitary conveniences shall not be suitable unless—...
(b) they and the rooms containing them are kept in a clean and orderly condition; and
(c) separate rooms containing conveniences are provided for men and women except where and so far as each convenience is in a separate room the door of which is capable of being secured from inside.[33]

Nevertheless, in conformity with the logic of *Void Where Prohibited*, the official replied: "'I believe that the requirements of the Workplace (Health, Safety and Welfare) Regulations 1992 to provide suitable and sufficient sanitary conveniences includes [sic] by implication reasonable opportunity to use these facilities.'"[34] Because "no systematic work" had previously been done on toilet access, the TGWU, according to the editor of *Hazards*, the British occupational health and safety magazine, began "pressing the health and safety authorities to do this research. All of this activity was stimulated by your US campaign."[35] Unfortunately for bus drivers, however, who sometimes, like their counterparts in the United States,[36] work five hours without a bathroom break on account of tight running times and lack of toilets on the road—16 percent of respondents surveyed by the union had to wait four hours or more—a definitional gap in the regulation left them for the time being uncovered because neither buses nor roads qualified

---

Director of Governmental and International Affairs of the UAW.

[32]Email from Kim Sunley to Marc Linder (Dec. 16, 1998).

[33]Workplace (Health, Safety and Welfare) Regulations 1992, sect. 20 (Stat. Instruments No. 3004).

[34]Email from Kim Sunley to Marc Linder (Dec. 16, 1998). The Director General was Jenny Bacon. The general duty clause provides: "It shall be the duty of every employer to ensure, so far as is reasonably practicable, the health, safety and welfare of all his employees." Health and Safety at Work Act 1974, ch. 37, sect. 2(1).

[35]Email from Rory O'Neill to Marc Linder (Dec. 10, 1998). However, because the government failed to close the "loophole" that OSHA closed in 1998, U.K. unions began a campaign to force it to entitle workers to go when they need to. Kevin Maguire, "Let People Go for Time Off in Loo, Says TUC," *The Guardian*, Feb. 21, 2003, at 6 (Lexis).

[36]See below chap. 14.

as workplaces,[37] and by the end of 2002, little progress had been made.[38]

The attention that the media paid both to the book and to the problem intensified in the beginning of 1998. To be sure, some initial press accounts failed to capture the subtlety of one of the book's central legal claims. In early January, for example, the *Wall Street Journal*, reporting on OSHA's plans to issue a standard interpretation mandating a reasonable time to use the toilet, asserted that the book "says no federal law assures time off to use restrooms."[39] The author corrected this mischaracterization in a letter to the editor:

> [T]he book explains that the Occupational Safety and Health Administration's 25-year-old sanitation standard requiring employers to "provide" toilets to their employees makes no sense whatsoever unless it is interpreted to include the obligation to let workers use those toilets. In other words, employers that do not let their employees use the toilet violate the regulation by failing to "provide" toilets. Unfortunately, although the law may theoretically give workers the right to use the bathroom, the agency that administers the law did not, until confronted with the information in this book, choose to enforce it.[40]

Later that month, Doug Henwood, the editor of the *Left Business Observer*, interviewed the authors on "Behind the News," his WBAI radio program broadcast in New York City. In early February the *Chicago Tribune* labor reporter published a long and sympathetic article about the problem and the book in the Sunday edition,[41] which was widely syndicated.[42] The author's op-ed pieces,

---

[37]Email Kim Sunley to Marc Linder (June 7, 1999) enclosing "Report on Toilet Facilities for Bus Drivers and Results of Survey."

[38]Email from Martin Mayer, TGWU national representative for UK bus drivers, to Marc Linder (Oct. 6, 2002).

[39]Al Karr, "Work Week," *Wall Street Journal*, Jan. 13, 1998, at 1, col. 5. See also Alis Valencia, "Facing Hard Truths: In Search of Dignity and Respect on the Job," *At Work—Creating a More Enlightened World of Business & Work*, July-Aug. 1998, at 21 ("companies have no legal obligation to let employees use the bathroom when they need to") (book review).

[40]Marc Linder, "The Commode and the Law," *Wall Street Journal*, Jan. 22, 1998, at A19, col. 1. A separate correction was published in the *Salt Lake Tribune*: "Author Marc Linder says employers that do not let their employees use the toilet violate an Occupational Safety and Health Administration regulation requiring employers to provide toilets. A Wall Street Journal item published in The Salt Lake Tribune implied otherwise." "Corrections and Clarifications," *Salt Lake Tribune*, Jan. 24, 1998, at B6 (Westlaw).

[41]Stephen Franklin, "Gimme a Break: A Minute or Two to Go to the Bathroom Isn't Necessarily a Worker's Right," *Chicago Tribune*, Feb. 8, 1998, sect. 6, at 5.

[42]E.g., Stephen Franklin, "Rest Breaks for U.S. Workers Are Not Guaranteed," *Kansas City Star*, Feb. 23, 1998; Stephen Franklin, "Forget Coffee...Some Don't Get Bathroom Break," (Fort Lauderdale) *Sun-Sentinel*, Feb.. 23, 1998.

which presented very compressed versions of the book, published later in February in *The National Law Journal*[43] and the business section of the Sunday *New York Times,*[44] were calculated to create maximum public embarrassment for OSHA for its absurd interpretation of its toilet standard and thus to put pressure on the agency finally to issue its new interpretation.

Public awareness of the scandal of lack of workplace toilet access swelled on March 13, 1998, when a huge number of newspapers carried a lengthy Associated Press article released the previous day as "Relief Is on the Way for Workers Who Need to Use the Bathroom."[45] While the most straightforwardly optimistic headline appeared in the *Chicago Tribune*: "U.S. to Order Worker Right to Use Toilet,"[46] a Canadian newspaper added the element of pathos: "U.S. Changing Humiliating Toilet-Break Regulations."[47] Written by Maggie Jackson, who had relied heavily on the author for her structure, orientation, and sources, the article noted that "for teachers, factory workers, telemarketers, farmworkers and others,

---

[43]Marc Linder, "'Let My People Go,' Cries Labor Union," *National Law Journal*, Feb. 2, 1998, at A19, col. 1-4. Bizarrely, Barbara Repa, "When the Boss Doubles as a Bathroom Monitor," http://www.nolo.com/Edit_Art/peepatrol.html (May, 29, 1998), characterized this article as a "glib jibe" in reaction to OSHA's citation of Hudson Foods.

[44]Marc Linder, "A Fight for Restroom Rights," *N.Y. Times*, Feb. 22, 1998, sect. 3, at 13, col. 4-5.

[45]Among the U.S. and Canadian newspapers publishing the article (under various titles) beginning on March 13 were: (Cedar Rapids) *Gazette*, *Detroit News*, (Riverside, California) *Press-Enterprise*, (Bergen County, NJ) *Record*, *Hartford Courant*, (Marshalltown, IA) *Times-Republican*, *Oelwein Register*, *Orange County Register*, *Toronto Star*, (Oklahoma City) *Journal Record*, *Times-Picayune*, *Chicago Tribune*, *San Francisco Chronicle*, *Fort Worth Star-Telegram*, *San Antonio Express-News*, *Richmond Times-Dispatch*, (Cleveland) *Plain Dealer*, *Roanoke Times*, *Vancouver Sun*, (Worcester, MA) *Telegram & Gazette*, *Calgary Herald*, (Allentown, PA) *Morning Call*, *Indianapolis Star*, *Columbus Dispatch*, (Spokane) *Spokesman-Review*, *Seattle Times*, *Des Moines Register*, *Salt Lake Tribune*, *Greensboro News & Record*, *San Diego Union and Tribune*, *Florida Times Union*, *Albany Times Union*, *Charleston Gazette & Daily Mail,* (Quincy, MA) *Patriot Ledger*, (Springfield, IL) *State Journal-Register*, *Fresno Bee*, *Cincinnati Enquirer*, *Houston Chronicle*, *Florida Today*, *Grand Rapids Press*, *Fort Dodge Messenger*, *Muscatine Journal*., *Waterloo Courier*, *Sioux City Journal*, *Mason City Globe-Gazette*. See also Mike Augspurger, "Bathroom Breaks Will Be Enforced by New Law," *Burlington Hawk Eye*, Mar. 15, 1998 (stating that local employers in Burlington, Iowa, let workers go to the bathroom "'as needed'"). Also the website of CNN Interactive carried the article.

[46]"U.S. to Order Worker Right to Use Toilet," *Chicago Tribune*, Mar. 15, 1998, sect. 1, at 18.

[47]"U.S. Changing Humiliating Toilet-Break Regulations," *Vancouver Sun*, Mar. 14, 1998.

meeting this simple need can mean humiliating pleas for permission and even a risk of losing their job." Jackson then quoted the author as being "'horrified to learn that employers can get away with this'" and saying: "'This isn't a problem in every workplace, but it's much more widespread than we had originally believed'"[48]—and, as it turned out, than even he himself knew at the time.[49]

Outraged at the author's naivete, that very day, a reader of the Greenville (South Carolina) *News* wrote him that the fact that he had been "'surprised'" that "employers can get away with refusing employees leave to use bathrooms" suggested that "[t]he University of Iowa must be an ivory tower even farther removed from reality than the average Ivy League campus. Employers in the U.S. can get away with every goddamned thing they want to get away with."[50]

On March 15, the book ("It's a pretty great title"), the phenomenon of access denied ("incredibly sexist"), and the possibility of government intervention ("The Bathroom Access Act of 1998") were discussed seriously on the national television program "ABC Good Morning America Sunday," by free-lance journalist Katherine Davis and others.[51] The following morning the author was interviewed on a Phoenix news-talk radio station for the program "Wake Up Arizona"[52] and on other radio stations. The following week the new national consciousness of employers' autocratic denial of toilet access found its way into the "Dilbert" cartoon strip, which featured the evil human resources director announcing a new vacation policy to deal with the rule that all vacation time has to be used in the year it was earned: "I realize this is not always convenient. So I've decided to be flexible. From now on, any time you spend in the restroom will count as vacation."[53] In the last week of March the author was interviewed about the book and the problem on "Employment & Labor Lawcast," an audiotape legal news service listened to by thousands of lawyers largely while traveling or exercising.[54]

More significantly, on April 2 the author appeared at noon for an hour on "The Tom Pope Show," a radio program of the Dudley Broadcasting Network targeting a black audience, which was broadcast on stations from Columbia,

---

[48]Maggie Jackson, "Relief Is in the Way for Workers Who Need to Use the Bathroom," Associated Press, Mar. 12, 1998 (Westlaw).

[49]See below ch. 8.

[50]Letter from W.J.D. Wells, Greer, S.C., to Marc Linder (Mar. 13, 1998).

[51]"ABC Good Morning America Sunday" (Mar. 15, 1998, 10:00 am ET), Transcript # 98031508-J02 (Lexis). The editor-in-chief of the press that published the book called contributor Chris Cuomo's comments "asinine." Email from Fran Benson to Marc Linder (Mar. 16, 1998).

[52]KFYI 910 AM, 8:20 a.m., Mar. 16, 1998.

[53]"Dilbert," *Des Moines Register*, Mar. 22, 1998.

[54]"Employment & Labor Lawcast" 4(6) (week of Mar. 30, 1998).

South Carolina to Ontario, Canada.[55] During this live broadcast, workers from auto factories in the Detroit area called in to complain about lack of toilet access at work, symbolized by waits of upwards of an hour. When the author asked about the impact of the much-vaunted relief system that the United Automobile Workers had negotiated with the major automobile manufacturers over the years, a relief worker called in to protest that relief was the last thing he was allowed to give: because General Motors had done little or no hiring in years in order to increase productivity through attrition, relief workers had been assigned their own independent production tasks, which left them with little time to provide relief. The author also used the opportunity to give out the telephone number of and to urge listeners to call Emily Sheketoff, the Deputy Assistant Secretary of Labor for OSHA, the Clinton administration's political commissar at the agency, who was reputed to be the ultimate decisionmaker, demanding that OSHA issue a broad toilet standard interpretation.

A week before OSHA was moved to act, the agency's new head, Charles Jeffress, revealed to a reporter its failure to understand the extent of the problem it was seeking to eliminate: "Jeffress thinks bathroom breaks are not a widespread problem. 'Potentially it's an issue where you have an assembly line operation,' he said, adding the problem tends to occur when a supervisor is 'being too harsh.'" It was unclear whether Jeffress was merely ignorant or whether the claim was part of his strategy "to mend fences with the Republican majority in Congress that unsuccessfully tried to abolish the agency in 1995." His understatement may, after all, have been designed to mollify Representative Cass Ballenger, the North Carolina Republican who chaired the subcommittee with oversight over OSHA and had "wondered if OSHA would be interfering with business operations with a toilet rule that's too specific. 'You could really make it awful difficult to run an operation,' he said. Ballenger suggested that a public hearing or comment period be arranged before the toilet rule takes effect."[56]

---

[55]Letter from Gwen Pope, Dudley Broadcasting Network, to Marc Linder (Mar. 23, 1997 [should be 1998]).

[56]Brian Tumulty, "OSHA Chief Works to Make Agency Relevant in Workplace," Gannett News Service, Apr. 1, 1998 (Westlaw).

3

Iowa OSHA Anticipates Federal OSHA

I wonder how Iowa survived 150 years of statehood without bathroom break regulation.[1]

Meanwhile, in Iowa, a state-plan state, Byron Orton, the Labor Commissioner, who has oversight over the state OSHA program, acted in advance of and more resolutely than federal OSHA. The immediate background of this action went back to 1995 when, following up on a telephone conversation, the author wrote a letter to Mary Bryant, the Iowa Occupational Safety and Health Administrator, asking for the (then) Department of Employment Service's "official position on whether employees in Iowa are entitled to use the bathroom at work." The letter went on to ask: "If, as you have already indicated, you take the position that OSHA's requirement that employers provide toilets does not imply that they also permit employees to use those toilets, could you please also explain what the purpose is of the toilet provision if the law does not require employers to permit workers to use them."[2] Bryant's official reply was only partially responsive, yet expressed the official position with all imaginable clarity: "Although the OSHA standards require that an employer provide toilet facilities, they are silent on the issue of allowing workers to use them. Therefore, we have no regulation to force them to do so and could not issue a citation for their not allowing it."[3] It was in no small part Deborah Berkowitz's submission of this letter to Federal OSHA in 1997 that prompted that agency to change its interpretation of the law.

In 1996, Charles Whisenand, who was working as an assistant manager of a Holiday "convenience" store in Ames, Iowa, was diagnosed with congestive heart failure requiring him to take medicine that made him urinate often. As he later described his situation: "'We were asked to work alone for eight to 10 hours a

---

[1]Letter from Iowa State Senator Steve King to Labor Commissioner Byron Orton (Apr. 26, 1998). Orton made available to the author a copy of his correspondence with King.

[2]Letter from Marc Linder to Mary Bryant (Oct. 31, 1995).

[3]Letter from Mary Bryant to Marc Linder (Nov. 17, 1995).

day without relief. When I asked about their policy and the scheduling that caused me such pain, they did not think it was such a problem. I was told that if I could not handle such a job then I had the privilege of working somewhere else.'" Instead of getting a relief worker, Whisenand was fired. The company purportedly provided breaks only in states that mandated them by law, of which Iowa was not one. Whisenand then contacted his state senator, Johnie Hammond,[4] who on March 3, 1997,[5] introduced a bill in the Iowa Senate, which resembled a unique law enacted in Minnesota in 1988.[6] Though inadequate in general and probably of limited help to Whisenand, the Iowa bill did provide that: "An employer shall allow an employee...a paid fifteen-minute rest break during every consecutive four-hour period of work. To implement this requirement, an employer shall provide an employee, at the worksite, with reasonable access to restroom facilities." Civil penalties ranged from $1,000 for violations to $3,000 for repeated violations demonstrating "a pattern of abusive employment practices."[7] Despite its relatively restrained intervention, the bill "was killed in the Business and Labor Committee...by, as Whisenand put it, 'my fellow Republicans who think it's overregulating businesses.'"[8]

In October 1997 Orton's involvement was triggered by a column about bathroom access published in the *Des Moines Register*, the state's leading newspaper, and widely syndicated throughout the United States.[9] Serendipitously, the columnist, Rekha Basu, had heard about *Void Where Prohibited* before its publication from her mother, who was friends with Dr. Bernard Lown,[10] who had written a blurb for the book. When she mentioned that the state OSHA administrator had written the author stating that the law imposed no obligation on employers to let workers use the toilet, but that an OSHA attorney whom Basu had interviewed stated that "'Our position would probably be that an employer needs

---

[4]Rekha Basu, "Progress on Bathroom Breaks," *Des Moines Register*, Apr. 12, 1998, at 6AA, col. 3-5.

[5]On http:www2.legis.state.ia.us/GA/77GA/BillHistory/SF/00200/SF00267.html.

[6]See Marc Linder and Ingrid Nygaard, *Void Where Prohibited: Rest Breaks and the Right to Urinate on Company Time* 63 (1998).

[7]Iowa Senate File 267, sects. 2 and 4, on http://www2.legis.state.ia.us/cgi-bin/Legislation/Bill.pl.

[8]Basu, "Progress on Bathroom Breaks."

[9]See, e.g., Rekha Basu, "Restroom Privileges Restricted," *Great Falls Tribune* (Nov. 7, 1997); idem, "Many Employees Forbidden to Answer Nature's Call," *Herald-Dispatch* (Huntington, WV), Nov. 8, 1997.

[10]A professor emeritus of cardiology at the Harvard School of Public Health, Lown had received the Nobel Peace Prize in 1985 on behalf of International Physicians for the Prevention of Nuclear War.

to make sure they allow the employee to use it,'" Basu asked: "So which is it?"[11]

In order to gauge the embarrassment engendered by this column, the author made numerous telephone calls to Iowa OSHA officials and asked Mark Smith, the president of the Iowa Federation of Labor, to speak to Orton about changing the agency's position. But when Smith called Orton, the commissioner told him that he had already met with his staff to change the policy, some of whom had informed the author that they still asserted that the employer's obligation to provide a toilet did not require the employer to let workers use it. This confusion prompted Byron Orton to intervene. In the version of his letter to the editor that the *Des Moines Register* published on November 2, 1997, Orton—who had been an administrative law judge in the state unemployment compensation system and Industrial Commissioner in charge of the state workers compensation system before being appointed Labor Commissioner in 1995[12]—observed that although OSHA standards "are silent on the issue of allowing employees to use" toilets, those standards "are to be construed with a view of promoting their intended purpose and achieving reasonable results." Consequently, "consistent with pronouncements from federal OSHA," Orton, who was doubtless alluding to the Hudson Foods case, stated that Iowa OSHA "takes the position that the employer's obligation to provide toilet facilities includes the obligation to allow employees to use the toilet facilities." Although he had no doubt that most employers in Iowa treated their employees "with respect and dignity" concerning toilet access, "based upon my 23 years in state government, I...know that some employers deny restroom privileges to employees except during designated breaks with the threat that an individual's employment will be terminated if he or she violates such a rule." Orton nevertheless then concluded:

> Such a draconian rule unnecessarily exposes employees to health hazards, callously tramples on the dignity of employees and jeopardizes the public policy of this state to encourage harmonious labor-management relations. [I]n those instances where employees are denied timely restroom privileges, Iowa OSHA is prepared to act by way of citations and penalties.[13]

---

[11]Rekha Basu, "Give Employees a Break," *Des Moines Register*, Oct. 19, 1997, at 6AA, col. 3. Although the Iowa OSHA administrator Mary Bryant wrote the letter to Marc Linder on Nov. 17, 1995, stating that the agency had no power to cite employers who denied employees bathroom breaks, as late as August 1998 Cynthia Tofflemire, an Iowa State Labor Commission attorney, denied that such a statement had ever been made. "Accommodating Bathroom Use Challenges Employers," *Successful Job Accommodation Strategies* 4(4) (Aug. 1998) (Lexis).

[12]Byron Orton, Iowa Labor Commissioner, "Biographical Information" (undated).

[13]Byron Orton, "OSHA Silent on Potty Breaks," *Des Moines Register*, Nov. 2, 1997, at AA7:1.

Why, after stating that the employer's obligation to provide toilets included the obligation to allow employees to use them, Orton nevertheless referred to workers' "privileges" rather than rights to void, is unclear. But in a section of the letter that the newspaper cut, Orton made the important point that: "The call of nature is not controlled by the time clock and does not always wait for coffee and lunch breaks. I am aware of situations where employees have remained at their work stations and unceremoniously soiled themselves, even vomited, because they had been told that to leave their work station to answer a call of nature would result in discharge from employment."[14]

On January 21, 1998, Orton issued a memorandum to all Division of Labor staff, the Iowa Federation of Labor, the UAW, the Iowa Association of Business and Industry, and the National Federation of Independent Business, on the "Interpretation of 1910.141(c) Toilet Facilities." Orton's memorandum in part tracked the federal OSHA draft of July 1997, which will be discussed later.[15] Noting that it was important for Iowa OSHA to take a uniform approach with regard to the general industry sanitation standard and especially regarding employees' use of employer-provided toilets, he requested that the new "guideline" be used in dealing with this issue. Of the standard's requirement that employers provide a sufficient number of toilets, Orton's memorandum stated: "If an employer failed to have in the workplace the necessary number of toilets, or if the employer kept them locked, it would be a violation of the standard. Similarly, an employer restricting the employees' use of toilets in the workplace is not providing toilets and thus is in violation of the sanitation standard." After pointing to one dictionary definition of "provide" as to "'supply or make available (something wanted to needed),'" Orton formulated the new guideline:

> Use of toilet facilities is, therefore, a necessary component to the "providing toilet requirement." To comply with the sanitation standard, employers must allow employees to use toilet facilities when employees need to use the facilities. Exactly how employers insure employees' access to the toilet facilities will depend on the conditions at each place of employment and may vary from workplace to workplace. For example, in assembly line production settings, many employers establish some kind of signaling and relief system for workers. Employers who have not made such arrangements, or who have made inadequate arrangements, are not in compliance with the standard. Failure to comply shall result in citations and penalties.[16]

---

[14]Letter from Byron Orton to Des Moines Register, Oct. 22, 1997.

[15]See below ch. 6.

[16]Byron Orton, Memorandum Re: Interpretation of 1910.141(c) Toilet Facilities (Jan. 21, 1998); published in Iowa Federation of Labor, *Political Action Update*, 98-06 (Feb. 20, 1998).

Orton's approach was somewhat more rigorous—or, as he himself put it, "a bit stronger"[17]—than that of the July 1997 Federal OSHA draft or the final interpretation of April 6, 1998,[18] because it was not hedged in by repeated references to a reasonableness criterion, which might eventually be interpreted by pro-employer administrative or judicial officials as authorizing firms to require workers to wait extended periods of time. In the draft of his letter to the editor of the *Des Moines Register*, Orton had threatened OSHA action "where employees are exposed to health hazards associated with being denied timely restroom privileges,"[19] but no such condition appeared in his memorandum because the author had, through Smith of the Iowa Federation of Labor, persuaded him to drop it lest employers contest the legitimacy of the policy on the grounds that there was no medical evidence of a health hazard linked to a limited number of denials of the right to void. As the Associated Press noted: "In part because of Linder's research, Iowa in January became only the second state to explicitly protect workers' legal rights in this area."[20]

In response to Iowa OSHA's interpretive action, Steve King. the vice chair of the Business and Labor Committee of the Iowa state Senate and well-known anti-labor right-wing Republican, inaugurated a belligerent correspondence with Orton. One week after Orton had issued his memorandum, King opined that "reasonable people ought to be able to work out reasonable policies without federal or state intervention," adding that he did "not believe that a case can be made that the regulatory intent was consistent with your interpretation."[21] In reply, Orton declared that he was "at a complete loss as to how there could be any reasonable interpretation of the sanitation standard other than the interpretation I announced. ... The position that the requirement to provide toilet facilities does not encompass the requirement to allow employees to use toilet facilities is unconvincing to me. OSHA standards are to be construed with a view of promoting their intended purpose and achieving reasonable results. [A]ny other interpretation would not only defeat the intended purpose of the OSHA sanitation standard, it would clearly fly in the face of reasonableness and common sense."[22]

---

[17]Telephone interview with Byron Orton (Oct. 18, 2002).

[18]See below ch. 6.

[19]Letter from Orton to Des Moines Register (Oct. 22, 1998).

[20]"Bathroom Access Causes Work Troubles," *Des Moines Register*, Mar. 13, 1998, at 2M, col. 2 (State Edition; NewsBank). Minnesota had enacted a statute in 1988 that offered workers inadequate protection by requiring employers to let them go to the restroom once every four hours. See Linder and Nygaard, *Void Where Prohibited* at 63-64.

[21]Letter from Steve King to Byron Orton (Jan. 28, 1998). In 2002 King was elected to the U.S. House of Representatives.

[22]Letter from Byron Orton to Steve King (Feb. 9, 1998).

Undaunted—and taking umbrage at Orton's "insinuation that my interpretation...was unreasonable and lacked the mark of common sense"—King asked Orton what "the standard for Iowa workplaces concerning restroom usage" should be and specifically: "How one would deal with a person who habitually abuses the privilege of having a restroom facility which they are permitted to use at their leisure?"[23]

Apparently having forgotten that he had himself used the very same term in his letter to the *Des Moines Register*, Orton took issue with King's "characterization of a restroom facility as a 'privilege' for working Iowans. Pursuant to OSHA law, employees have a legal right to restroom facilities in the workplace; pursuant to the laws of biology and a civilized society, restroom facilities are a human necessity." Recasting King's question in terms of employees' "abus[ing] their rights...by utilizing restroom facilities for reasons other than their intended purpose, e.g., sleeping, illegal activity, etc.," Orton observed that "employers already have the ability and authority to discipline workers.... My interpretation provides them no protection. Rather, current OSHA law would...shield employers from citations and penalties. As with any OSHA standard, OSHA law provides the affirmative defense of 'employee misconduct.'" Of overriding practical significance was Orton's statement: "Please understand that the frequency of use, standing alone, does not establish 'employee misconduct' so long as the employee is using the restroom facility for its intended purpose." Orton also bolstered his view that it would serve no purpose to require employers to provide toilets "if employees are not allowed to use them when needed" by reference to two similarly structured OSHA provisions that require employers whose employees are exposed to potentially injurious corrosives shall "provide suitable eye wash facilities for quick flushing or washing" and that require them to "provide" respirators under certain conditions. Though neither says anything about allowing workers to use eye wash or respirators, it would be unreasonable to interpret them as conferring no "legally protected right" to use them. Orton was even able to boast that the Iowa Association of Business and Industry had expressed to him "their appreciation for clarifying an issue in a fashion that now provides solid guidance for their members."[24]

Soon the corporate animal slaughter plants in Iowa would have little for which to express their gratitude to Orton.

---

[23]Letter from Steve King to Byron Orton (Apr. 6, 1998).

[24]Letter from Byron Orton to Steve King (Apr. 21, 1998).

Part II

An Historical-Exclusionary Interlude

Toilet facilities are an essential part of the worker's everyday life. They are as important to him or her during working hours as they are at home. It is management's responsibility to supply toilet facilities in sufficient number for the size of the employee group and in locations convenient for those using them.[1]

---

[1]Arkansas Department of Labor, Safety Code # 6: Safety Code for Industrial Sanitation, sect. 6, at 10 ("Explanatory Comments"), on http://www.state.ar.us/labor/pdf/code6_industrial_sanitation.pdf.

4

How OSHA Almost Revoked Its Toilet Standard in 1978

To develop the correct attitude among employees in respecting the management's investment in sanitation facilities, responsibility...rests with the employer. Not only should he seek actively and constantly the worker's cooperation in maintaining the facilities in a sanitary condition, he must also demonstrate regular and thorough upkeep in the toilets.... When the employer does this then a share of the responsibility for adequate sanitation also belongs to the employee. It should be his obligation to use rather than abuse toilet...facilities. Adequate toilets and lavatories are provided for his use because they are essential to his well being and health while at work as they are at home. By the same token, the facilities should be used by him and left usable for the person who follows him just as he would wish it done in his home.[1]

With the narrative having reached April 6, 1998, the date on which OSHA, animated by the combined impact of the UFCW's lobbying campaign and the adverse publicity surrounding the publication of *Void Where Prohibited*, finally issued its long-awaited interpretation of its toilet standard, it is time to exhume the astounding irony and long-forgotten fact that 20 years earlier OSHA had proposed and come very close to revoking that toilet standard altogether. Had OSHA succeeded in deleting the standard in 1978, the agency would have been very hard pressed to identify a basis on which to construct employers' obligation to let workers stop work to use the toilet in 1998.

Since OSHA's spokesman barely a week prior to April 6 had admitted that "'a clarification that merely providing restrooms is not enough will be a big change,'"[2] and since it is implausible that OSHA had received no complaints about access during the preceding 27 years, its claim that it had always viewed

---

[1]Arkansas Department of Labor, Safety Code # 6: Safety Code for Industrial Sanitation, sect. 6, at 10 (Explanatory Comment) (n.d. [after 1951]), on http://www.accessarkansas.org/labor/pdf/code6_industrial_sanitation.pdf.

[2]Simon Nadel, "Restricting Regular Bathroom Access Can Spur Negative Workplace Repercussions," *U.S. Law Week* 66(37):2579-80 at 2579 (Mar. 31, 1998) (quoting OSHA communications director Stephen Gaskill).

the requirement as implicit[3] lacks credibility.

Such skepticism is reinforced by the declaration of George Guenther, OSHA's first administrator, in 1972 that standards dealing with restroom facilities "have 'little direct relationship to occupational safety and health.'"[4] Guenther's view was in no way idiosyncratic. The fact that in 1977-78 the Carter administration OSHA under Eula Bingham, a self-professed "advocate of worker rights to health and safety,"[5] could have contemplated destroying the hinge that implicitly anchored workers' right to void when they need to may seem preposterous; yet the only plausible alternative interpretation is that OSHA never considered the link at all.

The course of this obscure development can be traced back to efforts by the Carter administration to propitiate employers that had been vigorously complaining during the Ford interregnum that OSHA had "forced business to make unjustified profit-reducing changes in factories."[6] In 1975 a conference committee of the Democratic-controlled Congress issued a fiscal year 1976 appropriations bill report directing the Labor Department to undertake "[r]eview and simplification of existing OSHA standards and elimination of so-called 'nuisance standards' or standards which do not deal with workplace conditions that are clearly hazardous to the health or safety of workers or are more properly under the jurisdiction of State Departments of Public Health."[7] Pressed for the Republican nomination by the even more openly pro-employer Ronald Reagan,[8] President Ford had told a group of businessmen in 1976 that "they would like 'to throw OSHA into the ocean.'"[9]

By 1978, Douglas Soutar, vice president for industrial relations at the American Smelting and Refining Company and for many years a central figure in articulating big business's national labor law and labor relations agenda, who had lobbied against the enactment of OSHA, still feared that it would become "a Trojan Horse, useful in accelerating the eventual nationalization of our business system." The passage of eight years had vindicated his initial opposition to what

---

[3]See below ch. 6.

[4]"Easing of New Job-Safety Law Provision for Small Firms Backed by Labor Agency," *Wall Street Journal*, June 23, 1972, at 2, col. 3-4.

[5]Charles Noble, *Liberalism at Work: The Rise and Fall of OSHA* 188 (1986).

[6]David Burnham, "Ford Termed Cool to 3 Key Agencies," *N.Y. Times*, Jan. 16, 1976, at 1, col. 1, at 30, col. 6 (Proquest).

[7]*Department of HEW and Related Agencies Appropriation Bill, Fiscal Year 1976*, 94th Cong., 1st Sess. 9 (1975).

[8]On Reagan's view of OSHA during that presidential campaign, see "Why Nobody Wants to Listen to OSHA," *Business Week*, June 14, 1976, at 64 (Lexis).

[9]David Burnham, "Worker Safety Agency, Under Fire, Has Little Impact But Big Potential," *N.Y. Times*, Dec. 20, 1976, at A1, at B6, col. 2 (Proquest).

he had foreseen "as a super agency which would drastically impair industry's prerogatives and freedom to manage, its ability to generate capital, to borrow, to earn an adequate return on investment." By displacing labor-management law, OSHA was enabling workers to attain what they could not through collective bargaining.[10] Nevertheless, not even a tough, cynical, and realistic capitalist spokesman like Soutar would have predicted in 1978 that 20 years later OSHA's Trojan Horse would insinuate itself into the workplace via the bathroom through the agency's interpretation of what had been an American National Standards Institute (ANSI) voluntary toilet standard and impose at-will voiding breaks on employers. After all, employers had favored such "national consensus" standards[11] of ANSI, whose committees were "composed primarily of corporate safety engineers, with a light sprinkling of labor and government representatives."[12]

Bingham's predecessor in the Ford Administration, Morton Corn, had testified before both congressional appropriations committees in 1976 that OSHA was busy eliminating "unessential regulations"; the majority of those "cited as trivial" having been derived from the national consensus standards that had been adopted during the agency's "formative months as a baseline for its compliance actions. ... Under those consensus standards, a large number of irrelevant ingredients got on our books."[13] In rushing to assure House members eager to learn whether OSHA had followed their directions to eliminate "nuisance standards,"[14] Corn insisted that he did not "think the agency should be concerned with the esthetic qualities of the work environment" such as prohibiting compounds with "a horrible odor."[15]

---

[10]Douglas Soutar, "A Management Viewpoint," *Labor Law Journal* 29(8):492-98 at 493 (Aug. 1978).

[11]*Occupational Safety and Health Act, 1970: Hearings Before the Subcommittee on Labor of the Committee on Labor and Public Welfare of the United States Senate*, 91st Cong., 1st and 1d Sess. 328 (1970) (statement of J. Sharpe Queener, Safety Director, Du Pont Co. and representing U.S. Chamber of Commerce).

[12]John Mendeloff, *Regulating Safety: An Economic and Political Analysis of Occupational Safety and Health Policy* 36 (1979).

[13]*Departments of Labor and Health, Education, and Welfare Appropriations for 1977: Hearings Before a Subcommittee of the Committee on Appropriations House of Representatives: Part 1: Department of Labor Related Agencies*, 94th Cong., 2d Sess. 551, 562 (1976).

[14]*Departments of Labor and Health, Education, and Welfare Appropriations for 1977: Hearings Before a Subcommittee of the Committee on Appropriations House of Representatives* at 538, 563. See also *Departments of Labor and Health, Education, and Welfare Appropriations for 1977: Hearings Before a Subcommittee of the Committee on Appropriations United States Senate: Part 7*, 94th Cong., 2d Sess. 4336 (1976).

[15]*Departments of Labor and Health, Education, and Welfare Appropriations for*

In February 1977, just two weeks after having become president, Carter, in a "Report to the American People," declared that his administration would remove unnecessary regulations.[16] By midyear he was expressing the radical anti-prevention belief that "the Federal Government ought to get out of those kinds of detailed safety precautions when the worker can observe with his or her own eyes that a danger exists, and then have the safety regulations covered perhaps by increases in the payment of workmen's compensation if an employer does have a dangerous place for the employees to work. ... In the safety area, I don't think many of them [i.e. regulations] are necessary. But we are doing what we can to now to simplify the whole system...."[17]

Bingham and Secretary of Labor Ray Marshall, who was closely identified with the labor movement,

> developed a good idea of the themes that would be the basis of their OSHA policy. One of the initial concerns that Bingham had expressed was that "OSHA was about some silly things" and was widely perceived as being preoccupied with "frivolous, irrelevant rules and regulations." President Carter and Ray Marshall shared this concern, and all three knew that OSHA was in deep trouble with Congress, with the small business community, and with organized labor. ... These ideas formed the basis of a bold public relations stroke engineered by Bingham's public affairs officer Frank Greer which tied together OSHA's new programs and policies under the rubric of "common sense." At a press conference on May 19, 1977, Marshall and Bingham announced with great fanfare that OSHA was being redirected to follow new "Common Sense Priorities" and that from now on it would focus its energies on the really serious problems of the workplace. There were three main parts to the program: to "get serious about serious dangers"; to help the small businesses comply with OSHA rules; and to clarify and simplify safety rules. ... Under its third priority—simplifying safety rules—OSHA had already begun combing through over a thousand consensus standards to revise unclear ones and eliminate unnecessary or irrelevant ones.[18]

1977: Hearings Before a Subcommittee of the Committee on Appropriations House of Representatives at 557.

[16]James Carter, "Report to the American People," in *Public Papers of the Presidents of the United States: James Carter: 1977*, Book I, at 69-77 at 74 (1977 [Feb. 2, 1977]).

[17]"Remarks and a Question-and-Answer Session at a Public Meeting in Yazoo City, MS]," in *Public Papers of the Presidents of the United States: James Carter: 1977*, Book II, at 1316-34 at 1331 (1978 [July 21, 1977]).

[18]Judson MacLaury, "The Occupational Safety and Health Administration: A History of its First Thirteen Years, 1971-1984: 4. Eula Bingham Administration, 1977-1981: Of minnows, whales and 'common sense'" on http://www.dol.gov/asp/programs/history/osha13bingham.htm (quotations from interview with Bingham). For the text of the Marshall-Bingham announcements, see *Congressional Record* 123:15926 (1977).

Bingham, who declared that "OSHA does not exist to mediate between labor and management on health and safety issues.... We exist to limit human suffering and to protect working men and women,"[19] agreed in March 1978 with employers' criticism of OSHA for "over-enforcement of nuisance standards." Of the more than 5,000 safety standards that OSHA had adopted at its inception, "many of them trivial and unrelated to actual workplace hazards," Bingham mentioned those that had been issued by ANSI (such as the toilet standard) as "never" having been "intended to be enforceable." It was for that very reason that she had proposed revoking more than 1,100 standards that were out of date and/or not related to safety or health.[20]

In September 1977 BNA's *Occupational Safety & Health Reporter* reported that a draft list of standards to be revoked was being compiled by two different OSHA work groups that had been formed in June: one from OSHA's safety standards office and the other from the Labor Department's solicitor's office and other OSHA divisions. The career civil servant in charge of the project, John Proctor,[21] the deputy director of safety standards, stated that there was not full agreement between the two groups about the deletion of 1910.141(c)(1)(i) and (ii). Proctor shed considerable light on what may have been the opportunistic rather than principled reason underlying the drive to eliminate government supervision of workplace bathrooms: "These deletions were included tentatively however, 'as means for OSHA to clearly indicate that its future enforcement efforts will be directed at high hazard industries.... Proctor also noted that the revocation of 1910.141(c) and table J-1 [which specifies how many toilets must be provided] would remove OSHA regulation of toilets which has 'subjected our compliance safety and health officers and other officials to ridicule and much criticism.'"[22]

---

[19]Eula Bingham, "The Responsibilities of OSHA," *Labor Law Journal* 29(8):487-92 at 487 (Aug. 1978). See also Bingham's congressional testimony in *Oversight Hearings on the Occupational Safety and Health Act: Hearings Before the Subcommittee on Compensation, Health and Safety of the Committee on Labor of the House of Representatives*, 95th Cong., 1st Sess. 294-320 (1977).

[20]Eula Bingham, "OSHA: Only Beginning," *Labor Law Journal* 29(3):131-36 at 131, 133 (Mar. 1978).

[21]When asked 24 years later, Proctor had no recall of the details of the toilet standard, which was a health standard, whereas he had been in charge of safety standards. Telephone interview with Barbara Belaski, OSHA Standards Office (Nov. 21, 2002). Belaski, who worked for Proctor in 1978, spoke to Proctor and transmitted this message from him. Belaski herself stressed that many of the standards had been revoked because they should never have been adopted in the first place as part of national consensus standards since they were covered by other agencies or dealt with public health issues.

[22]"Revocation of Laundry, Bakery Rules Among Suggestions by OSHA Work Groups," *Occupational Safety & Health Reporter* 7(14):420 (Sept. 1, 1977).

Finally, on December 5, 1977, Bingham formally announced OSHA's proposal to eliminate a large number of general industry health and safety standards and gave the public the opportunity to comment on it. Secretary of Labor Marshall called the initiative "'another significant step' in the redirection of OSHA enforcement toward serious workplace hazards."[23] Using the open-front toilet seat requirement as an example of the unnecessary regulations being eliminated, Marshall declared that of the thousands of accidents and illnesses suffered by workers annually, "'to the best of our knowledge, none...has been caused by the shape of a toilet seat....'"[24] In support of the plan to revoke the national consensus standards derived chiefly from ANSI standards, Bingham stressed that

> Congressional committees in the exercise of their oversight authority, have recommended that OSHA review its standards and revise those which are unclear or irrelevant to employee safety and health. [I]n the appropriations bill for fiscal year 1977 Congress directed OSHA to eliminate "so-called nuisance standards" or standards which do not deal with workplace conditions that are clearly hazardous to the health or safety or workers or are more properly under the jurisdiction of State Departments of Public Health."
>
> Both the Secretary of Labor and the Assistant Secretary for Occupational Safety and Health have indicated their commitment...to redirect its enforcement efforts towards more significant safety and health hazards. To initially implement this commitment, OSHA proposes to revoke, as expeditiously as possible under the Act, those regulations which most clearly have no direct or immediate relationship to the safety or health of employees.[25]

Among the criteria for selecting regulations meriting revocation mentioned by OSHA were: obsolete or inconsequential; concerned with comfort or convenience; directed toward public safety or property protection; subject to enforcement by other regulatory agencies; encumbered by unnecessary detail; and adequately covered by other general standards. In particular the penultimate deletion criterion, on which OSHA would draw in revoking part of the toilet standard, was justified as "permit[ting] employers greater flexibility in selecting the specific methods to abate these workplace hazards, including the development of new technology."[26] In addition, OSHA, exercising its power under the statute to issue de minimis notices rather than citations where a violation of a standard has "no

---

[23]"Revocation of 'Irrelevant' Standards Proposed by OSHA: Comment Period Set," *Occupational Safety and Health Reporter* 7(28):947 (Dec. 8, 1977).

[24]Richard Madden, "Safety Unit Would Drop 1,100 Rules," *N.Y. Times*, Dec.6, 1977, at 20, col. 1 (Proquest).

[25]*Federal Register* 42:62734 (Dec. 13, 1977).

[26]*Federal Register* 42:62735.

direct or immediate relationship to safety or health,"[27] announced that for the duration of the rulemaking proceedings it would not cite employers for violations of standards proposed for revocation.[28] As OSHA's Program Directive to field and national offices explained, "this policy will significantly relieve the concerns of the regulated employers with regard to those trivial workplace conditions which have no direct or immediate impact on safety or health."[29]

OSHA then proposed revoking the entire toilet standard, including the provision of toilet paper,[30] with the exception of: (1) the mandate that "toilet facilities...shall be provided in all places of employment," but stripped of its specification of how many toilets were required;[31] and (2) the injunction that "[t]he sewage disposal method shall not endanger the health of employees."[32] Significantly, this radically stripped-down provision would have conferred on employers more discretion than some had dared suggest in 1972 at the time of public hearings and comment on an earlier major revision of the sanitation standard, when no employer even raised the issue of sloughing off the obligation to furnish toilet paper.[33] At that time, for example, the National Association of Manufacturers, denying "that it was the intent of Congress to include the general sanitation standards in" OSHA, had argued that because they "are more appropriately matters of public health and, as such, have been traditionally regulated and controlled by local, municipal and state laws,...there is no desireability for pre-emption by the Federal Government."[34] Even more far-reaching had been the testimony of another employers association that one of its members' medical doctor had taken the position that: "'We do not know of any health or safety justification or basis for specifying the...number of toilet facilities....'"[35] To be

---

[27]29 USC sect 658(a).

[28]*Federal Register* 42:62735. See also OSHA Program Directive #200-68 (Dec. 2, 1977), in *Employment Safety and Health Guide*, ¶11,132 at 12,137-38 (Transfer Binder: Developments 1977-1978).

[29]OSHA Program Directive #200-67 (Dec. 1, 1977), in *Employment Safety and Health Guide*, ¶11,133 at 12,138-40 at 12,139 (Transfer Binder: Developments 1977-1978).

[30]29 CFR sect. 1910.141(c)(1(v).

[31]29 CFR sect. 1910.141(c)(1)(i).

[32]29 CFR 1910.141(c)(1)(iii); *Federal Register* 42:62737, 62801.

[33]On the hearing and comments, see Marc Linder and Ingrid Nygaard, *Void Where Prohibited: Rest Breaks and The Right to Urinate on Company Times* 57-61 (1998).

[34]United States Department of Labor, Occupational Safety and Health Administration, "Hearing on Proposed Revision of Sanitation Standards" at 39-40 (Nov. 8, 1972, Washington, D.C.) (testimony of Kenneth Schweiger, Director, Employee Relations, NAM).

[35]United States Department of Labor, Occupational Safety and Health Administration, "Hearing on Proposed Revision of Sanitation Standards" at 19 (testimony of Alvin

sure, other employers that had suggested that "no specific numbers...be used with reference to number of toilet facilities required," added the constraint "that the standard should require only 'adequate' or 'sufficient' toilet facilities."[36] And in 1972 the medical director of Eastman Kodak Company had advocated adoption of the following "performance standard," which many employers today might find uncongenial: "'Every place of employment shall be provided with a sufficient number of toilet facilities which shall be readily accessible to employees without unreasonable delay that would create an unsanitary condition or health hazard.'"[37] The fact of overriding significance about the 1972 hearings is that, despite these calls by some employers for elimination of OSHA's regulation of the number of toilets, no employer argued that firms were not obligated to let workers go to the bathroom; on the contrary, many of them detailed practices and proposals that seem downright pro-labor three decades later.[38]

After an extension, the public comment period ran until May 1978. Before the period expired, OSHA had received requests from labor organizations to discuss the revocations, and the agency did hold such meetings between March 13 and 23;[39] it offered the same opportunity to other interested groups, but none made such a request. OSHA officials presented its position on specific proposals for revocation at the meetings and a summary of them was entered into the public record on April 11. OSHA received a total of 230 comments, including 58 from trade associations, 89 from management, 22 from labor unions, 40 from government, and 21 from other interested persons. Based on these comments, OSHA decided to revoke 607 of about 700 general industry standards proposed for revocation.[40]

Mardon, Director, Public Affairs and Public Relations, Associated Industries of New York State, Inc.). To be sure, "without unreasonable delay" could also be interpreted as meaning that the number of toilets must be adequate to enable workers in fact to get to use them during scheduled mass breaks rather than requiring employers to let workers go individually whenever they need to.

[36]*Federal Register* 38:10931 (1973).

[37]United States Department of Labor, Occupational Safety and Health Administration, "Hearing on Proposed Revision of Sanitation Standards" at 228 (testimony of Norman Ashenburg).

[38]For a detailed exposition, see Linder and Nygaard, *Void Where Prohibited* at 57-61.

[39]Unfortunately, the very brief account in "OSHA, Unions Meet on Project: Final Decision to Be Based on Record," *Occupational Safety & Health Reporter* 7(4):1638 (Mar. 30, 1978), was skimpy, lacking any useful detail.

[40]*Federal Register* 43:49726-28 (Oct. 24, 1978). After considerable searching, OSHA's Docket Office located Docket No. S250 (containing the records of the standards deletion project including the hearings) at the National Archives in Suitland, MD. Unfortunately the materials there are stored in 67 boxes labelled only with arbitrary accession

While the distinctly pro-labor Representative George Miller (Democrat of California) was praising Bingham at a congressional OSHA oversight hearing for having "made a dramatic start in getting rid of frivolous regulation,"[41] labor unions, unsurprisingly, opposed the revocation initiative, AFL-CIO President George Meany even urging withdrawal of the whole project.[42] The UAW considered it "'something of an overreaction to a well-orchestrated attack by industry on the enforcement activities of OSHA.'"[43] Unassailable was the union's critique of OSHA's suggestion that the Act's general duty clause could be the basis for citing hazards no longer covered by individual standards: since the general duty clause applies only to hazards "that are causing or are likely to cause death or serious physical harm,"[44] the standard of proof would be "much higher." The United Paperworkers International Union objected that failure to retain the sanitation standards "would result in 'a very unhealthy condition for thousands of workers.'" The outgoing director of the National Institute for Occupational Safety and Health, Dr. John Finklea, stated the obvious in explaining NIOSH's opposition to the elimination of the sanitation standards in section 1910.141 on the grounds that they were "'designed to promote personal hygiene and to prevent the spread of communicable disease.'"[45] The director of industrial safety of the District of Columbia Minimum Wage and Industrial Safety Board echoed the sentiments of labor unions in arguing that the revocations were "'an arbitrary decision to offer to opponents a sign that OSHA intends to deregulate and be less offensive.'" Despite being recipients of these alleged deregulatory gifts, employers downplayed the revocations as "'essentially cosmetic.'" The Motor Vehicle Manufacturers Association, for example, urged a more comprehensive revision of Part 1910, recommending "the development of performance rather than specification standards."[46]

Why and how a performance standard would have been superior to specifying

---

numbers; Vanessa Reeves of the Docket Office and a co-worker travelled to Suitland and made a heroic effort to identify the materials on 1910.141(c), but discovered that the papers in the individual boxes lack any order, appearing as if they had been dumped into the boxes at random. Telephone interview with Vanessa Reeves, OSHA Docket Office (Nov. 18 and 21 and Dec. 13, 2002).

[41] *Congressional Record* 124:16672 (1978).

[42] "Meany Urges OSH to Withdraw Proposal: Unions Question Criteria Used by Agency," *Occupational Safety & Health Reporter* 7(51):1893 (May 18, 1978).

[43] "Auto Workers Union Urges Retention of Standards, Use of De Minimis Notice," *Occupational Safety & Health Reporter* 7(38):1393 (Feb. 16, 1978).

[44] 29 USC sect. 654(a)(1).

[45] "Auto Workers Union Urges Retention of Standards" at 1393, 1394.

[46] "OSHA Should Appoint Advisory Group on Revisions, Steelworkers Union Says," *Occupational Safety & Health Reporter* 7(4):1639-40 (Mar. 30, 1978).

the appropriate number of toilets, the employers organization did not explain. But the fact that two decades later employers complained that the OSHA Memorandum's performance (or reasonableness) standard was too vague suggests that the objection was and remains opportunistically fungible.

Without any substantive public explanation of its reasons, OSHA, "[u]pon reexamination of the evidence...determined that, in the best interest of worker protection," the following provisions "should not be revoked": section 1910.141 (c)(1)(i) (toilet facilities—general); Table J-1 (minimum number of water closets); 1910.141(c)(1)(ii) (mobile crews); and 1910.141(c)(2)(i) (requirement of doors, walls, or partitions high enough "to assure privacy").[47] The remaining provisions slated for revocation were revoked, including 1910.141(c)(1)(iv), which required an increase in the number of toilets according to the number of non-employees allowed to use them:

> Although a number of commentators opposed revocation of this provision, OSHA has concluded that agency authority in this area should extend only to requiring minimum numbers of facilities for emplyees [sic]. Use of such facilities in a place of employment by the public would be more generally a problem of public health, and any additional requirements for such shared facilities beyond the OSHA provisions would appropriately fall under the purview of local or state building codes or public safety and health authorities. Further, this particular provision is in large part inconsequential in that the provision lacks substantive requirements appropriate for mandatory enforcement by this agency with respect to determining the relative usage of the facilities by the public and the number of additional water closets that would there be necessary.[48]

The subsections requiring toilet paper and covered receptacles (for feminine hygiene products in women's bathrooms) were revoked without further explanation because OSHA "as a matter of policy has adjudged them to encumber this section with unnecessary detail"[49]—the revocation criterion must frequently used by OSHA.[50] Finally, OSHA revoked the provision that had prompted the greatest clamor and media comment[51]—the requirement of a split (open-front) toilet seat in all toilets installed or replaced after June 4, 1973.[52] The proliferation of toilet seats with an opening in the front had begun in the 1960s in an effort to prevent

---

[47]*Federal Register* 43:49737.

[48]*Federal Register* 43:49736.

[49]*Federal Register* 43:49736.

[50]*Federal Register* 43:49728.

[51]E.g., "OSHA Drops 928 'Nuisance' Rules," Facts on File World News Digest, Nov. 3, 1978, at 833 C3 (Lexis); "Why It's So Hard to End Regulations," *U.S. News & World Report*, Nov. 6, 1978, at 72 (Lexis).

[52]1910.141(c)(3)(ii); *Federal Register* 43:49736.

a gonorrhea epidemic that had spread from this source. This "concern seems somewhat warranted" from a biological standpoint because the gonococcal bacterium can remain viable in secretions on wood for more than three days, but several studies cast doubt on toilet seats as a source of infection.[53] In this context the seat mandate was neither one of OSHA's "absurdities" nor "meant only to promote comfort and convenience...."[54] Moreover, the source of the OSHA standard, the ANSI national consensus standard (the *American Standard Safety Code for Industrial Sanitation in Manufacturing Establishments*), had included it since at least 1935.[55]

A partial explanation of OSHA's revocation of the toilet standard came from a long-time OSHA industrial hygienist, Robert Manware, who worked on the revocation project in 1977-78. Within the larger framework of the regulatory deletions, he observed that, although toilet paper was "a nice thing to have," OSHA officials believed that providing it "was going to happen without federal intervention."[56] To be sure, Manware conceded that if OSHA had subsequently discovered that empirically significant numbers of employers were not providing toilet paper, perhaps it would have re-regulated the subject. Alternatively, he speculated that OSHA, even without a standard, might still be able to cite employers on the basis of the statutory general duty clause. Since he was willing to go so far as to conjecture that lack of toilet paper might cause the "serious physical harm"[57] required to trigger the application of the general duty clause, it would be ironic that OSHA had revoked such a standard as part of a campaign designed to eliminate regulations having no immediate relationship to safety or health.

Manware could not recall what labor unions must have said at the public meetings to persuade OSHA to rescind the revocation of section 1910.141 (c)(1)(i), but he suspected that one reason underlying the original revocation was the agency's belief that, in the light of a 1973 federal appeals court decision invalidating Table J-2, which specified the number of lavatories employers were

---

[53]Ingrid Nygaard and Jeffrey Lynch, "Public Restrooms: Past, Present and Future" 7-8 (unpub. MS, University of Iowa, College of Medicine, Environmental Health 175:197 (Fall 2001)).

[54]Timothy Clark, "The 'Facts' About OSHA's 1,100 Revoked Regulations," *National Journal* 10(32):1298 (Aug. 12, 1978) (Lexis).

[55]American Standards Association, *American Standard Safety Code for Industrial Sanitation in Manufacturing Establishments* sect. 3.13(c) (ASA Z4.1-1935, Apr. 1, 1935).

[56]Telephone interview with Robert Manware, OSHA, Washington, D.C. (Nov. 22, 2002).

[57]According to 29 USC sect 654(a)(1): "Each employer shall furnish to each of his employees employment and a place of employment which are free from recognized hazards that are causing or are likely to cause death or serious physical harm to his employees."

required to provide, on the grounds that OSHA had failed to provide substantial evidence supporting the proposal,[58] it would also not be able to persuade a court that it could empirically justify specifying a specific number of toilets.[59] However, such apprehension would have been inapposite in 1978 since the time period within which an employer could have judicially challenged the validity of the toilet standard had already expired in 1973.[60]

At the March 1978 hearings considerable opposition was voiced to the toilet standard, in particular, the requirement of a split seat, on the grounds that such matters were questions of public health, which state and local health agencies were competent to regulate. Many in OSHA perceived such complaints as pretexts for dismantling the agency altogether, just as they viewed the revocation project as an effort to blunt that attack. Indeed, at the time some officials at OSHA who were concerned about the revocations suspected that employers might try to contest citations by state or local health agencies on the grounds of federal preemption.[61]

Even before the revocation had gone into effect, an immensely amused President Carter could not refrain from praising the action: "In 1 day this year Eula Bingham terminated the application of 1,100 OSHA regulations. And although a few of those are still on the books—she has to go through a procedure to eliminate them—they are not being enforced. And I know Billy [Carter], at his service station, says that the OSHA inspections and regulations are much less onerous than they were before. [*Laughter*]"[62]

Jerry Purswell, who came to OSHA in June 1978 as the Director of the Directorate of Safety Standards Programs, found on his arrival a "tense mood" surrounding the deletion of "nitpicking" standards, which had "stalled" as a result of the opposition of industrial unions, which did not want standards revoked for which the unions would then have to bargain collectively with employers. Once this "impasse" had been reached, OSHA Administrator Bingham "dropped" the

---

[58]Associated Industries of New York State, Inc. v. United States Department of Labor, 487 F.2d 342 (2d Cir. 1973).

[59]Telephone interview with Manware.

[60]"Any person who may be adversely affected by a standard issued under this section may at any time prior to the sixtieth day after such standard is promulgated file a petition challenging the validity of such standard with the United States court of appeals...." 29 USC sect. 655(f). Rob Swain, OSHA Counsel U.S. Dept. of Labor Solicitor's Office, Washington, D.C. , agreed with the view expressed in the text concerning the time-barred preclusion of a challenge to the validity of the standard. Telephone interview with Rob Swain (Jan. 6, 2003).

[61]Telephone interview with Tom Seymour, former OSHA official (Nov. 18, 2002).

[62]"Interview," in *Public Papers of the Presidents of the United States: James Carter: 1978*, Book II, at 1588-96 at 1594 (1978 [Sept. 23, 1978]).

project in Purswell's "lap" and left it to him to "move it along." Purswell then met privately with the unions and explained to them that after all the swipes that the business press such as the *Wall Street Journal* and *Business Week* had made at OSHA, eliminating the standards with detailed specifications had become politically very important to Bingham. Purswell then succeeded in negotiating a settlement with the unions, which surprised Bingham.[63] (Bingham did "remember that Purswell sat with the unions and they were in agreement with what was published.")[64] Although he had no specific recall of how the negotiations with the unions over the toilet standard proceeded, Purswell did remember that it was one of the standards discussed and that if the provision prescribing the number of toilets was ultimately restored in the final rule, the reason was the unions' complaint that otherwise they would be forced to negotiate with employers over the subject as well as their fear that if the general industry standard were deleted, it would become that much more difficult to induce OSHA ever to promulgate a field sanitation standard for migrant farmworkers (which in fact took 14 years to issue). When asked how OSHA viewed future enforcement of such requirements as the provision of toilet paper following their revocation, Purswell—after mentioning, like Manware, the general duty clause and then quickly retracting the thought on realizing that it could not meet the "serious physical harm" standard—emphasized that Secretary of Labor Ray Marshall, not foreseeing unions' precipitous decline, had strongly insisted at the time that these sorts of standards were really part of labor relations and collective bargaining and not safety and health issues for OSHA to regulate.[65]

Sheldon Samuels, the director of safety and health at the Industrial Union Department of the AFL-CIO—whom Purswell identified as the coordinator of union representatives negotiating with him—while vehemently rejecting Purswell's characterization of unions' approach as trying to avoid having to bargain collectively over terms and conditions that OSHA had imposed by means of standards, did confirm that unions advocated adoption by OSHA of commonplace, community standards, such as those that a county health department would enforce, which would have included a specified number of toilets, toilet paper, and the other elements of the ANSI-OSHA standard. He also maintained that the unions and OSHA had an implicit understanding that, in spite of the revocation, the agency would continue to cite employers that failed, for example, to provide

---

[63]Telephone interview with Jerry Purswell (Nov. 23, 2002).

[64]Email from Eula Bingham to Marc Linder (Nov. 25, 2002).

[65]Telephone interview with Purswell. Though private, Purswell's "quiet talks with unions that have been offended by [OSHA's] deletion project," were not secret. Timothy Clark, "What's All the Uproar over OSHA's 'Nit-Picking' Rules?" *National Journal* 10(40):1594 (Oct. 7, 1978) (Lexis).

toilet paper. To be sure, Samuels could offer no legal basis for a regulatory agency's continuing enforcement of a regulation that it had expressly revoked,[66] let alone a real-world explanation as to how inspectors and supervisors in the twenty-first century who never knew that OSHA had briefly mandated toilet paper from 1971 to 1978 might become executors of this implicit understanding. Samuels, who, in addition to joining ranks with Manware and Purswell in suggesting the general duty clause as a proper vehicle for enforcing the revoked standard, found a silver lining even in an OSHA inspector's concluding that he or she lacked the power to cite an employer for failing to furnish toilet paper: somehow the fact of the firm's omission would find its way into the public domain and embarrass the employer.[67]

Peg Seminario, who is today the head of occupational safety and health at the AFL-CIO and worked on the standards deletion project in 1978, was able to "recall that our main approach to dealing with the standard' [sic] deletion project was to assess each of the proposed changes to determine if they would remove important safety and health protections, or were simply redundant or unnecessary requirements. The toilet standard would have been one that we would have argued should be kept."[68]

It is true that state and local governments have adopted the Uniform Plumbing Code and Uniform Building Code, which mandate various minimum numbers of water closets for many types of buildings, including some where people are employed.[69] How stringent enforcement of these requirements would be is another matter. In Iowa City, for example, the Building Inspection Division noted that its enforcement activities with regard to the number of toilets were focused on new construction and renovation; when asked about toilets in existing buildings, an official at first stated that the agency had no such jurisdiction, but then, after reflection, recalled that on one occasion the city had written a letter to a restaurant requiring that the toilets be unlocked following a complaint that, in order to prevent vandalism, it had locked its toilets at night while it was still

---

[66]Nevertheless, in 2002 OSHA in North Carolina, where the state legislature has adopted all Federal OSHA standards for the state's OSHA, N.C. Gen. Stat. sect. 95-131(a) (2002), cited an employer for having violated 1910.141(c)(1)(i)—"Toilet facilities were not provided in accordance with TABLE J-1"—because "no toilet paper was provided for the men's rest room." Asheville Metal Finishing, Inc., Insp. No. 305186124 (issued May 10, 2002).

[67]Telephone interview with Sheldon Samuels, Solomons, MD (Nov. 23, 2002).

[68]Email from Peg Seminario to Marc Linder (Nov. 26, 2002).

[69]International Association of Plumbing and Mechanical Officials, *Uniform Plumbing Code: 2000 Edition,* table 4-1 at 31-33 (2000); International Conference of Building Officials, *Uniform Building Code*, vol. 1: *Administrative, Fire- and Life-Safety, and Field Inspection Provisions* table A-29-A at 1-397-98 (1994).

open. With regard to the non-functioning of toilets (or even of the only toilet), the official observed that such a condition would have to have persisted for a considerable period of time (at least weeks) before the agency would intervene, and even then only on a case by case basis.[70]

Had this regulatory function been vacated by OSHA, it is difficult to imagine that any such public health agencies would ever have required an employer that maintained toilets in compliance with the codes to let workers (outside of the food-service industry and other public-health-related employments) stop working and use those toilets whenever they needed to void. Enforcing such rules at the intersection of health and employment is uniquely the province of an occupational safety and health agency. Under the headline, "U.S. Agency on Job Safety and Health Discards 928 'Nuisance' Rules," *The New York Times* ran an Associated Press article in 1978 alleging that OSHA "was glad to see the rules go. They never saved any lives or warded off any injuries or illnesses, as far as agency officials know." Although the paper quoted Purswell as stating that OSHA "had been unable to find a single case 'in which a woman became ill because of exposure to a sanitary napkin,'"[71] it apparently did not seek confirmation of the health-neutral consequences of non-use of toilet paper after defecation.

To be sure, some jurisdictions also require that toilet paper be provided for toilets,[72] which OSHA ceased to mandate in 1978. But, again, for example, in Iowa City, Inspection Services stated that it had no jurisdiction over the matter, while the Johnson County Public Health Department declared that it could require the provision of toilet paper only in the businesses it licenses—restaurants, bars, grocery stores, motels and hotels; in other buildings or workplaces it lacks jurisdiction.[73] When asked whether the agency would tell an employee who com-

---

[70]Telephone interview with Jan Ream, Inspection Services, Iowa City (Dec. 3, 2002).

[71]*N.Y. Times*, Nov. 25, 1978, at 24, col. 1 (Proquest).

[72]E.g., Connecticut Public Health Code, Regulations of Connecticut State Agencies 19-13-B108 (1985), requires toilet paper in public buildings, restaurants, large stores, and shopping centers. Similarly, in Alaska toilet paper must be provided in public facilities and facilities open to the public. Alaska Adm. Code 18:30.640 (2002).

[73]Telephone interview with Albert Moonsammy, environmental health specialist, Johnson County Department of Public Health, Iowa City (Dec. 3, 2002); telephone interview with Ralph Wilmer, Director, Johnson County Iowa Public Health Dept. , Iowa City (Dec. 4, 2002). The director identified the Food and Drug Administration's Model Food Code, which the Iowa legislature had adopted, as the source of this authority. The Model Food Code does state: "A supply of toilet tissue shall be available at each toilet." FDA, 2001 Food Code, sect. 6-302.11, on http://vm.cfsan.fda.gov/~dms/fc01-toc.html. The Illinois Food Safety Sanitation Code adopted this provision; 77 Ill. Adm. Code sect. 750.1110(d) (2002). However, neither the Iowa Code nor the regulations issued pursuant to them adopted this provision. Iowa Code sect. 137F.14 (1999); Iowa Adm. Code sect.

plained that he had himself been turned into a public health hazard by virtue of his employer's failure to provide toilet paper that unfortunately there was nothing the Public Health Department could do to correct the problem, its director acknowledged that the county general health nuisances regulation might be available for that purpose, but hastened to add that he was certain that the county attorney would refuse to prosecute on the grounds that the nuisance was too petty. He also strongly suspected that a similar gap in public health enforcement was typical throughout the United States.[74] The Peoria (Illinois) City/County Health Department confirmed that suspicion, adding that when it receives such "labor-management" complaints, it refers workers to OSHA.[75]

Failure to provide toilet paper in workplace bathrooms is not some hypothetical fabrication of a playful academic mind, though *The New York Times* seems to suspect that it is peculiar to one of the cheapest and "most dangerous employers in America...."[76] OSHA itself issued employers (including numerous large and nationally recognized firms) well over 1,100 citations for violating this standard between 1972 and 1978, when it was revoked.[77] Some state-plan OSHA

481.31.9 (1999). Informed of the absence of express statutory or regulatory authority for the toilet paper requirement, the surprised director looked up the codes himself and confirmed this fact, insisting commonsensically, however, that regardless of whether toilet paper was expressly required by the statute or regulation, it was part and parcel of providing a toilet. Telephone interview with Wilmer.

[74]Telephone interview with Wilmer. Under the regulation: "Health nuisance means any act, omission to act or condition which injures or threatens the health and safety of one or more persons and shall include but shall not be limited to a) Any business trade, manufacture or other operation or condition of property, which gives rise to noxious gases, vapors...or fumes which injure or threaten the health or safety of individuals or the public. b) The storage, collection, discharge, or depositing of any offal, filth, refuse or contaminated material in any place...so as to threaten the health or safety of individual or public, or to be conducive to the breeding and harborage of flies, rats, or other vermin." If the person fails to abate the nuisance, the board of health can cause the abatement of the nuisance with "[a]ll expenses incurred thereby...paid by the owner...or occupant of...property and the same shall be a lien upon the property." Johnson County Board of Health Regulation of Health Nuisances, sects. 1.3 and 5 (Aug. 22, 1968; amended June 20, 1974). The abatement provision would presumably mean that the county would buy the toilet paper and charge the owner in perpetuity.

[75]Telephone interview with unidentified 26-year veteran of Peoria City/County Health Department (Dec. 4, 2002).

[76]David Barstow and Lowell Bergman, "At a Texas Foundry, An Indifference to Life," *N.Y. Times*, Jan. 8, 2003, at A1, col. 1-2, at A14, col. 2 (nat. ed.) (reporting on Tyler Pipe).

[77]The search ("19100141 c01 v") conducted in Lexis-Nexis OSHAIR file (Dec. 25, 2002) identified 1,179 citations, which by and large ceased by September 1978; one from

agencies, in California[78] and Washington,[79] for example, have retained the toilet paper requirement in their OSHA programs; and from 1990 to 2002 Cal/OSHA issued 257 citations to employers for having failed to provide an adequate supply of toilet paper,[80] while the Washington State program issued 29 citations.[81] If even Cal/OSHA-imposed penalties in the range of $110-150[82] are efficacious in discouraging smaller employers from repeating such an act of disrespect, it must be presumed that imposition of a $1,200 fine in 2002 taught the offending employer a lesson.[83]

Nor is failure to provide toilet paper a feature exclusively of workplaces in the United States. According to a recent survey, one third of teachers, nurses, call center staff, home workers, and managers in Great Britain reported that they deferred defecating at work because of the condition of the toilets; six percent specified lack of toilet paper as the reason. As the result of a vicious circle, six percent also revealed that that very act of deferring made them constipated and thus unable to go easily.[84]

Nevertheless, in 1978 the *Washington Post* editorially applauded the death

---

1988 was presumably the result of a typo, while two from early 1979 may have been issued by inspectors who did not know that the standard had been revoked. The larger employers cited by OSHA included A&P, Albertson's, American Airlines, Armour, Chrysler, Coca Cola Bottling, Food Fair, Hartford Fire Insurance, Holiday Inn, J. C. Penney, Kroger, Levi Strauss, Newport News Shipbuilding, Penn Central Railroad, Pepsi Cola Bottling, Phillips Petroleum, Shop-Rite, Safeway (eight citations), U.S. Steel, United Parcel Service, and Youngstown Sheet and Tube.

[78]Cal. Adm. Code tit. 8, sect. 3364(d) (2002).

[79]Wash. Adm. Code sect. 296-24-12007(d) (2001).

[80]Search ("California and 3364(d)") conducted in Lexis-Nexis OSHAIR file (Dec. 2, 2002).

[81]Search ("Washington and 0241200701 d") conducted in Lexis-Nexis OSHAIR file (Dec. 5, 2002).

[82]E.g., Nice Art, Inc., Insp. No. 119944114 (Apr. 6, 2001); San Raphael Citrus Corp., Insp. No. 126213313 (Feb. 4, 1999); Sing Sing Fashion, Inc., Insp. No. 120309745 (May 4, 2001).

[83]JLH Industries, Insp. No.120166939 (Aug. 16, 2002). The firm made wooden kitchen cabinets. Of the 29 citations issued by Washington OSHA (WISHA) only two imposed monetary penalties, but they amounted to $540 and $1,500; both of these companies manufactured concrete products and the latter company, which violated numerous standards, was out of business before the abatement plan had been completed. Yakima Precast, Inc. Insp. No. 111208203 (Nov. 5, 1999); Pre-Mix Products of the Northwest, Insp. No. 115471229 (June 28, 1994).

[84]Christine Norton, "Loo Breaks at Work—Is There a Problem?" *European Health and Safety Magazine* 2(7):26-27 (Nov. 2002).

of "bad rules" and "nonsensical rules" "that should never have been born."[85] Looking back in 2003, an official of the Portable Sanitation Association International, the organization designated by ANSI as the standard developer of its Minimum Requirements for Sanitation in Places of Employment standard, when informed that OSHA had dropped the provision of toilet paper in the 1970s, was not at all surprised: she speculated that it had been deemed a "frill" or an "option," like air conditioning in an automobile.[86] That OSHA itself knew and knows better is obvious from the definition it crafted in 1987 for the field sanitation standard for farmworkers: "Toilet facility means a fixed or portable facility designed for the purpose of adequate collection and containment of the products of both defecation and urination which is supplied with toilet paper adequate to employee needs."[87]

Despite revocation of the requirement, "managements concerned about whether employees are loafing" may have developed an opportunistic interest in providing toilet paper after a "very large corporation in Manhattan" discovered that the "cylinder in the center of the plastic roller, which holds the tension spring for the roller, has a hollow space slightly more than three quarters of an inch wide and more than four inches long," which can house a self-contained miniature microphone transmitter making it possible for a "'security' aide at a remote post to listen in...."[88]

---

[85]"OSHA: Cleaning Up the Rules," *Washington Post*, Dec. 2, 1978, at A18 (Lexis).

[86]Telephone interview with Millicent Carroll, Portable Sanitation Association International, Bloomington, MN (Jan 17, 2003).

[87]29 CFR sect. 1928.110(b) (2002). Because the construction sanitation, which was issued in 1993, incorporated the general industry sanitation standard, from which the toilet paper provision had already been removed, it lacked the toilet paper requirement. *Federal Register* 58:35076 (1993).

[88]Vance Packard, *The Naked Society* 79 (1964).

5

Groups of Workers Not Even Covered by OSHA's Toilet Standard

It's remarkable how basic and persistent and extreme some deprivations of human rights are, while most of us rant about far more rarefied details of law and life.[1]

The applicability of OSHA's toilet access regulation is very broad. As the agency itself declares in an interpretive regulation: "The legislative history...clearly shows that every amendment or other proposal which would have resulted in any employee's being left outside the protection afforded by the Act was rejected. The reason for excluding no employee, either by exemption or limitation on coverage, lies in the most fundamental of social purposes of this legislation, which is to protect the lives and health of human beings in the context of their employment."[2] Nevertheless, OSHA's reach in general and that of its toilet standard in particular (and hence of the right to void at work) is not universal. It does not extend to: domestic household workers (who are without protection);[3] with some exceptions, state and local government workers in the 26 states in which Federal OSHA (which excludes public employees) has jurisdiction;[4] federal workers (except congressional employees covered by the Congressional Accountability Act of 1995[5] and postal workers); agricultural workers (who are covered by a set of field sanitation standards);[6] railroad train crews; or construction workers (who are covered by a less stringent set of standards).[7]

---

[1]Email from Prof. Michael Saks, Arizona State University College of Law, to Marc Linder (Mar. 24, 1998).

[2]29 CFR sect. 1975.3(a) (2002).

[3]29 CFR sect. 1975.6.

[4]29 USC sect. 652(5) (2000).

[5]2 USC sect. 1341.

[6]29 CFR sect. 1928.110. For the text, see below App. III.

[7]29 CFR sect. 1926.51. For the text, see below App. IV.

State and Local Government

Occupational safety and health obligations in the public sector are, as in other areas of the law in the United States, dysfunctionally balkanized. When Congress enacted OSHA in 1970, it excluded all Federal, state, and local governments from the definition of covered "employer."[8] At the same time, Congress authorized the Secretary of Labor to approve state plans for the development and enforcement of occupational safety and health standards that would "preempt applicable Federal standards."[9] One of the central conditions for approval of such state plans is that the standards themselves as well as their enforcement "be at least as effective in providing safe and healthful employment and places of employment" as Federal OSHA's standards.[10] In one respect, however, Congress required as a condition of approval that the state plans be more "effective" than Federal OSHA by mandating that such a plan in the judgment of the Secretary of Labor "contains satisfactory assurances that such State will, to the extent permitted by its law, establish and maintain an effective and comprehensive occupational safety and health program applicable to all employees of public agencies of the State and its political subdivisions, which program is as effective as the standards contained in an approved plan."[11] Consequently, while the 21 state-plan states (Alaska, Arizona, California, Hawaii, Indiana, Iowa, Kentucky, Maryland, Michigan, Minnesota, Nevada, New Mexico, North Carolina, Oregon, South Carolina, Tennessee, Utah, Vermont, Virginia, Washington, and Wyoming)[12] and the three states (Connecticut, New Jersey, and New York) with approved public-sector programs only,[13] cover their government employees, the 26 states under Federal OSHA's jurisdiction—which excludes the governmental sector at all levels—are not required to have such coverage and many do not.

An evaluation performed in 2000 by the Office of Audit of the Office of Inspector General of the U.S. Department of Labor of this "hodgepodge system"[14] underscored that "the public sector poses the same or even greater overall risk of

---

[8]29 USC sect. 652(5) (2000). On unsuccessful attempts to amend OSHA to cover state and local government employees, see *Occupational Safety and Health Law* 66-67 (2d ed.; Randy Rabinowitz ed. 2002).

[9]29 USC sect. 667(b).

[10]29 USC sect. 667(c)(2). See also 29 CFR sect. 1902.

[11]29 USC sect. 667(c)(6).

[12]29 CFR sect. 1952.

[13]29 CFR sect. 1956.

[14]U.S. Dept. of Labor, Office of Inspector General, Office of Audit, "Evaluating the Status of Occupational Safety and Health Coverage of State and Local Government Workers in Federal OSHA States" at 8 (Rep. No. 05-00-001-10-001, Feb. 9, 2000), on http://www.oig.dol.gov/public/reports/oa/2000/05-00-001-10-001.pdf.

workplace injury and illness as the private sector." The reason is that, contrary to popular misperception, the public sector does not consist only of desk jobs; large numbers of public employees work in hospitals, as police and fire fighters, corrections officers, vehicle mechanics, and wildlife workers, exposed to many hazards.[15] Whether this structural similarity in exposure to health and safety hazards also applies to toilet access is not clear. On a semi-anecdotal level, the chief legal counsel to the Arkansas Department of Labor reported that she has received about one call per week for 20 years from workers in the private sector wanting to know whether they have a right to go to the bathroom, but did not recall ever having received such an inquiry from a public-sector employee.[16]

The "Audit" concluded that nine of the Federal OSHA states (Arkansas, Illinois, Kansas, Maine, New Hampshire, Ohio, Oklahoma, Rhode Island, and Wisconsin), "had in place the staffing and legislation which contained the basic elements for a viable OSHA program...."[17] To be sure, in its response OSHA

---

[15]U.S. Dept. of Labor, Office of Inspector General, Office of Audit, "Evaluating the Status of Occupational Safety and Health Coverage of State and Local Government Workers in Federal OSHA States" at 11.

[16]Telephone interview with Denise Oxley, Chief Legal Counsel, Ark. Dept. of Labor (Oct. 25, 2002).

[17]U.S. Dept. of Labor, Office of Inspector General, Office of Audit, "Evaluating the Status of Occupational Safety and Health Coverage of State and Local Government Workers in Federal OSHA States" at 7. The audit in fact referred to 12 states, including Florida, New Jersey, and Guam, which are disregarded here because Guam is not a state, in the interim Florida abolished safety inspections for public employers (1999 Fla. Laws ch. 240, sect. 14 at 2148, 2165; 2001 Fla. Laws ch. 65, sect. 9 at 610, 612), and New Jersey's public sector program has been approved by Federal OSHA. The inclusion of Arkansas is inappropriate in the present context. First, according to the agency's chief legal counsel, the Arkansas Department of Labor's public employee safety and health program lacks a toilet standard that it could apply to county and local government employees (including teachers); and second, even with regard to state employees, the department is limited to enabling aggrieved workers to file grievances within a state grievance system. The state's non-codified industrial sanitation code for manufacturing does, however, contain potentially expansive access provisions. Telephone interview with Denise Oxley (Oct. 31, 2002); Ark. Code Ann. sect. 11-2-110 (2002); Ark. Dept. of Labor, Safety Code #6: Safety Code for Industrial Sanitation, sect. 6 (n.d.), on http://www.state.ar.us/labor/pdf/code6_industrial_sanitation.pdf. Finally, an older workplace safety and health law, empowering the department to make inspections, does not apply to the section on the provision of toilets, which is focused on separate facilities for men and women, not on the number of toilets, let alone access. Ark. Code Ann. sects. 11-5-107, 11-5-112. The Supervisor of Arkansas Occupational Safety and Health (AOSH) conceded that since AOSH lacks the power to cite employers for violating federal OSHA standards, since the American National Standards Institute toilet standard has not adopted

commented that it was "very concerned" that the audit's use of the word "acceptable" to describe the programs in these states "that may not fully meet or even come close to meeting the State plan approval requirements is inappropriate. Although there may be some awareness and some formal attention directed to the hazards these workers face in such States, the word 'acceptable' gives an impression that the programs are better than they may in fact be."[18] Unlike these nine states, of the remaining 17 Federal OSHA jurisdictions, Alabama and Delaware had no program, while the others (Colorado, the District of Columbia, Georgia, Idaho, Louisiana, Massachusetts. Mississippi, Missouri, Montana, Nebraska, North Dakota, Pennsylvania, South Dakota, Texas, and West Virginia), were deficient in one or more of the following areas: legislative or gubernatorial authority for a program; occupational safety and health standards equivalent to those of OSHA; methods for compelling compliance with those standards; a review system for contested cases; and/or adequate staffing.[19] In addition, seven states in this latter group (Colorado, Georgia, Louisiana, Mississippi, Missouri, Texas, and West Virginia) cover only state employees, but not local government workers.[20]

Among the nine Federal OSHA states classified by the "Audit" as maintaining OSHA-viable public employee programs, Wisconsin has enacted a statute that gives public employees rights equivalent to those provided by OSHA[21] and requires the Department of Workforce Development to adopt standards that provide protection at least equal to that provided to private-sector employees by OSHA.[22] The Ohio Public Employment Risk Reduction statute mandates adoption of Federal OSHA standards;[23] the statute does permit exceptions, but the

---

OSHA's Memorandum, and since the state's "antiquated" industrial sanitation code does not apply to state workers, he would lack the power to cite, for example, school districts for not letting teachers go to the bathroom when they need to, in spite of his personal knowledge that many an elementary school teacher is a "prisoner" in her classroom. Telephone interview with Michael Watson (Oct. 29, 2002).

[18]U.S. Dept. of Labor, Office of Inspector General, Office of Audit, "Evaluating the Status of Occupational Safety and Health Coverage of State and Local Government Workers in Federal OSHA States" at 17.

[19]U.S. Dept. of Labor, Office of Inspector General, Office of Audit, "Evaluating the Status of Occupational Safety and Health Coverage of State and Local Government Workers in Federal OSHA States" at 9, 25-27.

[20]U.S. Dept. of Labor, Office of Inspector General, Office of Audit, "Evaluating the Status of Occupational Safety and Health Coverage of State and Local Government Workers in Federal OSHA States" at 13-14.

[21]Wisconsin Statutes Annotated sect. 101.055(1 (2001).

[22]Wisconsin Statutes Annotated sect. 101.055(3).

[23]Ohio Revised Code ch. 4167.07 (2001).

state agency has in fact adopted OSHA's section 1910.141 (with one modification not pertinent here).[24] In Illinois, pursuant to the state occupational health and safety law applicable to state employees,[25] the state agency has adopted OSHA's general health and safety standards including Part 1910.[26] Maine,[27] Oklahoma,[28] and Rhode Island[29] have adopted or incorporated by reference Part 1910. In Rhode Island each state and local subdivision agency head is responsible for establishing and maintaining an "effective and comprehensive occupational safety and health program" and providing safe and healthful places and conditions of employment.[30] New Hampshire has adopted a code provision essentially identical to Federal OSHA's general industry sanitation standard.[31]

Finally, under a Kansas state law empowering the secretary of human resources to enter any workplace, including state agencies and public works, to investigate sanitary conditions and methods of protecting workers from dangers, the secretary can order the employer to make changes to protect workers endangered by conditions injurious to workers' health.[32] Although the law provides for assessment of penalties, the agency has not adopted rules for invoking them, apparently because, according to the head of the Industrial Safety and Health Unit of the Division of Workers Compensation, which is charged with ensuring public employees with a safe and healthy work environment: "Public sector employers have always voluntarily abated identified hazards." He also stated that, although he had never heard of OSHA's 1998 toilet standard Memorandum: "If a public sector employee (school teacher) were to bring a complaint (inadequate toilet facilities or breaks) to this section, an investigation would be conducted which would include employee and management interviews. 29 CFR 1910.141(c)(1)(i) could be cited and abatement required. This may or may not include the provision which you reference. Verification of the correction would be required and the employee could be contacted to assess the adequacy of the fix. No monetary penalty would be assessed." He suggested that a school could set up a system under which the principal or assistant principal or some other person would enter

---

[24]Ohio Adm. Code ch. 4167-3-01(A) and (L) (2001-2002).

[25]820 Illinois Comp. Stat. Ann. sect. 225/2 (1999).

[26]Code of Illinois Rules tit. 56, pt 350, sect. 280 (2002).

[27]Maine Rev. Stat. Ann. Tit. 26 sect. 561, 563 (1988); Code of Maine Rules 12 179 002-2 (2001).

[28]40 Okla. Stat. Tit. 40 sect. 401-25 (2001); Okla. Adm. Code sect. 380:40-1-2(a) (1999).

[29]Code of R.I. Rules sect. 16 050 002-4 (1999).

[30]R.I. Gen Laws sect. 28-20-10 (2000).

[31]N.H. Rev. Stat. Ann. ch. 277 (1999); Code of N.H. Rules, Labor 1403.76(a) (1998).

[32]Kansas Stat. sect. 44-636(a) (2000).

the classroom for a teacher who had to void, just as there are systems in place to substitute for a teacher who has to leave for any emergency.[33]

Although their programs are deficient in important ways, West Virginia[34] and the District of Columbia[35] have also adopted or incorporated by reference Part 1910, while Nebraska has created a Workplace Safety Consultation Law, which covers public and private employers subject to the state workers compensation statute, which empowers the state Department of Labor to "conduct workplace inspections...to determine whether employers are complying with" Federal OSHA standards.[36] However, as the manager of the program commented, the department lacks the power to fine an employer for violation of an OSHA standard; all it can do is attempt to persuade the employer to comply and, if the latter refuses, in the case of a public employer, to repeat the effort at persuasion at the next higher political level. Since state and local public employers are not subject to Federal OSHA jurisdiction, the department cannot even, as it can with private employers, refer the case to OSHA. Although he had never heard of OSHA's 1998 toilet standard Memorandum and repeatedly stated that he found it unbelievable that schoolteachers had toilet access problems, he declared that the next time a school was inspected—which does not occur often since inspections tend to focus on more hazardous workplaces—toilet access would be checked into.[37]

This reference to schoolteachers, some of whom may be required to remain with their elementary school pupils as long as three and a half hours without a break,[38] is important because they face special access problems and their medical consequences. A study by the authors of *Void Where Prohibited* of teachers in two school districts (Des Moines and Iowa City) revealed that 21.2 percent of those who made a conscious effort at work to drink less to decrease their voiding frequency reported a urinary tract infection during the preceding 12 months compared to only 9.9 percent among those who did not restrict their fluid intake.[39]

---

[33]Telephone interview with Rudy Leutzinger, head of Industrial Safety and Health Unit, Kansas Department of Human Resources, Topeka (Oct. 17, 2002); the quotations are taken from emails from Leutzinger to Marc Linder (Oct. 21 and 22, 2002).

[34]W.Va. Code Ann. sect. 21-3A-7 (1996); W.Va. Code of State Rules sect. 42-15-4.1 (1998).

[35]D.C. Code Ann. Sect. 32-1101 (2001); Code of D.C. Municipal Regulations tit. 7 sect. 2003.1(d) (1999).

[36]Neb. Rev. Stat. sect. 48-443(1), 446(2) (2001).

[37]Telephone interview with Bill Taylor, Workplace Safety Consultation Program Manager, Lincoln, NE (Oct. 17, 2002).

[38]Brian Tumulty, "Teachers Wrestle With Issue of Bathroom Breaks: OSHA Might Come to Rescue," Gannett News Service, Apr. 1, 1998 (Westlaw).

[39]I. Nygaard and M. Linder, "Thirst at Work—an Occupational Hazard?" *International Urogynecology Journal* 8:340-43 (1997).

This linkage was also confirmed by the president of the Houston Federation of Teachers, who has had to defend teachers disciplined for leaving their class to use the bathroom.[40] Public school teachers nevertheless have less legal protection than most workers because of the aforementioned exclusion of the public sector from Federal OSHA.[41]

Federal Government

Congress excluded the United States government from the definition of "employer" under OSHA,[42] but it did charge the head of every Federal agency (except the Postal Service)[43] with establishing and maintaining "an effective and comprehensive occupational safety and health program" and providing "safe and healthful places and conditions of employment, consistent with the standards set under section 655,"[44] which include OSHA's toilet standard. OSHA has itself promulgated regulations reinforcing that agency heads "shall comply with all occupational safety and health standards issued under section 6[55]...."[45] Moreover, a presidential Executive Order from 1980 requires agency heads to "[a]ssure prompt abatement of unsafe or unhealthy working conditions,"[46] but, while empowering the Department of Labor to conduct inspections,[47] it fails to provide for any enforcement such as monetary penalties or judicially enforcible abatement orders. Nevertheless, despite the lack of such enforcement procedures, according to Tom Marple, Director of OSHA's Office of Federal Agency Programs, since "everyone's working for the same boss," OSHA has other tools at its disposal; although theoretically this alternative enforcement path leads all the way to presidential intervention, such recourse has never proved necessary, and discussions at the deputy assistant secretary level of any federal department have

---

[40]L.M. Sixel, "Raise Your hand If You Need Relief," *Houston Chronicle*, Sept. 11, 1998, at 1 (Westlaw).

[41]Carrie Mason-Draffen, "Bad News: Bathroom Breaks Can Be Withheld," *Newsday*, Aug. 13, 2000, at F10 (Lexis), erroneously informed public school teachers complaining that they sometimes could not go to the bathroom for four hours that since New York State does not regulate public employees and defers to federal regulations, teachers have no rights.

[42]29 USC sect. 652(5).

[43]On the recent extension OSHA coverage to the U.S. Postal Service, see below ch.. 16.

[44]29 USC sect. 668(a)(1).

[45]29 CFR sect. 1960.16.

[46]Exec. Order No. 12196, sect. 1-201(c) (1980).

[47]Exec. Order No. 12196, sect. 1-401(i).

always proved sufficient to secure compliance.[48]

Agriculture

After what the D.C. Circuit Court of Appeals called OSHA's "disgraceful chapter of legal neglect"—14 years of "intractable...resistance to issuing" sanitation rules to protect farmworkers[49]—OSHA finally issued a field sanitation standard in 1987.[50] Because this standard is codified separately from the general industry sanitation standard, it is not covered by the 1998 interpretive Memorandum. And although it contains certain protective elements lacking in the industry toilet standard,[51] it is also subject to two serious limitations: it does not require employers to provide toilets to workers who work fewer than three hours per day[52] (two hours under Cal/OSHA[53] and without limits under Washington OSHA[54]) and applies only to an "agricultural establishment where eleven...or more employees are engaged on any given day in hand-labor operations in the field"[55] (five or more workers in California[56] and one or more in Washington[57]). Congress and OSHA withheld protection from these workers not because they were deemed invulnerable to health risks associated with urinary or fecal retention, but merely to benefit certain agricultural employers at the expense of their workers.

In the words of the Reagan administration OSHA:

> The requirement to provide toilets and handwashing facilities applies to all farmworkers covered under the scope of this standard, with one notable exception. The sanitation facilities need not be provided, under paragraph (c)(2)(v), to "employees who perform field work for a period of 3 hours or less (including transportation time to and from the field) during the workday." This exemption is limited to the provision of toilets and handwashing facilities and does not extend to drinking water. (Employers must provide drinking water, as specified in paragraph (c)(1), to all employees regardless of the length of their workday).

---

[48]Telephone interview with Tom Marple, Washington, D.C. (Oct. 21, 2002).

[49]Farmworker Justice Fund, Inc. v. Brock, 811 F.2d 613, 614 (D.C. Cir. 1987).

[50]29 CFR sect. 1928.110. For the full text, see below Appendix III.

[51]See below ch. 6.

[52]29 CFR sect. 1928.110(c)(2)(v).

[53]8 Cal. Code of Regulations sect. 3457(c)(2)(A) (2002).

[54]Wash. Adm. Code sect. 296-307-095 (2002).

[55]29 CFR sect. 1928.110(a).

[56]8 Cal. Code of Regulations sect. 3457(c)(2)(A).

[57]Wash. Adm. Code sect. 296-307-09503.

The Migrant Legal Action Program, asserting that workers can be seriously infected from human feces or poisoned by pesticide residues even if they work for only one or two hours per day, opposed the proposed, or any exemption based on the number of hours worked.... Another farmworker representative pointed out that if an exemption based on time were allowed, it should include travel time to and from the fields, especially as the travel almost always is over rural, isolated roads that do not have rest stations.... Agricultural trade associations generally support this provision, arguing that requiring employers to provide toilet and handwashing facilities for part-time employees would be burdensome and unnecessary.... Moreover, OSHA's expert witness on urinary tract infections, Dr. Anemias [sic; should be Ananias] C. Diokno testified that, from a medical point of view, 3 hours was about the average safe amount of time urine could be withheld before bacteria multiply and blood flow decreases due to pressure from the overstretched bladder....

Consequently, OSHA has concluded that an exemption for part-time work is appropriate, because the risks are less to employees and because, under the circumstances, requiring provision of the facilities would be unnecessarily burdensome to employers. Nevertheless, the Agency believes that the time should be strictly limited in order to adequately protect farmworkers. OSHA's 1976 field sanitation proposal exempted field work of two hours or less (41 FR 17576). The 1984 proposal raises the period to three hours per work day, but adds the requirement that travel time to and from the field be included in the total (49 FR 7605), which assures farmworkers that they will not be away from facilities for more than three hours.[58]

In fact, OSHA misstated the history of the standard. In 1976, when OSHA Administrator Morton Corn was animated by the recognition both that farm workers "have the same physiological and hygienic needs, and are exposed to similar health hazard risks as are their industrial counterparts" and that Congress intended "to bring agricultural employees into the mainstream of the American labor force,"[59] he proposed exempting field work lasting less than two hours including travel time to and from the workplace.[60] The claim by the Reagan administration OSHA in 1984 that exempting work lasting three hours or less was appropriate "since there is little need for most workers to relieve themselves at the worksite when the work is for very short time periods and when facilities presumably are available before and after work"[61] was and is simply factually incorrect. Even if there is no reliable medical evidence that waiting three hours to urinate causes increased bacterial loads (and thus urinary tract infections), it may still cause pain, painful urgency, or urinary incontinence.[62] Indeed, even

---

[58]*Federal Register* 52:16050-96 at 16090 (1987).

[59]*Federal Register* 41:17576-79 at 17577 (1976).

[60]*Federal Register* 41:17578 (29 CFR sect. 1928.110(d)(iv)).

[61]*Federal Register* 49:7589, 7600 (1984).

[62]Email from Dr. Ingrid Nygaard to Marc Linder (Jan 10 and 13, 2003). For medical

OSHA's expert witness from the 1980s, Dr. Ananias Diokno, agreed in retrospect that, despite his testimony about bacterial infections,[63] if workers have to go more often than every three hours, "we should not corral them...they should be allowed to go."[64] And although expensive medications with side effects could possibly reduce voiding frequency, it is unclear why workers should be forced to bear that double burden instead of being permitted to urinate when they need to.[65]

Moreover, OSHA statements from the 1980s to the contrary are suspect because they were made in connection with the Reagan administration's systematic refusal to issue any field sanitation standard at all. Typical of the reasons or excuses that OSHA offered up in its last-ditch effort in 1985 were the following: "whatever risk that is present from lack of adequate sanitation in fields where people work (with the exception of heat stroke) is extremely difficult to separate from the risk present from non worksite conditions of these same workers"; "[i]t is injudicious to take inspectors out of high-hazard worksites and put them in farmer's [sic] fields":"private-sector market incentives also operate to promote improved public health."[66]

The efficacy of market incentives has apparently left something to be desired: OSHA has cited agricultural employers 1,836 times for failing to provide toilet and/or handwashing facilities since the field sanitation standard went into effect,[67]

---

sources, see below ch. 13.

[63]Diokno's testified at the OSHA hearing in the mid-1980s that "[i]t is expected that a normal person in general should void approximately four to six times a day to avoid [bladder] overdistension." Ananias Diokno, "Urinary Tract Infection" at 6 (n.d. [ca. 1984]), in OSHA, Docket No. H-308, Doc. No. 23 (copy furnished by OSHA). Dr. Nygaard testified at the Jim Beam OSHA hearing in 2002 that she was not familiar with any study showing as few as three to six voids per day. See below ch. 13.

[64]Telephone interview with Dr. Ananias Diokno, Chief of Urology, William Beaumont Hospital, Royal Oak, MI (Jan. 16, 2003).

[65]Email from Dr. Ingrid Nygaard to Marc Linder (Jan. 17, 2003). Diokno—who in 2003 was not familiar with the latest and most reliable urinary frequency study by FitzGerald et al. (see below ch. 13), which challenges the received wisdom—expressed the opinion that urinating every 3-4 hours is normal except among people with very small bladders (under 300 cc); he regarded more frequent urination as "disruptive," mentioning in particular problems that arise when people are out shopping and have to know where all the bathrooms are and especially for women, who, with all the "contraptions," might take 15 minutes to urinate. He recommended that such frequent urinators see a doctor for drugs or behavioral therapy to reduce their frequency; although he conceded that the medication has side effects, he did not believe that it was doing people a favor to let them continue to urinate frequently.

[66]*Federal Register* 50:15086-92 at 15090, 15092 (1985).

[67]Lexis-Nexis OSHAIR file search ("19280110 c02 i") conducted on Dec. 31, 2002.

with an additional 728 citations issued by Cal/OSHA since 1992.[68]

Another legal disability to which farmworkers are subjected is time and distance. The field sanitation standard provides: "Toilet and handwashing facilities shall be accessibly located and in close proximity to each other. The facilities shall be located within a one-quarter-mile walk of each hand laborer's place of work in the field."[69] California[70] and Washington[71] have created the alternative of five minutes. Federal OSHA had originally proposed a 5-minute rule as well, but eventually dropped it in favor of the quarter-mile rule.[72]

OSHA was well aware of the negative impact of distance on use. In promulgating the field sanitation standard in 1987 that established the quarter-mile walk standard, the Reagan administration OSHA expressly stated that distance was "a major concern of both employers, who often deal with perishable crops, and farmworkers, who usually are paid on a piece-work basis."[73] This lumping together of employers and farmworkers is curious since a maximum legal distance of a quarter-mile imposes no restriction on employers, which are free to buy or rent more toilet facilities and to space them closer than a quarter-mile away if they wish. In contrast, farmworkers who fear losing piece wages if they have to take a long walk to urinate are powerless to engage in self-help by acquiring additional toilet facilities and positioning them closer.

Although it might seem plausible that agricultural employers paying hand harvesters on a piece rate might be indifferent to how often workers stop to void, New Mexico OSHA and Cal/OSHA reported that farm labor contractors often do not permit workers to walk across the field to a port-a-potty because they insist on finishing the harvesting in that field by a certain time.[74] According to the litigation director of California Rural Legal Assistance: "As with many of these farm labor issues, lack of access to the toilets is usually less a function of direct refusal to allow a worker to use them (when they are there) and more an issue of the pressures of piece rate, etc creating a situation where workers fear they will lose pay or preferential work assignments if they take breaks for anything."[75]

---

[68]Lexis-Nexis OSHAIR file search ("3457 c02 a") conducted on Dec. 31, 2002.

[69]29 CFR sect. 1928.110(c)(2)(iii) (2002).

[70]8 Cal. Code of Regulations sect. 3457(c)(2)(D) (2002) (whichever is shorter).

[71]Washington Adm. Code sect. 296-307-09506 and 296-307-09518(8) (2002).

[72]*Federal Register* 41:17576-79 at 17578 (29 CFR sect. 1928.110(d)(2)(i)); *Federal Register* 52:16050-96 at 16090 (1987).

[73]*Federal Register* 52:16090.

[74]Telephone interview with George Vigil, program manager for statistics section, New Mexico OSHA, Albuquerque (Sept. 20, 2002); telephone interview with Susana Freund, Cal-OSHA, Ventura (Oct. 30, 2002).

[75]Email from Cynthia Rice, CRLA, San Francisco, to Marc Linder (Dec. 12, 2002).

Toilet access is a problem, for example, in California's Central Valley, where employers "often don't provide enough toilets for farm workers" or ones that are so dirty that workers try not to void during nine-hour shifts. As one female field worker put it: "'It was either cover your nose and mouth and go in there, or have people looking at me sitting in the field.'"[76] The lack of toilet facilities for farmworkers is common in West Texas, too.[77] During a United Farm Workers organizing campaign at a mushroom farm in Quincy, Florida, the employer instituted a procedure of counting bathroom breaks weekly as a disciplinary measure designed to discourage workers from backing the union.[78]

In Washington State one problem for farmworkers, according to one of their public interest lawyers, is the practical unavailability of toilets. Whereas some farmers provide portable toilets that they move within a field or apple orchard, thus complying with state OSHA's five-minute walk or quarter-mile rule, others have toilets that are in one point in the field or orchard. If, as the farmworkers work, they move farther from the toilet, often more than five minutes or a quarter-mile, they "merely relieve themselves in the fields." This problem is exacerbated when workers are doing piece work: because "a break is on their dime,...few will use facilities if they are too far away. I noted in [the] harvest this year that when workers are not making minimum wage while working piece rate, the employer fires them. Once farmworkers are aware...that they will be discharged under such circumstances, how far/close a facility is may be moot as they will relieve themselves in the orchard/field to make minimum wage and save their jobs."[79]

To be sure, farmworkers are not the only group of employees whose voiding

---

[76]"Bathroom Access Causes Work Troubles," *Des Moines Register*, Mar. 13, 1998, at 2M, col. 2.

[77]Email from Polly Bone, Branch Manager, Texas Rural Legal Aid, Plainview, TX, to Marc Linder (Dec. 17, 2002).

[78]Rebecca Schleifer, Review of *Void Where Prohibited*, in *University of Pennsylvania Journal of Labor and Employment Law* 2:603-608 at 607 (2000).

[79]Email from Patrick Pleas, Northwest Justice Project, Wenatchee, to Marc Linder (Dec. 11, 2002). Under the Fair Labor Standards Act it is unlawful to dock a worker's wages for breaks less than 20 minutes—but only to the extent that the worker's wage is pressed below the minimum wage or time and a half for overtime over an entire pay period. Although in general it would be lawful for an employer to fire an at-will employee who failed to reach the minimum wage on a piece rate (provided that the employer paid him the minimum wage for however long he had worked), it would be unlawful retaliation if the reason for not having achieved the minimum wage were the worker's having exercised his right under OSHA to go to the toilet (just as it would be unlawful to press the wage below the minimum by docking wages). 29 USC sect. 660(c)(1); see below ch 17.

frequency is adversely affected by the piece-rate form of compensation. As a recent study of women workers at a large pottery manufacturing plant revealed, even where 85 percent of respondents reported that the bathroom was easily accessible, 83 percent that there was no time limit on bathroom breaks, and 87 percent that they did not have to ask permission to take a break to go to the toilet, nevertheless 87 percent also reported "waiting until the last minute to go to the bathroom," and only 78 percent went to the bathroom even three or four times—that is, at least one time in addition to the scheduled 20-minute lunch and 10-minute break—during an eight hour day The authors conjectured that the piecework-bonus incentive wage system "may have contributed to a reluctance to take rest breaks in order to increase production...."[80] In other words, some firms may be in a position to undermine the effectiveness of OSHA's imposition on employers of an obligation to let workers go when they need to go by making workers internalize the compulsions of uninterrupted production and by shifting to them the responsibility for not exercising their right to void.

Railroads

Special voiding problems of railway train crews surfaced after *Void Where Prohibited* went to press. Most prominent was the "'disgusting and barbaric'"[81] toilet system that the Norfolk Southern Railway maintained in its locomotives. In 1997 a class action was filed on behalf of 5,000 train crew members (over whom OSHA does not exercise jurisdiction)[82] seeking monetary damages and an

---

[80]Sheila Fitzgerald, Mary Palmer, Victoria Kirkland, and Leslie Robinson, "The Impact of Urinary Incontinence in Working Women: A Study in a Production Facility," *Women & Health* 35(1):1-16 at 5, 8-9 (quote), 13 (quote) (2002). On especially hot summer days the workers were given one additional 10-minute rest break. *Id.* at 5. Although the published article does not specify this detail, Fitzgerald stated that the survey asked only whether respondents voided 1, 2, 3, or 4 times daily; 68.1 percent of those who answered voided 3 or 4 times. Email from Sheila Fitzgerald to Marc Linder (Oct. 3, 2002). Although the questionnaire asked respondents how often they voided, the article referred to the number of times; when asked how the authors converted the data, Fitzgerald replied: "We did not do any conversions and now that I think about it the information is probably useless. The better question is 'how many times do you use the BR at work?'" Email from Sheila Fitzgerald to Marc Linder (Oct. 4, 2002).

[81]Christopher Dinsmore, "Norfolk Southern Crew Members Sue the Railroad Over Commodes," (Norfolk) *Virginian-Pilot*, Aug. 27, 1997, at D1 (Lexis) (quoting lawsuit).

[82]OSHA does not "apply to working conditions of employees with respect to which other Federal agencies...exercise statutory authority to prescribe or enforce standards or regulations affecting occupational safety or health." 29 USC 653(b)(1). The Federal

end to the use of bags that line a commode and that are afterwards placed in a bucket, to be disposed of later. The Norfolk Southern is the only railroad that does not provide chemical or flush toilets on long-haul service locomotives, which were not required by federal law to have toilets[83] until the Federal Railroad Administration of the Department of Transportation promulgated a regulation in 2002 phasing in conforming toilets.[84]

Construction

Because the construction industry is not covered by the general industry sanitation standard, it is also not subject to Federal OSHA's Memorandum embodying workers' right to void at-will. Many construction workers complain that they "are subjected to the lack of facilities, or disgusting portable toilets that they are compelled to use, often without toilet paper, and almost always the stench is unbearable."[85] The special OSHA standard for the construction industry provides that toilets at construction jobsites "shall be provided for employees" according to a table mandating at least one toilet for 20 or fewer employees, one toilet seat and one urinal per 40 workers for 20 or more workers, and one toilet seat and one urinal per 50 workers for 200 or more workers. Jobsites lacking a sanitary sewer "shall be provided with one of the following toilet facilities unless prohibited by local codes": privies or chemical, recirculating, or combustion toilets. However, these requirements "shall not apply to mobile crews having transportation readily available to nearby toilet facilities."[86]

Railroad Administration (FRA) has such statutory authority, 49 USC sect. 20101 (2002), and in 1978 it published a policy statement declaring that it was exercising such authority, thus preempting OSHA. However, the FRA also stated that since OSHA's general industry sanitation standard was generally applicable to railroad workplaces, a certain overlap existed, as a result of which consultation between the agencies concerning sanitation in rolling stock was called for. 43 *Federal Register* 10583-90 at 10588-89 (1978). It is this state of affairs that prompted OSHA's jurisdiction expert to characterize the likelihood of preemption as a "resounding maybe." Telephone interview with John Solheim, Washington, D.C. (Dec. 12, 2002). Nevertheless, under a recent U.S. Supreme Court decision, the FRA appears to have validly triggered the statutory preemption of OSHA. Chao v. Mallard Bay Drilling, Inc., 534 U.S. 235 (2002).

[83]Dinsmore, "Norfolk Southern Crew Members Sue the Railroad Over Commodes."

[84]67 *Federal Register* 16032 (2002); 49 CFR 229.137 (2002).

[85]Susan Lippman, Book Review of *Void Where Prohibited*, *NY Hard Hat News*, Apr. 2001, at 7.

[86]29 CFR sect. 1926.51(c)(1), (3), and (4) (2001). For the full text, see below Appendix IV.

According to labor union officials, the problem that construction workers face is not—in the union or nonunion sector for that matter—so much denial of permission to stop work to urinate, as is the case in manufacturing or some service industries. Rather, it is the lack of a toilet altogether and the lack of handwashing facilities, especially after defecation. According to Stephen Cooper, who was international safety director of the Ironworkers Union and a member of OSHA's Advisory Committee on Construction Safety and Health (ACCSH) from 1976 until his retirement in 2001, most of the larger construction employers do provide adequate port-a-potties, but almost none provides handwashing facilities—which OSHA does not require on-site except for workers applying "paints, coating, herbicides, or insecticides, or in other operations where contaminants may be harmful to the employees"[87]—which he estimated would cost about $300

---

[87]29 CFR sect. 1926.51(f)(1). Subsection (f)(3) does state that "[l]avatories shall be made available in all places of employment. The requirements of this subdivision do not apply to mobile crews or to normally unattended work locations if employees working at these locations have transportation readily available to nearby washing facilities which meet the other requirements of this paragraph." It then goes on to provide that "[e]ach lavatory shall be provided with hot and cold running water, or tepid running water." In fact, however, OSHA does not enforce this provision; instead it in effect applies the "transportation readily available to nearby washing facilities" exception to all construction sites, despite the fact that in a standard interpretation antedating the adoption of the lavatory provision OSHA had already expressly found that, with regard to the provision of toilets, "'mobile crews' job functions require continual or frequent movement from jobsite to jobsite on a daily or hourly basis. Such is not the normal situation for work crews involved in housing construction." OSHA, Standard Interpretation, 1926, Letter from Roy Gurnham, Director, Office of Construction and Maritime Compliance Assistance, to William Carroll (Apr. 19, 1993), on http://www.osha.gov. OSHA created this nonenforcement mode by means of a standard interpretation declaring that subsections (f)(2)-(4) "only apply to permanent places of employment. The general scope statement (1910.14(a)(1)) [of the general industry sanitation standard] limiting the application of these provisions was inadvertently omitted in the June 30 [1993] Federal Register publication. The Office of Construction and Civil Engineering Safety Standards is in the process of correcting this statement." OSHA, Standard Interpretation 1926.51 (Feb.10,1994), on http://www.osha.gov/pls/oshaweb/owadisp.show_document?p_table=INTERPRETATIONS&p. In fact, OSHA has never corrected the situation, in part because correction would require amending the standard by opening it for notice and comment. Because new construction sites are, virtually by definition, nonpermanent, OSHA now treats virtually all new construction sites as nonpermanent places of employment, and the lavatory standard is a nullity. Telephone interview with Calvin Branch, Occupational Safety and Health Specialist, Construction Standards and Guidance, OSHA, Washington, D.C. (Jan. 13, 2003). OSHA has issued only 19 citations for violations of section 1926.51(f)(3)(ii), of which only three were issued by Federal OSHA after 1993. Search ("19260052 f03 ii")

per week for a self-contained trailer with its own (non-running) water supply and hot water. Among smaller construction employers, including those in the union sector, failure to provide the required number (or even any) toilets is common. Since many construction workers defecate during the workday and the increasing number of female workers change sanitary napkins without being able to wash their hands, feces- and blood-borne pathogens pose a significant health risk for workers when they eat or smoke. These circumstances do give rise to the one instance in which employers do deny permission to stop work to void: if workers needing to defecate have no toilet altogether and decide to drive (as far as 10 to 25 miles and back) to a gas station or convenience store, employers may fire them.[88]

Yet here, too, strong unions and worker militance can make a difference. The business manager of Local 948 of the International Brotherhood of Electrical Workers in Flint, Michigan reported that if the on-site port-a-potties are filthy, some members will tell the employer that they are not going to put up with the conditions and then drive to the closest public bathroom, even though they are paid by the hour, and the employer will not even attempt to dock their wages. Conversely, he observed that workers paid on a piece-rate in the nonunion construction sector would definitely be discouraged from voiding by the loss of wages.[89]

Cooper's view notwithstanding, however, when the ACCSH—with which OSHA, by its own procedural rule, is required to consult in issuing, modifying, or revoking a construction industry standard[90]—discussed this very issue on

---

in Lexis-Nexis OSHAIR file (Jan. 13, 2003). In late 2002 the Cal/OSHA Standards Board adopted a proposal to require at least one washing station per 20 employees in construction; the stations have to be located so that workers can readily wash in connection with using a toilet; mobile work crews having available transportation to such facilities are excluded. "An Ironworker's Success: Construction Hand Washing Proposal Adopted by Cal-OSHA Standards Board," *Cal-OSHA Reporter* 29(46) (Nov. 29, 2002) (Lexis).

[88]Telephone interview with Steve Cooper (Sept. 18, 2002).

[89]Telephone interview with Charles Marshall, business manager, IBEW Local 948, Flint, MI (Oct. 29, 2002).

[90]29 CFR sect. 1911.10(a) (2002). Pursuant to OSHA: "An advisory committee may be appointed by the Secretary to assist him in his standard-setting functions under section 6 of this Act. Each such committee shall consist of not more than fifteen members and shall include as a member one or more designees of the Secretary of Health, Education, and Welfare, and shall include among its members an equal number of persons qualified by experience and affiliation to present the viewpoint of the employers involved, and of persons similarly qualified to present the viewpoint of the workers involved, as well as one or more representatives of health and safety agencies of the States. An advisory

January 29, 1999, it decided to include a provision requiring employers to let workers go to the toilet "on an as-needed basis" in the draft proposal to revise OSHA's sanitation standard for construction. The crucial colloquy took place between two members, Larry Edginton, the Director of Safety and Health of the International Union of Operating Engineers, and the chairman of the ACCSH, Stuart Burkhammer, Vice President and Manager of Safety and Health Services, Bechtel Corporation:

MR. EDGINTON: [T]he other issue I had again has to do with access to the facilities. And I don't think we need to be talking about access to facilities under limited access work sites. We need to be thinking about it in terms of all work sites.... And perhaps, under table D-2, we should add language that says that employees shall be allowed to use the facilities on an as-needed basis. We should not restrict this question of use to simply mobile sites. It should be treated as all sites.
CHAIRMAN BURKHAMMER: Maybe, I'm naive and I haven't been in this business very long. But I have never, ever, ever heard an employer tell an employee they can't go to the bathroom. If you got to go, you've got to go, you know. I mean, I --
MR. EDGINTON: I used to think that, too, Mr. Chairman, until I had people start telling me it was so.
CHAIRMAN BURKHAMMER: It's a sad world we live in. All right. Where would you like to put that, Larry?
MR. EDGINTON: Under table D-2. Perhaps, this could be a 1926.51(c)(2). ...
CHAIRMAN BURKHAMMER: And what's your wording?
MR. EDGINTON: Employees shall be allowed to use toilet facilities on an as-needed

---

committee may also include such other persons as the Secretary may appoint who are qualified by knowledge and experience to make a useful contribution to the work of such committee, including one or more representatives of professional organizations of technicians or professionals specializing in occupational safety or health, and one or more representatives of nationally recognized standards-producing organizations, but the number of persons so appointed to any such advisory committee shall not exceed the number appointed to such committee as representatives of Federal and State agencies." 29 USC sect. 656(b). Furthermore, 29 CFR sect. 1912.3(a) provides: "This part applies to the Advisory Committee on Construction Safety and Health which has been established under section 107 of the Contract Work Hours and Safety Standards Act (40 U.S.C. 333), commonly known as the Construction Safety Act. The aforesaid section 107 requires the Secretary of Labor to seek the advice of the Advisory Committee in formulating construction standards thereunder. The standards which have been issued under section 107 are published in Part 1926 of this chapter. In view of the far-reaching coverage of the Construction Safety Act, the myriad of standards which may be issued thereunder, and the fact that the Construction Safety Act would also apply to much of the work which is covered by the Williams-Steiger Occupational Safety and Health Act of 1970, whenever occupational safety or health standards for construction activities are proposed, the Assistant Secretary shall consult the Advisory Committee."

basis.
CHAIRMAN BURKHAMMER: Shall be allowed. Is that what you said, shall be allowed? Steve, do you accept that recommendation in the motion?
MR. COOPER: Yes.[91]

The well-informed *Occupational Safety and Health Reporter* quoted a National Institute for Occupational Safety and Health representative as saying that "'unbelievable as it may be,' there are still some employers who have toilet facilities for their employees but are unwilling to let their employees take time out to use them. This practice is 'unacceptable and must be stopped,' ACCSH members contended."[92]

Despite the fact that ascertaining whether a construction employer has a toilet on the construction site or not would appear to be the easiest imaginable inspection action, advocates such as Jane Williams assert that OSHA refuses to enforce the standard and has failed to issue any citations for violations for ten years. It seeks, in her view, to justify its inaction on the grounds that it has higher enforcement priorities in the construction industry devoted to avoiding serious injuries and deaths, neither of which results from a lack of toilets.[93] Contrary to Williams's claim, however, OSHA has issued 845 citations in the years after 1992 for violations of the construction sanitation standard requiring the provision of a prescribed number of toilets.[94]

For a number of years, the OSHA ACCSH has been proposing stricter sanitation standards. In 1997 it set up a subcommittee to consider the need for restrooms and washing facilities, the absence of which, according to Cooper, its chair, bordered on third-world conditions.[95] It called for a revision of the sanitation standard "because too many construction workers are denied basic sanitation

---

[91]Transcript, U.S. Department of Labor, OSHA Advisory Committee on Construction Safety and Health , Vol. 2, January 29, 1999, on http://www.osha.gov/doc/accsh/transcripts/jan29-99.html. Although the Comments on Draft Sanitation Standard occupy pages 304-85 according to the index, the website version contains no pagination.

[92]"Workers to Use Toilet Facilities as Needed Under New Proposed OSHA Sanitation Rule," *Occupational Safety and Health Reporter* 28(34):1076 (Feb. 3, 1999).

[93]Telephone interview with Jane Williams, OSHA Advisory Committee on Construction Safety and Health, Scottsdale, AZ (Sept. 18, 2002). In contrast, it is true, as noted above, that OSHA has effectively nullified its lavatory standard for the construction industry.

[94]Search ("date is aft 1992 and 19260051 c01") in Lexis-Nexis OSHAIR file (Dec. 29, 2002). A small proportion of these cases was opened before 1993 and perhaps some of the citations were issued then as well, though the cases were not closed until after 1992.

[95]"OSHA Group Mulls Restrooms, Decon," *Impact on Construction Safety and Health* 15(2):2 (Sept. 1997).

rights."[96] On Oct. 7, 1998, the ACCSH unanimously voted to "recommend to OSHA that revision of 29 CFR 1926.51 be placed on the Agency's regulatory agenda for rule making in 1999."[97]

The ACCSH draft sanitation standard of January 1999, which was approved by a vote of 11-1 and which the OSHA Directorate of Construction agreed to "take...and put...through the normal process, and come back to the ACCSH with the OSHA version,"[98] would have required one toilet facility for the first nine employees and one toilet seat and one urinal for each additional 10 workers. In addition, it would have included the aforementioned provision, which was similar to, but not quite so expansive as the OSHA Memorandum of April 6, 1998: "Employees shall be allowed to use the toilet facilities on an as needed basis." Finally, for limited access worksites, it would have required employers to "ensure transportation is available to toilet/wash facilities when facilities are unavailable at the site." The ACCSH would also have required for the first time that hand washing facilities (including hot potable water where practicable) be provided in near proximity to the toilets.[99]

From November 1999 through May 2001, the Department of Labor in the Clinton and Bush administrations stated four times in its semi-annual regulatory agendas that it believed that the proposal recommended by the ACCSH on October 7, 1998 "raises important issues regarding the type of sanitation facilities needed for construction workers. OSHA intends to issue an ANPRM [Advance Notice of Proposed Rulemaking] to consider revisions to the sanitation standard that would include washing facilities, gender-separate and lockable toilet facilities, and (where other OSHA standards require change rooms), gender-separate and lockable change facilities."[100] At the end of 1999, when OSHA placed the proposed sanitation standard on its regulatory agenda for long-term action (meaning that publication of a proposed or final rule would not occur for at least a year), Williams and Cooper, the co-chairs of the ACCSH, told OSHA administrator Charles Jeffress that providing sanitary washing facilities that are accessible to construction workers was "more important than any ergonomics, noise, or

---

[96]"Workers to Use Toilet Facilities as Needed."

[97]Transcript, OSHA, Advisory Committee on Construction Safety and Health, October 7, 1998, on http://www.osha.gov/doc/accsh/transcripts/oct7-98.html; ACCSH, Minutes of October 7-8, 1998, Meeting, on http://www.osha.gov/doc/accsh/meeting minutes/oct98.html.

[98]ACCSH, Minutes of Jan. 28-29, 1999, Meeting, on http://www.osha.gov/doc/accsh/ meetingminutes/jan99.html.

[99]ACCSH, OSHA Regulations (Standards - 29 CFR 1926.51 - Sanitation) (Jan. 1999) (copy furnished by Jane Williams).

[100]*Fed. Reg.* 66:25724 (2001). See also *Fed. Reg.* 64:64669 (1999);*Fed. Reg.* 65:23061, 74109 (2000).

chromium standard" and that none of the other standards mattered if OSHA was not able to resolve this one quickly.[101] Already by the end of the Clinton administration, however, progress toward codification had been halted because, as Cooper observed, OSHA's "'attitude seems to be that it's not a priority because it doesn't save lives'"[102]—a position that, as seen earlier, OSHA officials use to justify the higher priority they accord the prevention of falls, dismemberments, and electrocutions over the enforcement of the general industry sanitation standard.

By the latter part of 2001, the Bush administration announced that "OSHA is withdrawing this entry from the agenda at this time due to resource constraints and other priorities."[103] And a revised sanitation standard "probably won't make the regulatory agenda," as the *Occupational Safety and Health Reporter* observed, since OSHA itself had stated that the focus of the new Bush appointee as administrator would be "on putting things on the regulatory agenda that can be accomplished quickly."[104] *The Building Tradesman*, the organ of the Michigan construction unions, added that with the proposed standard "dead in the water," building trades workers "have the unfortunate distinction among nearly all U.S. industries...that their employers are not required to provide them with even the most rudimentary method to wash their hands. On some smaller construction sites, employers are not required to provide any bathroom facilities at all. The closest fast food restaurant, or more likely, the nearest ditch or behind the nearest tree, sometimes are the only facilities available."[105]

When OSHA finally took a step toward creating parity with workers covered by the general industry sanitation standard, it was obviously guided by the principle enunciated in the April 6, 1998 Memorandum, from which entire paragraphs were taken verbatim. On June 7, 2002 it issued an opinion letter in response to a request for an interpretation of "nearby" in the aforementioned exception for mobile crews "having transportation readily available to nearby toilet facilities." OSHA's response first defined mobile crews as consisting of "[w]orkers who continually or frequently move from jobsite to jobsite on a daily or

---

[101]"OSHA Should Make Sanitation Standard Top Industry Priority, ACCSH Members Say," *Daily Labor Report*, Dec. 13, 1999, at A-8 (Lexis).

[102]Marty Mulcahy, :Sanitation Standard Moves Slowly, But, Now, Not-So-Surely," *The Building Tradesman*, Dec. 22, 2000, on http://www.detroitbuildingtrades.org/newspapr/dec222000.html.

[103]*Fed. Reg.* 66:61884 (2001).

[104]Taashi Rowe,"Lack of Modern Field Sanitation Rule Affects Health of Women in Building Trades," *Occupational Safety and Health Reporter* 32(330:787-88 (Aug. 15, 2002).

[105]Marty Mulcahy, "Could Michigan OK a Better Sanitation Rule in 2002?" *The Building Tradesman* 51(12):1 (June 7, 2002).

hourly basis," but excluding those "who report to a conventional construction project, where they work for more extended periods of time (days, weeks, or longer)...." The opinion letter then went on to focus on the key issue common to both standards: "for purposes of this standard, 'nearby' means prompt access--sufficiently close so that employees can use them when they need to do so." The only relevant change that the opinion letter made to the language of the April 6, 1998 Memorandum reflected the chief difference in obstacles to access between general industry and construction: instead of focusing on the unavailability as a function of the length of time during which workers are not allowed to use the toilets, the construction standard declared: "Toilets that take too long to get to are not 'available.'"[106] Whereas the general industry sanitation standard mandates that members of mobile crews "must have transportation immediately available to nearby toilet facilities,"[107] the June 7, 2002 opinion letter concludes that "in general, toilets would be considered 'nearby' if it would take less than 10 minutes to get to them,"[108] without explaining how it arrived at this fixed standard.

Jane Williams has argued that "[t]he flaw in this letter is captured by '10 minutes or less'; specifically it does not state that public facilities, such as those in a Burger King cannot be substituted for the employer's duty to provide same as 'permission' for substitution is not obtained...."[109] Although Williams's criticism concerning the deficiencies of non-dedicated public toilets is well-taken, it bears no logical relationship to the adequacy of the 10-minute rule.[110]

Women construction workers have special access problems. One especially poignant case involved Karen Olson, who worked as a flagger on outdoor construction projects, standing for long hours directing traffic without access to a toilet, although the employer, Propheter Construction, was required under the union contract to provide toilet facilities or to make arrangements for bathroom breaks. When the employer told flaggers to void outdoors or in the bushes, Olson, the company's only female flagger, "refused to use anything but a bathroom." The bursting of her bladder on June 25, 1999 caused Olson to leave the

---

[106]Letter from Russell Swanson, Director, Directorate of Construction, to Nicholas Mertz, Re: sanitation, mobile crews, 1926.51(c)(4) (June 7, 2002) (furnished by Noah Connell, director of OSHA's Office of Construction Standards and Compliance Assistance, OSHA, who stated that the letter was in the process of being posted on the OSHA website).

[107]29 CFR sect. 1910.141(c)(1)(ii).

[108]Letter from Swanson to Mertz.

[109]Fax from Jane Williams to Marc Linder (Oct. 2, 2002).

[110]For an overview of Williams' critique of the history of OSHA's failure to "promulgate effective language or enforcement directives so as to accomplish their charter as regards [construction] Sanitation," see [Jane Williams?], "A10-25: Sanitation in Construction Activity Report Index" at 3 (July 2002) .

jobsite to go to the emergency room. Despite Olson's having notified her employer that she had to leave for emergency medical treatment, Propheter fired her that very day for insubordination. Her lawsuit was grounded on the claim that the injury to her bladder "was related to the infrequent use of the bathroom and...'holding' her urine for extended periods of time."[111]

Although the Seventh Circuit Court of Appeal's earlier ruling in 2000 in *DeClue v. Central Illinois Light Co.*[112] prompted a federal trial judge in Illinois to dismiss Olson's hostile workplace environment claim, the lower court did refuse to dismiss her disparate impact claim, eventually prompting a settlement. In rejecting Propheter's motion for summary judgment on Olson's disparate-impact claim, the judge declared:

> Defendant seems to believe that providing some restroom facility somewhere on the construction site, approximately 4.5 miles from plaintiff, is per se sufficient to satisfy *DeClue*. However, this court believes that a reasonable factfinder could determine that defendant's conditions were the sort that could deter women, but not men, from seeking construction jobs with defendant. Walking 9 miles (4.5 miles to the restroom facilities and 4.5 miles back to the job position) would certainly deter most people from using any restroom facilities which were technically available, such that a reasonable jury could find that there was an "absence"of facilities at defendant's site. Furthermore, plaintiff has come forward with evidence suggesting that cars were not readily available to take flaggers to and from restroom facilities.[113]

The employer's cavalier attitude toward its workers' hygiene was underscored by its lawyer's statement that "the company makes it clear to all employees that it generally does not provide portable bathrooms. Instead, Propheter directs its employees who need a bathroom break to ask a co-worker to step in for them while the employee uses a nearby restroom."[114]

---

[111]Olson v. Propheter Construction Equipment Inc., Complaint, 00 C 839 (N.D. Ill, 2000). See generally, Molly McDonough, "Title VII and the Bathroom Break," *National Law Journal,* Sept. 25, 2000, at B5 (Lexis).

[112]DeClue v. Central Illinois Light Co., 223 F.3d 434 (7th Cir. 2000). For a discussion, see below ch. 8.

[113]Olson v. Propheter Construction Co., Denial of Defendant's Motion for Summary Judgment (N.D. Ill., Apr. 18, 2001).

[114]Ted Gregory, "Lawsuit Raises a Flag About Roadwork," *Chicago Tribune*, Sept. 5, 2000, at 1 (Lexis).

Part III

Promulgation: Voiding on the Man's Time

> [W]hat I remember about the shithouses at the General Dynamics shipyard is that that's where we did most of our organizing.... For a long time, the women were better organized than the men, but then GD...hired a female foreman so they had someone to raid the women's shithouse. Your work will help people spend more time hanging out in bathrooms talking trouble.[1]

---

[1]Email from Prof. James Pope, Rutgers-Newark Law School, to Marc Linder (Apr. 27, 1998).

6

OSHA Comes to Its Senses

Marc Linder is eagerly awaiting what Uncle Sam plans to do in the bathroom.[1]

Present Federal law requires that the employers provide bathrooms, but they don't actually have to let employees go to the bathroom. No shit. (So to speak.)[2]

Finally, on April 6, 1998, in the wake of the avalanche of embarrassing media coverage of OSHA's scandalous administrative failure to uphold and enforce workers' right to void at work unearthed by *Void Where Prohibited*,[3] John B. Miles, Jr., the Director of the Directorate of Compliance Programs in OSHA's national office, issued a Memorandum to regional OSHA administrators and state-plan OSHA agencies explaining its

interpretation that this standard requires employers to make toilet facilities available so that employees can use them when they need to do so. The employer may not impose unreasonable restrictions on employee use of the facilities. OSHA believes this requirement is implicit in the language of the standard and has not previously seen a need to address it more explicitly. Recently, however, OSHA has received requests for clarification of this point and has decided to issue this memorandum to explain its position clearly.[4]

This approach contrasts with that adopted for the field sanitation standard for agricultural worker by OSHA during the Reagan administration. Whereas the

---

[1]"Bathroom Access Causes Work Troubles," *Des Moines Register*, Mar. 13, 1998, at 2M, col. 2 (State Edition; NewsBank).

[2]David Stevens, "The Right to Pee Lazy: Class Struggle in the WC" (Mar. 22, 1998), on http://x10.dejanews.com/getdoc.xp?...NTEXT=892134249.116654116&hitnum=1.

[3]"[A]fter the book was issued, a wave of press interest ensued that brought pressure on OSHA." Book Review, [National Employment Law Project,] *Update*, Summer 1998, at 14.

[4]OSHA, Interpretation of 29 CFR 1910.141(c)(1)(i): Toilet Facilities. For the full text, see below Appendix II.

Memorandum takes as its point of departure the employer's obligation to let workers go when they need to go and applies a reasonableness requirement as a negative constraint on the restrictions that employers are permitted to place on workers' freedom to stop work immediately, the field sanitation standard is based on a reasonableness approach paired with a quasi-paternalistic regime under which employers are assigned the task of introducing their charges to the precepts of modern personal hygiene:

Reasonable use. The employer shall notify each employee of the location of the sanitation facilities and water and shall allow each employee reasonable opportunities during the workday to use them. The employer also shall inform each employee of the importance of each of the following good hygiene practices to minimize exposure to the hazards in the field of heat, communicable diseases, retention of urine and agrichemical residues:

(i) Use the water and facilities provided for drinking, handwashing and elimination;

...

(iii) Urinate as frequently as necessary.[5]

The 1998 Memorandum then went on to offer the following medical rationale for OSHA's interpretation:

The sanitation standard is intended to ensure that employers provide employees with sanitary and available toilet facilities, so that employees will not suffer the adverse health effects that can result if toilets are not available when employees need them. Individuals vary significantly in the frequency with which they need to urinate and defecate, with pregnant women, women with stress incontinence, and men with prostatic hypertrophy needing to urinate more frequently. Increased frequency of voiding may also be caused by various medications, by environmental factors such as cold, and by high fluid intake, which may be necessary for individuals working in a hot environment. Diet, medication use, and medical condition are among the factors that can affect the frequency of defecation.

Medical studies show the importance of regular urination, with women generally needing to void more frequently than men. Adverse health effects that may result from voluntary urinary retention include increased frequency of urinary tract infections (UTIs), which can lead to more serious infections and, in rare situations, renal damage (see, e.g., Nielsen, A. Waite, W. [sic; should be Walter, S.], "Epidemiology of Infrequent Voiding and Associated Symptoms," Scand J Urol Nephrol Supplement 157). UTIs during pregnancy have been associated with low birthweight babies, who are at risk for additional health problems compared to normal weight infants (see, Naeye, R.L., "Causes of the Excess [sic; should be Excessive] Rates of Perinatal Mortality and the [sic] Prematurity in Pregnancies Complicated by Maternity [sic; should be Maternal] Urinary[-]Tract Infections," New England J. Medicine 1979; 300(15); 819-823). Medical evidence also

---

[5] 29 CFR 1928.110(c)(4) (2002).

shows that health problems, including constipation, abdominal pain, diverticuli, and hemorrhoids, can result if individuals delay defecation (see National Institutes of Health (NJH) Publication No. 95-2754, July 1995).

OSHA's field sanitation standard for Agriculture, 29 CFR 1928.110, based its requirement that toilets for farmworkers be located no more than a quarter mile from the location where employees are working on similar findings. This is particularly significant because the field sanitation standard arose out of the only OSHA rulemaking to address explicitly the question of worker need for prompt access to toilet facilities.[6]

OSHA then bolstered its new interpretation with these semantic and policy arguments:

The language and structure of the general industry sanitation standard reflect the Agency's intent that employees be able to use toilet facilities promptly. The standard requires that toilet facilities be "provided" in every workplace. The most basic meaning of "provide" is "make available." See Webster's New World Dictionary, Third College Edition, 1988, defining "provide" as "to make available; to supply (someone with something);" Borton Inc. V. OSHRC, 734 F.2d 508, 510 (l0th Cir. 1984) (usual meaning of provide is "to furnish, supply, or make available"); Usery v. Kennecott Copper Corp., 577 F.2d 1113, 1119 (10th Cir, 1978) (same); Secretary v. Baker Concrete Constr. Co., 17 OSH Cas. (BNA) 1236, 1239 (concurring opinion; collecting cases); Contractors Welding of Western New York, Inc., 15 OSH Cas. (BNA) 1249, 1250 (same). Toilets that employees are not allowed to use for extended periods cannot be said to be "available" to those employees. Similarly, a clear intent of the requirement in Table J-1 that adequate numbers of toilets be provided for the size of the workforce is to assure that employees will not have to wait in long lines to use those facilities. Timely access is the goal of the standard.

The quoted provision of the standard is followed immediately by a paragraph stating that the toilet provision does not apply to mobile work crews or to locations that are normally unattended, "provided the employees working at these locations have transportation immediately available to nearby toilet facilities which meet the other requirements" of the standard (29 CFR 1910.141(c)(1)(ii)[]) (emphasis supplied). Thus employees who are members of mobile crews, or who work at normally unattended locations must be able to leave their work location "immediately" for a "nearby" toilet facility. This provision was obviously intended to provide these employees with protection equivalent to that the general provision provides to to [sic] employees at fixed worksites. Read together, the two provisions make clear that all employees must have prompt access to toilet facilities.[7]

OSHA was able to furnish only limited precedent to support its new position by reference to a couple of its older opinion letters that dealt with the related physical-spatial issue of "unobstructed free access":

---

[6]OSHA, Interpretation of 29 CFR 1910.141(c)(1)(i): Toilet Facilities.

[7]OSHA, Interpretation of 29 CFR 1910.141(c)(1)(i): Toilet Facilities.

OSHA has also made this point clear in a number of letters it has issued since the standard was promulgated. For example, in March 1976, OSHA explained to Aeroil Products Company that it would not necessarily violate the standard by having a small single-story building with no toilet facilities separated by 90 feet of pavement from a building that had the required facilities, so long as the employees in the smaller building had "unobstructed free access to the toilet facilities." Later that year, it explained again, in response to a question about toilet facilities at a U-Haul site, "reasonableness in evaluating the availability of sanitary facilities will be the rule." Again in 1983, OSHA responded to a request for a clarification of the standard by stating, "[i]f an employer provides the required toilet facilities...and provides unobstructed free access to them, it appears the intent of the standard would be met."[8]

OSHA then concluded its discussion by observing:

In light of the standard's purpose of protecting employees from the hazards created when toilets are not available, it is clear that the standard requires employers to allow employees prompt access to sanitary facilities. Restrictions on access must be reasonable, and may not cause extended delays. For example, a number of employers have instituted signal or relief worker systems for employees working on assembly lines or in other jobs where any employee's absence, even for the brief time it takes to go to the bathroom, would be disruptive. Under these systems, an employee who needs to use the bathroom gives some sort of a signal so that another employee may provide relief while the first employee is away from the work station. As long as there are sufficient relief workers to assure that employees need not wait an unreasonably long time to use the bathroom, OSHA believes that these systems comply with the standard.[9]

The regime established by the Memorandum is thus a hybrid. On the one hand, in principle it creates the right to take voiding breaks at will. (The notion of an "at-will voiding regime" had been introduced in *Void Where Prohibited*.[10] At the Jim Beam Brands Kentucky OSHA appeals hearing in 2002, the book's coauthor, Dr. Ingrid Nygaard, called it "the policy of choice" and "the most medically tenable policy for voiding,"[11] prompting the employer's lawyer, to whom it was anathema, to add that the Secretary of Labor had "taken the same posi-

---

[8]OSHA, Interpretation of 29 CFR 1910.141(c)(1)(i): Toilet Facilities.

[9]OSHA, Interpretation of 29 CFR 1910.141(c)(1)(i): Toilet Facilities.

[10]Marc Linder and Ingrid Nygaard, *Void Where Prohibited: Rest Breaks and the Right to Urinate on Company Time* 160 (1998).

[11]Commonwealth of Kentucky Occupational Safety and Health Review Commission, Administrative Action No. 02-KOSH-0015, Secretary of Labor Commonwealth of Kentucky v. Jim Beam Brands Co., KOSHRC #3681-01, "Transcript of Administrative Hearing" at 73, 86 (Aug. 28, 2002) ("Transcript of Hearing").

tion"—that "a void-at-will policy is the right policy."[12]) Thus the Memorandum states that "this standard requires employers to make toilet facilities available so that employees can use them *when* they need to do so."[13] Indeed, the Memorandum does not even explicitly require workers to ask for, let alone obtain, the employer's permission to go to the bathroom, though employees may need to inform employers of medical conditions requiring especially frequent and/or lengthy bathroom breaks.[14] Such permission would, arguably, be required only where the employee's absence would be "disruptive"; and even where disruption is possible, once the relief worker appears, the sole condition for taking the break would be fulfilled and permission would not be required.

On the other hand, although the Memorandum confers no power on an employer to test the veracity of a worker's statement that he or she has to void, let alone to test the genuineness of the physical need before or after the fact, OSHA does craft a reasonableness criterion that immunizes employers from being cited for (or even committing) a violation of the standard for briefly delaying the beginning of the worker's bathroom break. (In addition, the Memorandum may leave intact whatever a particular employer's otherwise lawful powers may be to determine whether employees, instead of voiding, are engaged in other activities—such as smoking, telephoning, reading, talking, or resting—and to discipline them for such 'abuse'; but even in such cases, the employer's remedy might include termination, but not preemptive flat prohibitions on going to the bathroom.)[15] However, this reasonableness criterion is very narrow in the sense that, at the moment the worker needs to void, it is confined solely to the employer's objective assessment of the disruptive impact of the break on its operations. In turn, even this disruption is never permitted to trump the worker's need to void in any absolute sense, and the employer itself is impliedly required to temper the disruption by hiring additional relief workers or perhaps otherwise restructuring operations to enable workers to go to the bathroom, although this duty to avoid chronic understaffing presumably must have some economic limits. Finally, the disruption defense would be nullified altogether in cases in which employer-imposed waiting caused workers to void on themselves or to be otherwise harmed.

In sum, then, the Memorandum has fashioned the right to at-will bathroom breaks subject only to the possibility of the employer's brief, temporary veto, which the employer can justify, if at all, exclusively by reference to the necessity of preventing unacceptable disruption, which the employer itself is obligated to

---

[12]"Transcript of Hearing" at 103 (Aug. 29, 2002) (Rick Griffith, Jim Beam's attorney).

[13]See below Appendix II (italics added).

[14]See below ch. 13.

[15]On the issue of so-called abuse, see below ch. 17.

mitigate so as not to interfere unreasonably with the worker's right to void when he or she needs to. However, since the Memorandum gives the employer no means by which to gauge how urgently the employee has to void—and presumably there is no practical way for anyone other than the would-be voider to measure that urgency—this element of reasonableness is not within the employer's discretion to apply; rather, only OSHA, after the fact, may be able to determine whether any given delay was reasonable, and perhaps even then only in instances in which unreasonableness was palpably demonstrated by the worker's having voided in his or her clothing or on the floor. Thus a labor reporter fundamentally misconceived the structure of labor standards regulation in general and of the Memorandum in particular by asserting: "Federal work rules on the issue are murky.... What exactly timely access constitutes has largely been left to companies to interpret."[16] Because no labor department is omnipresent, employers may in the first instance "interpret" or even intentionally violate any regulation; but only a regulatory agency and/or court can authoritatively interpret, that is, enforce, a rule.

The reasonableness criterion thus has three dimensions, relating to: (1) the impact of the delay of the onset of the break on the worker; (2) the disruptiveness of the break to the employer; and (as a sub-element of (2)) (3) the burden to the employer of having to hire relief workers or to restructure its operations. If the Memorandum gives OSHA "too much wiggle room"[17] in declining to cite an employer for violating the standard, that discretion derives from these reasonableness criteria, which over time will have to be fought over by labor and capital at the workplace as well as in administrative and judicial processes. Nevertheless, despite the employer's limited temporal veto power, the decision to take a bathroom break in principle remains at the employee's will.[18]

Finally, OSHA informed regional and state enforcement agents that they were to use the following criteria in determining whether to issue a citation:

---

[16]Carol Kleiman, "Restroom Off Limits? Workers Merit a Break," *Chicago Tribune*, Nov. 10, 2002, Business Sect. At 5 (Lexis).

[17]"Ruling Makes Federal Case of Workers' Toilet Time," *Des Moines Register*, Apr. 9, 1998, at 3A, col. 2 (quoting Marc Linder).

[18]The legitimacy of retaining the term "at-will bathroom break" despite the fact that it is subject to a constraint is made plausible by the following analogy: Even in a pure at-will regime a worker would still be an at-will employee in spite of the fact that, though free to terminate her employment contract at any time, she might conceivably still be subject to the constraint of liability for damages (if only in tort) if she knowingly (albeit unintentionally) set in train the destruction of machinery or raw materials by quitting (without any provocation by the employer) in the midst of some production process without notifying any supervisor that she was walking out so that someone else would be assigned to take her place immediately.

Employee complaints of restrictions on toilet facility use should be evaluated on a case-by-case basis to determine whether the restrictions are reasonable. Careful consideration must be given to the nature of the restriction, including the length of time that employees are required to delay bathroom use, and the employer's explanation for the restriction. In addition, the investigation should examine whether restrictions are general policy or arise only in particular circumstances or with particular supervisors, whether the employer policy recognizes individual medical needs, whether employees have reported adverse health effects, and the frequency with which employees are denied permission to use the toilet facilities. Knowledge of these factors is important not only to determine whether a citation will be issued, but also to decide how any violation will be characterized.[19]

Initially, OSHA decided, presumably because of the potentially disruptive reaction of employers, but also to insure some consistency in interpreting a performance standard,[20] to centralize decisionmaking:

It is important that a uniform approach be taken by all OSHA offices with respect to the interpretation of OSHA's general industry sanitation standard, specifically with regard to the issue of employee use of toilet facilities. Proposed citations for violations of this standard must be forwarded to the Directorate of Compliance Programs (DCP) for review and approval. DCP will consult with the Office of Occupational Medicine. DCP will approve citations if the employer's restrictions are clearly unreasonable, or otherwise not in compliance with the standard.[21]

Nevertheless, "[s]hortly after the interpretation was issued, it was decided that review and approval was to be at the Regional Office level, but that copies of any citations issued based on the April 6, 1998 interpretation should still be sent to DCP. This topic continues to generate interest from the public. Early this year we had a Freedom of Information Act (FOIA) request for copies of citations issued. Therefore, please continue to send copies of any citations issued pursuant to the 1998 interpretation to the National Office."[22]

Finally, Federal OSHA, reinforcing one of the statutory conditions of Federal

---

[19]OSHA, Interpretation of 29 CFR 1910.141(c)(1)(i): Toilet Facilities

[20]Telephone interview with Dale Cavanaugh, safety engineer, OSHA Seattle Regional Office (Dec. 2, 2002).

[21]OSHA, Interpretation of 29 CFR 1910.141(c)(1)(i): Toilet Facilities.

[22]This memorandum was issued by Richard Fairfax, Miles's successor, on Aug. 11, 2000. OSHA, Interpretation of 29 CFR 1910.141(c)(1)(i): Toilet Facilities. Nevertheless, two years later Fairfax, in response to the author's request for copies of citations issued for denial of access, wrote that the OSHA National Office "does not maintain records that are responsive to your request." Letter from Richard Fairfax to Marc Linder (Oct. 29, 2002).

approval of a state OSHA plan,[23] instructed the state-plan states that they "are not required to issue their own interpretation in response to this policy, however they must ensure that State standards and their interpretations remain 'at least as effective' as the Federal standard."[24] By not requiring the state programs to inform Federal OSHA as to how they intended to go about enforcing the new interpretation—as, for example, the National Office does with regard to certain important matters[25]—the Directorate of Compliance Programs made thorough familiarity with the new obligation/right and vigorous state-plan enforcement less probable.

It is instructive to compare the April 6, 1998 Memorandum with the much briefer draft from July 1997, which had been written just before OSHA issued the first citation for lack of toilet access against Hudson Foods and in parts is verbatim identical with the Iowa OSHA Memorandum of January 21, 1998.[26] To begin with, it contained a straightforward derivation of the right to access from the requirement that a certain number of toilets be provided, which is lacking in the final version:

> It would be a clear violation of the standard if the employer failed to have in the workplace the necessary number of toilets, or, if having installed the necessary toilets, the employer kept them locked. If, however, an employer does not let the employees use the toilets that are in the workplace, the employer is also not "providing toilet facilities" as required by the standard. When an employer does not let the workers use the toilets, the effect is the same as if the employer had locked them.[27]

Second, whereas the published Memorandum merely refers to existing relief systems and states that they comply with the standard,[28] the draft conferred a quasi-prescriptive character on them:

---

[23]29 USC sect. 667(c)(2) (2000).

[24]OSHA, Interpretation of 29 CFR 1910.141(c)(1)(i): Toilet Facilities.

[25]For example, in a memorandum concerning congressional enforcement exemptions and limitations, the OSHA Administrator declared: "The States shall respond via the two-way memorandum to the Regional Office as soon as the State's intention regarding the enforcement activity limitations and exemptions is known...." OSHA, CPL 2-0.513 - Enforcement Exemptions and Limitations Under the Appropriations Act (Jan. 2, 2002), on http://www.osha.gov/pls/oshaweb/owadisp.show_document?p_table=DIRECTIVES &p_id=1.

[26]See above ch. 3. The Iowa OSHA version was derivative.

[27]John B. Miles, Jr., Memorandum (July 3, 1997). Stamped "DRAFT" and computer dated July 3, 1997, it bore the handwritten message "Confidential Do not circulate," but was nevertheless faxed to the UFCW.

[28]Miles, Memorandum (Apr. 6, 1998).

In order to accommodate various work situations in which workers are on continuous production/assembly lines, employers may have to establish some type of signaling and relief system for workers as well as provide backup workers to relieve the line workers as needed. If the employer in such a situation does not allow reasonable use, it is in effect "locking" the workers at their workstations."[29]

That OSHA ultimately refrained from mandating any specific system for insuring access may, on the macro-political level, have been rooted in trepidations similar to those expressed by a labor arbitrator who had been called on a quarter-century earlier to adjudicate a United Steelworkers' grievance over an assembly-line relief system under which workers had to wait 30 and 45 minutes to go to the bathroom. After sustaining the grievance on the grounds that the employer had violated its contractual duty to provide adequate relief, the arbitrator, noted labor law professor William Murphy, paused to air his own doubts as to how to fashion an appropriate remedy, especially since the union had declined to suggest one. In addition to "the difficulty that there is no one clear and obvious way of dealing with the problem, as for example, reinstatement with or without just cause," he observed that "any way of dealing with the problem is intimately related to personnel and production policy and the general exercise of supervisory authority. An arbitrator should be wary of an award which interferes with or unduly restricts the exercise of legitimate and necessary management authority." Like OSHA 25 years later, he confined himself to mentioning the various possibilities that the employer had already tried, including leadman and relief worker systems and combinations of the two, adding only that the company could "insist[ ], as it apparently has not heretofore insisted, that the leadmen diligently effectuate their relief duties," and/or build more toilets closer to the conveyor belt.[30]

Finally, the July 1997 OSHA draft contained much greater specificity about what is reasonable and unreasonable:

Reasonableness in allowing employees the use of the toilet facilities is the rule. When employers deny employee requests to use the bathroom, except during scheduled breaks, it may or may not be reasonable, depending upon the factual situation of each workplace. Every 30 minutes may not be [un?]reasonable, while every 90 minutes may. Some examples, however, of unreasonableness would be an employer requirement or suggestion that employees "hold it," or "don't drink fluids," or "wear adult diapers." These are not

---

[29]Miles, Memorandum (July 3, 1997).

[30]Mor-Flo Industries, Inc. 62 Labor Arbitration Reports (BNA) 398, 401 (1974). Murphy, who had previously worked at the U.S. Department of Labor and the NLRB, taught at the University of North Carolina. Association of American Law Schools, *Directory of Law Teachers 1973*, at 479.

reasonable solutions to a worker's need to use toilet facilities.[31]

The dilution of the interpretation on its way to promulgation may in part be a function of objections raised by poultry companies. When asked five years later whether OSHA had received protests from any employers when the Memorandum was released, John Miles stated that the poultry industry had become "concerned" following issuance of the citation to Hudson, but that he had "shared" the draft with the industry, which saw that it "could live with" the interpretation.[32]

As interventionist as OSHA's interpretation is, because it confers circumscribed discretion on employers to act as gatekeepers and, in that role, to place "reasonable" restrictions on the promptness with which they permit employees to stop working in order to go to the bathroom, as a pragmatic safety and health measure it lacks the broad and absolutist human rights approach of a 1996 French labor court ruling that as "a fundamental human freedom...the right to go to the toilet cannot be subject to authorization by a third person."[33]

---

[31]Miles, Memorandum (July 3, 1997).

[32]Telephone interview with John Miles, OSHA Dallas Regional Administrator (Nov. 12, 2002). In 1997 OSHA undertook a cooperative program with the Wage and Hour Division of the Labor Department to conduct a special study of 51, mainly southern, poultry slaughter plants. "Poultry Study Finds PSM Violations, Falls, Cuts Outweigh Repetitive Motion," *Occupational Health & Safety Letter* 28(8) (Apr. 13, 1998). In that connection Miles did inform management of the impending Memorandum. Telephone interview with Benjamin Ross, Assistant Regional Administrator for Compliance Programs, OSHA, Atlanta (Dec. 18, 2002). Senate Majority Leader Trent Lott and other senators had tried to pressure the Secretary of Labor not to initiate the project on the grounds that routine inspections had failed to document any need for it. Letter from Trent Lott et al. to Alexis Herman (May 21, 1997) (copy faxed to author by UFCW, June 20, 1997).

[33]Lamouroux v. SA Groupe Bigard, No. 9500433-436, slip op. at 13, 15 (Conseil de Prud'hommes de Quimper, Mar. 18, 1996). See below ch.10.

7

Is OSHA's Interpretation Valid as a Matter of Administrative Law?

> SHRM [Society of Human Resource Management]'s Lawrence lamented that something as basic as bathroom use needs to be federally regulated. "It's a little sad we even need to have guidance on that."[1]

The reason that OSHA did not consider issuing its April 6, 1998 Memorandum, which it insists is merely an interpretation of its pre-existing toilet regulation, as a standard in its own right, that is, as a so-called legislative regulation with the force of law, was the desire to avoid inordinate delay. As Helen Rogers, a safety specialist in OSHA's Directorate of Compliance Programs who had been involved in the drafting process and was listed as the contact person in the Memorandum, observed four years later: if OSHA had gone through the notice-and-comment process under the Administrative Procedure Act (APA), which requires agencies to give both general notice of proposed rule making in the *Federal Register* and an opportunity to interested parties "to participate in the rule making through submission of written data, views, or arguments,"[2] the standard would still not be in place today.[3] In an interview, John Miles, who as the

---

[1]Simon Nadel, "Restricting Regular Bathroom Access Can Spur Negative Workplace Repercussions," *U.S. Law Week* 66(37):2579-80 at 2580 (Mar. 31, 1998). This article also appeared in *Occupational Safety & Health Reporter* 27(44):1570 (Apr. 8, 1998), and *Occupational Safety & Health Daily* (Apr. 16, 1998).

[2]5 USC sect. 553(b) and (c) (2000). The Occupational Safety and Health Act itself contains a similar provision requiring the agency to "publish a proposed rule promulgating, modifying, or revoking an occupational safety or health standard in the Federal Register and...afford interested persons a period of thirty days after publication to submit written data or comments." 29 USC sect. 655(b)(2) (2000).

[3]Telephone interview with Helen Rogers, Washington, D.C. (Sept. 19, 2002). Rogers peremptorily refused to discuss changes that the Memorandum had undergone during the year it was being drafted on the grounds that the government does not discuss such internal policy formulation matters.

Director of OSHA's Directorate of Compliance Programs had been the official responsible for issuing the Memorandum, estimated that rulemaking would have taken eight to ten years.[4] And, as the Director of Legal Support at Virginia OSHA added, Federal OSHA does such "quick and dirty" interpretations both because promulgating legislative regulations takes so many years and the process often entails considerable litigation.[5]

As the federal courts have ruled repeatedly: "It is well-established that an agency may not escape the notice and comment requirements...by labeling a major substantive legal addition to a rule a mere interpretation."[6] If an employer challenges the validity of such a label, judges have to determine whether the putative interpretation "spells out a duty fairly encompassed within the regulation that the interpretation purports to construe"[7] or whether it significantly expand[s] the scope" of that regulation.[8]

Thus it is clear that OSHA cannot lawfully circumvent that process merely by dressing up what in fact is a legislative regulation as a mere interpretation. Yet it has some discretion in this area by virtue of the federal judiciary's repeated complaints that the distinction between legislative and interpretive regulations is "'enshrouded in considerable smog,'" "'fuzzy,'" "'tenuous,'" "'blurred,'" and "'baffling.'"[9] This "quite troublesome"[10] ambiguity derives, according to the leading administrative law treatise, from the real-world fact that "legislative rules often require many years and many thousands of staff hours to issue. ... It would be impossible for any agency to use the long and expensive process of issuing legislative rules to address definitively and in detail every issue that arises in the process of implementing a regulatory...program."[11] The question here, then, is whether OSHA's declaration that "this standard requires employers to make toilet facilities available so that employees can use them when they need to do so" properly fits within the "many thousands of pages of policy statements and interpretative rules that address myriad details that are not explicitly resolved by its

---

[4]Telephone interview with John Miles, OSHA Regional Administrator, Dallas Region (Nov. 12, 2002).

[5]Telephone interview with Jay Withrow, Director of Legal Support, Virginia OSHA (Oct. 31, 2002).

[6]Appalachian Power Co. Environmental Protection Agency, 208 F.3d 1015, 1024 (D.C. Cir. 2000).

[7]Paralyzed Veterans of America v. D.C. Arena, 117 F.3d 579, 588 (D.C. Cir. 1997).

[8]Appalachian Power Co. v. Environmental Protection Agency, 208 F.3d at 1024.

[9]American Mining Congress v. Mine Safety and Health Adm., 995 F.2d 1106, 1108-1109 (D.C. Cir. 1993) (citing numerous decisions).

[10]Kenneth Davis, *Administrative Law Text* 126 (3d ed. 1972).

[11]Richard Pierce, Jr. *Administrative Law Treatise* 1:305 (4th ed. 2002).

legislative rules," or whether what OSHA calls a mere interpretation is in fact one of the "legislative rules that describe the basic contours of the regulatory program" administered by OSHA.[12]

In a prominent decision from 1996, Richard Posner, the federal judiciary's most unconventional thinker, while admitting that distinguishing between interpretive and legislative regulations "is often very difficult,"[13] tried to avoid traditional pitfalls by offering a new approach.

> [W]hen a statute does not impose a duty on the persons subject to it but instead authorizes...an agency to impose a duty, the formulation of that duty becomes a legislative task entrusted to the agency. Provided that a rule promulgated pursuant to such a delegation is intended to bind, and not merely to be a tentative statement of the agency's view, which would make it just a policy statement, and not a rule at all, the rule would be the clearest possible example of a legislative rule, as to which the notice and comment procedure not here followed is mandatory, as distinct from an interpretive rule; for there would be nothing to interpret. [U]nless a statute or regulation is of crystalline transparency, the agency enforcing it cannot avoid interpreting it, and the agency would be stymied in its enforcement duties if every time it brought a case on a new theory it had to pause for a bout, possibly lasting several years, of notice and comment rulemaking. Besides being unavoidably continuous, statutory interpretation normally proceeds without the aid of elaborate factual inquiries. When it is an executive or administrative agency that is doing the interpreting it brings to the task a greater knowledge of the regulated activity than the judicial or legislative branches have, and this knowledge is to some extent a substitute for formal fact-gathering.
>
> At the other extreme from what might be called normal or routine interpretation is the making of reasonable but arbitrary (not in the "arbitrary or capricious" sense) rules that are consistent with the statute or regulation under which the rules are promulgated but not derived from it, because they represent an arbitrary choice among methods of implementation. A rule that turns on a number is likely to be arbitrary in this sense. ... Legislators have the democratic legitimacy to make choices among value judgments, choices based on hunch or guesswork or even the toss of a coin, and other arbitrary choices. When agencies base rules on arbitrary choices they are legislating, and so these rules are legislative or substantive and require notice and comment rulemaking, a procedure that is analogous to the procedure employed by legislatures in making statutes. The notice of proposed rulemaking corresponds to the bill and the reception of written comments to the hearing on the bill.[14]

The question within the framework limned by Posner then becomes whether there is any "process of cloistered, appellate-court type reasoning by which"

---

[12]Pierce, *Administrative Law Treatise* 1:305.

[13]Hoctor v. U.S. Dept. of Agriculture, 82 F.3d 165, 167 (7th Cir. 1996).

[14]Hoctor v. U.S. Dept. of Agriculture, 82 F.3d at 169-71.

OSHA "could have excogitated" the rule that employers have to let workers go to the bathroom when they need to go from the standard requiring employers to provide a certain number of toilets per certain number of workers: "The rule is arbitrary in the sense that it could well be different without significant impairment of any regulatory purpose."[15]

Employers' lawyers would argue that because the real point of OSHA's action was not merely to interpret the toilet standard, but to change it by imposing new obligations on employers and conferring new rights on employees, OSHA lacked the power to achieve this result by means of a mere interpretive regulation. Instead, it was required to comply with the APA's notice and comment procedures before publishing a legislative regulation[16] that would be binding on the public and courts.[17] To be sure, employers would be constrained to concede that OSHA had met one threshold prerequisite because it could show that the legislative regulation section 1910.141(c)(1)(i) does contain an ambiguous term ("provided") that requires interpreting.[18] But employers would argue that even if it were reasonable for OSHA to state that a standard requiring employers to provide toilets must also mean that employers must let workers use them, the interpretation that they have to let workers go whenever workers say they need to void (in contrast with, for example, at the beginning and end of the workday and during scheduled breaks, every two hours, or at whatever intervals a urologist hired by the company says it is normal to void) was not obviously implicit in, and therefore could not be merely an interpretation of, section 1910.141(c)(1)(i). Therefore, in this view, OSHA should have been required to solicit comment from the public before promulgating a new, stricter standard.

To be sure, the substance of the toilet standard itself imposes some constraints on logical interpretations. For example, it would be inappropriate to permit employers to designate a specific time interval such as 10 a.m. to 10:15 a.m. as the only bathroom break for all employees because 15 workers would not all be able to void with one toilet within a 15-minute period, especially if walking from the work station to the toilet and back and/or doffing and donning protective equipment absorbed some of time.

Rob Swain, OSHA Counsel in the Solicitor's Office of the Department of Labor, speculated that some judges might invalidate the Memorandum on the grounds that it was a legislative regulation in its own right and should have been subjected to notice and comment pursuant to the APA, whereas others would

---

[15]Hoctor v. U.S. Dept. of Agriculture, 82 F.3d at 171.

[16]5 USC sect. 553.

[17]Telephone interview with Baruch Fellner, attorney, Gibson, Dunn & Crutcher, Washington, D.C. (Sept. 17, 2002).

[18]Phillips Petroleum Co. v. Johnson, 22 F.3d 616, 619 (5th Cir. 1994).

uphold it as a mere interpretation. Nevertheless, when asked whether, if OSHA had an opportunity to repeat the process, it would (abstracting from the issue of the long delay) prefer to promulgate the rule as a legislative regulation, Swain unhesitatingly said no because, if OSHA had to go through that lengthy process for every minor development in each of its standards, the agency would never have time to do anything else but notice and comment proceedings.[19]

Although OSHA asserted in the Memorandum that the access "requirement is implicit in the language of the standard," if it is also true, as demonstrated in *Void Where Prohibited*,[20] that prior to 1998 OSHA had in fact taken the position that the standard did not require employers to let workers use the toilet, it would be difficult to deny that the Memorandum enlarged employers' obligations. Even if OSHA could plausibly argue that prior to 1998 it had never taken that enforcement position in refusing to issue a citation, but only in conversation with the author, it could still not refute the claim that in its first 27 years of existence it had never issued a citation for failure to make toilets available when workers needed to use them. In that sense the Memorandum could be viewed as having enlarged employers' obligations.

John Miles noted that casting the interpretation as a Memorandum rather than as an opinion letter to the UFCW—a method chosen because the agency had also received other inquiries on the subject—perhaps gave it "a little more weight," but probably did not lend it any more legal force.[21] The legal force of OSHA's Memorandum is thus more circumscribed than that of a legislative regulation, which the Supreme Court has described this way: "If Congress has explicitly left a gap for the agency to fill, there is an express delegation of authority to the agency to elucidate a specific provision of the statute by regulation. Such legislative regulations are given controlling weight unless they are arbitrary, capricious, or manifestly contrary to the statute."[22]

In contrast, a different analysis applies when "the Secretary regularly employs less formal means of interpreting regulations prior to issuing a citation. These include the promulgation of interpretive rules, and the publication of agency enforcement guidelines.... Although not entitled to the same deference as norms that derive from the exercise of the Secretary's delegated lawmaking powers, these informal interpretations are still entitled to some weight on judicial

---

[19]Telephone interview with Rob Swain, OSHA Counsel U.S. Dept. of Labor Solicitor's Office, Washington, D.C. (Oct. 28, 2002).

[20]Marc Linder and Ingrid Nygaard, *Void Where Prohibited: Rest Breaks and the Right to Urinate on Company Time* 55-56 (1998).

[21]Telephone interview with John Miles (Nov. 12, 2002).

[22]Chevron U.S.A., Inc. v. National Resources Defense Council, Inc. 467 U.S. 837, 843-44 (1984).

review. A reviewing court may certainly consult them to determine whether the Secretary has consistently applied the interpretation embodied in the citation, a factor bearing on the reasonableness of the Secretary's position."[23]

As an agency's interpretation of its own rule, the Memorandum is not binding on employers or courts, but the latter would afford it "substantial deference"[24] as an expression of the agency's expertise, unless it is "plainly erroneous or inconsistent" with the regulation.[25] As the U.S. Supreme Court explained the force of so-called interpretive rules in a case involving the Wage and Hour Administrator's interpretation of the Fair Labor Standards Act:

> Congress...did create the office of Administrator, impose upon him a variety of duties, endow him with powers to inform himself of conditions in industries and employments subject to the Act, and put on him the duties of bringing injunction actions to restrain violations. ... He has set forth his views of the application of the Act under different circumstances in an interpretative bulletin and in informal rulings. They provide a practical guide to employers and employees as to how the office representing the public interest in its enforcement will seek to apply it. ...
>
> The rulings of this Administrator are not reached as a result of hearing adversary proceedings in which he finds facts from evidence and reaches conclusions of law from findings of fact. They are not, of course, conclusive, even in the cases with which they directly deal, much less in those to which they apply only by analogy. They do not constitute an interpretation of the Act or a standard for judging factual situations which binds a district court's processes, as an authoritative pronouncement of a higher court might do. But the Administrator's policies are made in pursuance of official duty, based upon more specialized experience and broader investigations and information than is likely to come to a judge in a particular case. They do determine the policy which will guide applications for enforcement by injunction on behalf of the Government. Good administration of the Act and good judicial administration alike require that the standards of public enforcement and those for determining private rights shall be at variance only where justified by very good reasons. The fact that the Administrator's policies and standards are not reached by trial in adversary form does not mean that they are not entitled to respect. ... We consider that the rulings, interpretations and opinions of the Administrator under this Act, while not controlling upon the courts by reason of their authority, do constitute a body of experience and informed judgment to which courts and litigants may properly resort for guidance. The weight of such a judgment in a particular case will depend upon the thoroughness evident in its consideration, the validity of its reasoning, its consistency with earlier and later pronouncements, and all those factors which give it power to persuade, if lacking power to control.[26]

---

[23]Martin v. Occupational Safety and Health Rev. Comm'n, 499 U.S. 144, 157 (1991).

[24]Martin v. Occupational Safety and Health Rev. Comm'n, 499 U.S. at 150-51.

[25]Thomas Jefferson University v. Shalala, 512 U.S. 504, 512 (1994).

[26]Skidmore v. Swift, 323 U.S. 134, 137-40 (1944).

By the same token, although the Memorandum is nevertheless the law, or at least a species of law—even before any court has upheld it—until a court invalidates it in an action by an employer that contests a citation based on it, it is unclear whether a court would entertain a declaratory action to invalidate it by an employer that is covered by it (and thus has legal standing to challenge it) but that has not yet been adversely affected by it.[27]

OSHA may also have violated the APA by having failed, without explanation, to publish the interpretive Memorandum in the *Federal Register*, in which the APA requires an agency, "for the guidance of the public," to publish "statements of general policy or interpretations of general applicability formulated and adopted by the agency."[28] Instead, OSHA published it merely on its website together with its opinion letters addressed to individuals. OSHA Counsel Swain argued that OSHA was not obligated to publish it in the *Federal Register* because "it is not binding on anyone except OSHA," not even on the OSHA Review Commission, which could hold it invalid. To be sure, while noting that the Memorandum originated as a response to a request (from the UFCW) for a clarification of the toilet standard (in light of Iowa OSHA's declaration in a letter to the author that it could not require employers to let workers use the bathroom), Swain conceded that in form and content it is obviously not an opinion letter, both because it is directed to OSHA officials and instructs them in detail as to how to enforce the standard. Swain also stated that few federal agencies publish analogous materials in the *Federal Register*.[29]

Swain's argument that the Memorandum is not controlled by the APA publication requirement because it is not binding on the public is undercut by the fact that it is precisely the hallmark of an interpretive—as opposed to a legisla-

---

[27]Swain doubted whether a court would entertain such a declaratory judgment. Telephone interview with Swain (Oct. 28, 2002).

[28]5 USC sect. 552(a)(1)(D) (2000). According to one influential statement of the law: "Only the phrases 'of general applicability' and 'of general policy' qualify which interpretations and policy-statements must be published. 5 U.S.C. sect. 552 (a)(1)(D). Congress inserted these qualifiers where it previously had the phrase 'not...addressed to and served upon named persons.' See S.Rep. No. 813, 89th Cong., 1st Sess. 6 (1965). The Senate characterized this change as a 'technical' one, id., suggesting that it considered the phrases equivalent. In an earlier version of the FOIA, Congress distinguished rules of general applicability from those that were particularized in scope, offering rates as an example of the latter. S. Rep. No. 1219, 88th Cong., 2d Sess. 4 (1964). The legislative history thus indicates a rather obvious definition of 'general': that which is neither directed at specified persons nor limited to particular situations." Nguyen v. U.S., 824 F.2d 697, 700 (9th Cir. 1987).

[29]Telephone interview with Rob Swain, OSHA Counsel U.S. Dept. of Labor Solicitor's Office, Washington, D.C. (Oct. 24, 2002).

tive—regulation that it is not binding on the public. Nevertheless, the U.S. Department of Labor, for example, has published hundreds of pages of interpretive regulations under the Fair Labor Standards Act in the *Code of Federal Regulations*.

It is unclear whether OSHA failed to publish the Memorandum merely because it systematically follows the APA's publication mandates only with regard to legislative regulations, as Swain suggested,[30] or whether it was trying, for political reasons, to limit publicity—a rather quixotic hope in light of the massive media attention the Memorandum had unleashed.

---

[30]Telephone interview with Rob Swain (Oct. 28, 2002).

8

Reactions to OSHA's New Interpretation

> Policies Reflecting Sensitivity to Women's Needs. MMMA [Mitsubishi Motor Manufacturing of America] agrees that it shall create a nursing room where nursing mothers can express milk in private, and shall make certain that its practices with respect to personal and hygienic needs of its female employees are realistic, appropriate and fairly administered in accordance with the April 6, 1998 OSHA interpretation of 29 C.F.R. 1910.141(c)(1)(i).[1]

On the day following the issuance of the Memorandum, Deborah Berkowitz, who as the UFCW's occupational safety and health director was chiefly responsible for securing OSHA's attention on the issue of voiding rights, summarized the impact of public pressure on OSHA: "It's a great thing. You moved mountains...and all that publicity you got made it happen."[2]

What exactly "it" was emerged in a new light during an interview with Richard Fairfax, OSHA's deputy director of compliance (who was reputed to be one of the agency's most pro-labor officials) on April 9, 1998, to determine what constituted a "reasonable" delay in making toilets available to workers. In response to a question as to whether making assembly-line workers wait 30 minutes was reasonable, Fairfax, instead of using a bureaucrat's standard evasion that he could not reply without additional information about the circumstances, answered flat out, Yes—provided that: the workers did not have a medical problem; the employer stated that it was too expensive to hire more relief workers; and the company policy was general rather than a special policy of a particular supervisor. He was neither swayed by the response that making people wait 30 minutes might in the long run cause a medical problem nor in the slightest amused by the rhetorical question as to whether he had to wait for 30 minutes

[1]Equal Employment Opportunity Commission v. Mitsubishi Motor Manufacturing of America, Consent Decree, ¶44(e) (C.D. Ill. June 11, 1998), on http://www.eeoc.gov/docs/mmma.html (settling a class action sex discrimination suit).

[2]Email from Deborah Berkowitz to Marc Linder (Apr. 8, 1998).

with his hand raised until his boss gave him permission to go to the toilet.[3] (Almost five years later, John Miles, who as Director of Compliance had issued the Memorandum and been Fairfax's boss, went much farther, declaring in an interview not only that it was okay to require an assembly-line worker to wait 30 minutes, but that in general "what we had in mind back then was every 2-3 hours—that's what's normal.")[4] In the real world of industrial relations, a quarter-century earlier an arbitrator had sustained a grievance submitted by the United Steelworkers alleging that an employer had violated their collective bargaining agreement by failing to provide adequate relief to workers "tied to" an assembly line inasmuch as these employees "had to wait as long as 30 and 45 minutes after requesting relief before it was provided." Remarkably, the excessiveness of a 30-minute wait was so self-explanatory that the arbitrator did not even bother to justify this part of his decision.[5]

News coverage of OSHA's new interpretation was extensive locally and nationally, including, for example, National Public Radio's morning news program, though reporters were not always able to recognize that the notion of toilet "privileges" had become an anachronism.[6] Much of that reporting, not unexpectedly, was a mixture of amusement and bemusement, revealing less interest in the oppressive employer practices to be remedied than in the opportunity to joke about bodily waste elimination and government intervention in this semi-taboo area. On Friday April 10, according to a national labor reporter, "[t]he story was all over the place—on the AP wire and CBS radio news."[7] The Associated Press article of April 9[8]—which repeated some of the material in the AP

---

[3]Telephone interview with Richard Fairfax (Apr. 9, 1998). Deborah Berkowitz and Jackie Nowell, directors of the occupational safety and health department of the UFCW, characterized Fairfax as unusually prolabor.

[4]Telephone interview with John Miles, OSHA Dallas Regional Administrator (Nov. 12, 2002).

[5]Mor-Flo Industries, Inc., 62 Labor Arbitration Reports (BNA) 398, 399 (1974). Although the arbitrator wasted no time on the quantitative issue, he derived the employer's contractual responsibility to provide relief from a clause on incentive pay requiring the company to base its production standards on the full use of the employee's time minus various rest periods including "personal allowances." *Id.* at 399. The union itself had declined to suggest a remedy.

[6]Brian Tumulty, "Workers Get 'Timely Access' to Bathrooms," *Detroit News*, Apr. 10, 1998, at A7 (Westlaw); Brian Tumulty, "Ruling Makes Federal Case of Workers' Toilet Times," *Des Moines Register*, Apr. 9, 1998, at 3A, col. 2 at 4.

[7]Email from Brian Tumulty, Gannett workplace reporter, to Marc Linder (Apr. 10, 1998).

[8]The piece by Lawrence Knutson, "U.S. Backs Workers' Restroom Rights," went out on the AP wire on April 9 at 13:28 EST.

article of March 12—as it ran in the next day's *Wall Street Journal* under the title, "Employers Must Provide Workers Access to Restrooms," stated: "There's no problem for most of the nation's workers. But in some jobs, such as food processing, assembly lines and telemarketing, meeting a simple human need can involve pleading and even the risk of losing a job."[9] Even the press in New Zealand reported it.[10] The author was journalistically dubbed "the nation's leading authority on restroom access in the work-place...a.k.a. 'Dr. Toilet.'"[11] In a widely syndicated article distributed by the Gannett News Service, he was quoted as saying that "the interpretation was not as specific as he had hoped. 'I'm in no way saying it's bad, but they gave themselves too much wiggle room. Say you're on an assembly line and it takes a half-hour to get relief, are they going to cite the employer or not? They are just fudging here.'"[12] That the reference to 30 minutes was not hypothetical, but a real answer from OSHA's Compliance Directorate was, to be sure, lost, but the prospect of future disputes over subjective assessments of what "prompt access" means was nevertheless raised.

The labor reporter for *The New York Times*, who for months had been promising to write an article about the problem of workplace toilet access, but ultimately never did,[13] cautioned at one point early on that he might have problems getting the story past his editor for reasons of social propriety and squeamishness.[14] In the end, the national newspaper of record failed to report on the matter until it was forced to do so after OSHA had acted on April 6, 1998 and its journalistic competitors had covered the story. But the colleague to whom the labor reporter had hurriedly handed off the assignment was not familiar with the subject[15] and the uncritical article suspiciously resembled an OSHA press release. Since Katharine Seelye's only acknowledged source was in fact OSHA's spokesman—who unintentionally exposed OSHA's history of failure to enforce em-

---

[9]"Employers Must Provide Workers Access to Restrooms," *Wall Street Journal*, Apr. 10, 1998, at B8, col. 6.

[10]Email from Geoffrey Palmer, former New Zealand prime minister, to Marc Linder (Apr. 19, 1998).

[11]Michael Biesecker, "Hold It? Workplaces Are Required to Have Toilets, But Employee Access Is Sometimes Limited," *Winston-Salem Journal*, Apr. 26, 1998, at 19, on http://www.journalnow.com.

[12]Brian Tumulty, "Workers Get 'Timely Access' to Bathrooms," *Detroit News*, Apr. 10, 1998, at A7 (Westlaw).

[13]Email from Steven Greenhouse to Marc Linder (Jan. 5, 1998, Apr. 8, and Apr. 10, 1998).

[14]Various telephone interviews with Steven Greenhouse (1997).

[15]The questions she posed to the author while they interviewed each other late in the afternoon as her deadline loomed revealed her limited comprehension of the issues. Telephone interview with Katharine Seelye (Apr. 9, 1998).

ployers' obligation to provide toilets by conceding that he did not know how widespread complaints were and that denial of access was a subject which OSHA had been merely "hearing anecdotally...about"—it was hardly surprising that she neglected to provide any account of the responsibility that the agency bore for years of inaction and irrational regulatory interpretation and non-enforcement.[16]

Ironically, the closest the refined *Times* came to touching on the subject before the Memorandum had been issued was tangentially to its reporting on other newspapers' reporting on "Paula Jones and her failed lawsuit against the President of the United States": interviewing a schoolteacher visiting Washington on April 2 about her reactions to the sight of "dozens of morning newspapers from across the nation, all screaming a chorus of headlines," the journalist commented that she "zeroed in on her kind of story, under a far more modest headline, announcing Federal action on the problem teachers have in obtaining enough bathroom breaks." Taking heart from the teacher's observation, "'[n]ow there's a serious story,'" the reporter regarded her point as "such a morning-after refreshment today inside the Washington Beltway, that there is more to life than Paula Jones, come-hither looks and distinguishing characteristics."[17] (Ironically, public school teachers are excluded from the Federal OSHA program.)[18]

Reactions to OSHA's intervention were largely predictable, with employers "assail[ing] the interpretation as unnecessary and beyond the scope of OSHA's authority,"[19] and unions such as the United Automobile Workers applauding the action.[20] The UFCW, which had been far and away the dominant force in the labor movement pushing for workplace voiding rights, "hailed" OSHA's step as "'a victory for human dignity.'" The union's president, Doug Dority, conjured up the tension between profits and people, while nevertheless vastly underestimating the universe of the affected "people":

> Human dignity doesn't end at the door. Tens of thousands of poultry and other food processing workers are subjected to the indignity of being arbitrarily denied the use of

---

[16]Katharine Seelye, "Newest Right: A Restroom Break," *N.Y. Times*, Apr.10, 1998, A13, col. 1 (nat. ed.).

[17]Francis Clines, "Testing of a President: The Capital: On the Morning After, Springtime Inside the Beltway," *N.Y. Times*, Apr. 3, 1998, at A23, col. 1 (Lexis). The article to which the reporter was referring was presumably the Gannett News Service piece by Brian Tumulty, which was published in the Washington, D.C. area, without a byline, as "OSHA Fights for Teachers' Bathroom Rights," *Potomac News* (Woodbridge, VA), Apr. 2, 1998 (Luce Press Clippings).

[18]See above ch. 5.

[19]Powers, Kinder & Keeney, "Inspectors Must Check Access to Toilets," *Rhode Island Employment Law Letter* 3(9) (Oct. 1998) (Lexis).

[20]E.g., "Need to Go to the Bathroom?" *Solidarity*, Aug.-Sept. 1998, at 7.

bathroom facilities. ... Maintaining ever-increasing line speeds to fatten the already bloated profits of the industry leaves no room for consideration of the human beings who operate the lines and produce the profits. Denying workers the right to go to the bathroom is just part of the dehumanization of the poultry workforce.[21]

In contrast, an employer-side labor law firm, reminiscing in a newsletter about a quip by Ronald Reagan in 1978 to the effect that it was amazing that people had known how to climb ladders before OSHA had issued its "'144 rules and regulations on ladder-climbing,'" was typical in admitting that "it's hard not to stifle at least a chuckle at one of OSHA's latest pronouncements."[22] Another employment law newsletter cautioned: "Many employers have had complaints from their employees in the past, ranging from reasonable to harassing. It remains to be seen whether the new interpretation of reasonableness will ease the problems of employers and employees...or whether it will cause a nightmare of complaints to OSHA from employees."[23]

Baruch Fellner, a Washington, D.C. corporate lawyer and former OSHA attorney,[24] asserted that the agency was "stretching its interpretation of the standard." Moreover, he claimed, without revealing the empirical basis of his claim, that "[t]he denial of employee restroom breaks is not a commonplace occurrence and should not be such a public issue. 'In the millions of workplaces in American there are more ubiquitous and more important safety and health concerns.... It causes one to wonder about OSHA's priorities....'"[25] Asked four years later for the basis of his claim that denial of access was not commonplace, Fellner admitted that he had merely been engaged in "speculation" based on the fact that his employer-clients had never brought to his attention any complaints on this point by their employees, who, Fellner asserted, were not shy about making complaints; he asserted, however, that OSHA knew no more about the frequency of such denials than he did. Asked how to reconcile his assertion that employers do not stop workers from going to the bathroom when they need to with OSHA's finding

---

[21]"Union Charges at Poultry Plant Bring New Policy in Workplace Bathroom Rights," U.S. Newswire, Apr. 14, 1998 (Westlaw).

[22]Powers, Kinder & Keeney, "Inspectors Must Check Access to Toilets."

[23]Armstrong, Teasdale, Schlafly & Davis, "OSHA Update," *Missouri Employment Law Letter*, July 1998 (Lexis).

[24]"Fellner served for eight years in the Solicitor's Office of the Department of Labor as Counsel in the Occupational Safety and Health Division. In those capacities, he was in charge of all litigation before the Occupational Safety and Health Review Commission and before the Courts of Appeals." http://www.gdclaw.com/insidegdc/whoswho/bio/?contactId=129.

[25]"OSHA Issues Interpretation Letter for Standard on Bathroom Facilities," *Daily Labor Report*, Apr. 13, 1998, at 70 d10 (Lexis).

that access was a problem in the chicken slaughter industry,[26] Fellner allowed as how he was not familiar with those circumstances, but urged caution given the source of such allegations in connection with organizing efforts by the UFCW. He also argued that the standard did not imply any obligation to provide access, though like OSHA lawyers before 1998, he conceded that locking bathrooms and not permitting use would mean that those toilets had not been "provided" within the meaning of the regulation.[27]

Just days before OSHA acted, on April 1, 1998, an event occurred at the United Parcel Service facility in Winston-Salem, North Carolina, that gave added urgency to the need for implementation of the new approach, but at the same time demonstrated that progress would not be without setbacks. According to the grievance complaint that he filed the next day, James Jenkins, a part-time sorter on the 4 to 8 a.m. shift (who also worked part time as a driver),[28] working under a Teamsters Union contract that provides for no breaks for four-hour shifts:[29]

> was denied access to use of a restroom. When I finished sorting a trailer I told...supervisor...I was going to the restroom. She told me I could not go. I explained I couldn't wait any longer and started toward the restroom. I was met by [another supervisor] who also said I couldn't go to the restroom and falsely accused me of walking off the job. I at that time went to the restroom and returned to the sort afterward.
>
> On Thursday morning [April 2] at the end of the sort I was told by [the second supervisor] I was being issued a warning letter and a three day suspension for walking off the job. After that meeting I wasn't allowed time to talk with shop steward [ ]. I was told to clock out and so I did. Before leaving I stopped to ask [another] shop steward [ ] a few questions. At that time [the second supervisor] walked up and ordered me to leave the property. After exiting the building she sent a police officer to escort me off the property.[30]

Under the rubric "Remedy Requested" Jenkins wrote: "I'm filing against prejudice, harassment, discrimination, slander, libel and defamation of character.

---

[26]The problem is, moreover, not confined to the United States. In Britain, a caller to the Trades Union Congress "'bad bosses' hotline reported that her employer, a poultry producer selling to major high-street stores, required workers to put their names on a waiting list to go to the loo—making them wait up to two hours to be allowed to go." "So You Want to Go to the Loo? Tough, Say P-Taking Bosses," *T &G Workplace Record* (undated) (furnished by Transport and General Workers Union).

[27]Telephone interview with Baruch Fellner, Washington, D.C. (Sept. 17, 2002).

[28]Telephone interview with James Jenkins, Winston-Salem (Dec. 3, 2002).

[29]Telephone interview with Donny Brown, business agent, Teamsters Local 391, Goldsboro, NC (Oct. 30, 2002).

[30]Teamsters Local Union No. 391, "Complaint" (4-2-98) (copy furnished by Donny Brown, Local 391 business agent).

I'm seeking immediate termination for flagrant and damaging misuse of authority, compensation for lost time and damages suffered."[31] Jenkins, who had been working at UPS since 1990, reported later that before this incident there had never been any problems going to the bathroom; he conjectured that the new manager, like many in her position, was trying to show who was boss.[32]

In the wake of the publicity surrounding OSHA's about-face three weeks earlier and also as a result of the efforts of Jenkins' outraged co-workers who were seeking to bring his victimization to public attention,[33] in late April the editorial page of the Sunday edition of the *Winston-Salem Journal* carried a long and thoughtful piece ("Hold It?") on Jenkins' experience that added more context to the dispute:

> James Jenkins was about two hours into a five-hour [sic] shift sorting packages when nature called.
>
> Actually, nature had been calling for quite a while. But Jenkins, who'd worked for the United Parcel Service in Winston-Salem for nine years, had held out until he'd finished unloading a truck. When he had a moment to spare, he asked his supervisor if he could go relieve himself.
>
> She said no. She told him to hold it.
>
> Jenkins told her he'd been holding it for an hour already and that he really, really needed to go.
>
> She said no; he'd have to wait.
>
> Jenkins told her it was an emergency and that he'd be right back. He says he went to the restroom and was gone for no more than three or four minutes....
>
> Though there are often at least two accounts of an incident between an employer and an employee, the basics of Jenkins' account were confirmed by UPS....[34]

UPS's fanatic struggle to control its employees' time was vividly on display in the denouement: reinstated after his union grieved, Jenkins was fired on April 22 for "'stealing time'" after he had been put on an unfamiliar truck route and finished late because he had difficulty finding several addresses.[35] The journalist editorialized: "You'd think that along with life, liberty, and the pursuit of happiness someone would have made voiding your bladder an unalienable right. ... Adults should be able to go when they feel they need to go. It should be a basic

---

[31]Teamsters Local Union No. 391 "Complaint" (4-2-98).

[32]Telephone interview with Jenkins.

[33]Telephone interview with Jenkins.

[34]Michael Biesecker, "Hold It?" *Winston-Salem Journal*, Apr. 26, 1998, at 19 (http://www.journalnow.com).

[35]Telephone interview with Jenkins.

civil and human right."[36] Three days after the editorial appeared, the union requested Jenkins' reinstatement, which it secured, but without pay.[37]

A week earlier another North Carolina newspaper had reported that North Carolina OSHA—with which, in the wake of the fire at the Imperial Food Products, Inc. plant that had killed 25 workers, Federal OSHA had terminated its operational status agreement in 1991 and temporarily resumed concurrent jurisdiction[38]—which at the time of the issuance of the Federal OSHA Memorandum had "never cited an employer for failing to give employees access to restrooms," had also never received a formal complaint about the problem, though it had received anonymous phone calls. Presumably workers in one of the least organized and most antiunion states, no less than their counterparts elsewhere,[39] were apprehensive about losing their jobs after filing the signed complaints required to trigger on-site investigations.[40] The *Winston-Salem Journal*, too, had reported that North Carolina OSHA had "never issued a citation to an employer for not allowing workers to use the toilets the law says they must provide."[41] At the same time, another part-time UPS worker in Winston-Salem wrote a letter to the editor noting that after the Jenkins incident he saw "a supervisor writing something when someone asked to go to the restroom. When I asked, 'What are you doing?' I was told that names were recorded along with how long it took them to use the restroom."[42]

Teamsters Local 391 representative Claude Gray telephoned a complaint to North Carolina OSHA on April 24, 1998. Although four years later he recalled distinctly that his complaint dealt with UPS's treatment of Jenkins,[43] according to the agency's written version of the oral complaint that it sent to the complainant for his confirming signature, he alleged that there had been only three toilets for more than 100 employees and that the bathroom had been dirty because the toilets had overflowed. The extant complaint file contains, in addition to the returned and signed form, a separate sheet of paper with the names of two em-

---

[36]Biesecker, "Hold It?"

[37]Teamsters Local Union No. 391, Complaint (Apr. 29, 1998) (copy furnished by Donny Brown).

[38]*Federal Register* 56:55192 (Oct. 24, 1991).

[39]See below ch. 11.

[40]Shannon Buggs, "Your Innate Right: You Can't Be Stopped from Going," (Raleigh) *News & Observer*, Apr. 19, 1998, at E1 (Westlaw).

[41]Biesecker, "Hold It?"

[42]Glenn C. Fields, "Restroom Time Checked" (undated newspaper clipping, later than April 11; copy faxed by Donny Brown).

[43]Telephone interview with Claude Gray, vice president, Teamsters Local 391 (Oct. 28 and Dec. 4, 2002).

ployees stating: "These two employees have been disciplined for going to the bathroom."[44] The Health Compliance Officer visited the workplace on May 21 and substantiated the complaint about an inadequate number of toilets; although UPS asserted to OSHA that these workers could also use other toilets if the ones in their area were being used or dirty, in fact five workers complained that they had not been allowed to use any others and one employee had received a warning letter for using a bathroom other than the designated one; UPS sought to justify the letter on the grounds that the toilet had not been dirty. Interestingly, despite the inspector's failure even to raise the issue of Jenkins' having been suspended for going to the bathroom, at the closing conference on June 8 the official asked UPS's District Health and Safety Manager whether he had been "aware of the recent interpretation issued by OSHA concerning bathroom availability," and he responded that "he had read a letter concerning this issue."[45] Four days later OSHA issued a citation (without imposing a monetary penalty) for failure to provide the proper number of toilets.[46]

Oddly, however, despite the complaint concerning suspension of the two employees for having gone to the bathroom, the OSHA citation said nothing about

---

[44]This information comes from Complaint No. 200073849, which was still on file at the North Carolina OSHA Winston-Salem office, although the complete case file had been transferred to the state headquarters in Raleigh. The complaint was signed on May 7 and received on May 11. A supervisor in the office read the relevant information from the complaint to the author over the telephone without disclosing any names. Since these names would be redacted even if the agency made the complaint available under a Freedom of Information Act request, the identity of the second suspended worker—Jenkins being the first—would still remain unknown. Telephone interview with Nelson Edwards, North Carolina OSHA, Winston-Salem (Dec. 9, 2002). Whereas Edwards had assumed that the complainant Gray had submitted this additional sheet, after being informed of OSHA's final letter, alluded to below, which does not mention the suspension, another compliance officer and former supervisor opined that it was equally plausible that the OSHA intake officer had created that sheet based on information obtained from Gray. Asked why the officer would have done that, the official speculated that the sheet meant that the complaint was to be referred to the North Carolina Workplace Retaliation Discrimination Office, which she suggested contacting. Telephone interview with Roseanne Morgan (Dec. 13, 2002). However, the administrator of that agency stated that neither Jenkins' nor Gray's name appeared on its computer and that in the ten years of the agency's existence he had never received such a complaint for retaliation for bathroom use. Telephone interview with Skip Easterly, Administrator, Employment Discrimination Bureau, Raleigh (Dec. 13, 2002).

[45]United Parcel Service, Inc., Insp. No. 301931424, Narrative, OSHA-1A (June 9, 1998), and Worksheet, OSHA-1B (June 11, 1998) (quote at 2).

[46]United Parcel Service, Inc., Insp. No. 301931424, on http://www.155.103.6.10/cgi-bin/est/est1xp?i=301931424.

denial of access. Instead, it merely stated that toilets had not been provided in accord with table J1 "for the male employees who worked in preload and reload, and for the male mechanics who were required to use the men's bathroom in the mechanics area." It then went on to note that the violation could be abated by marking the bathrooms in that area unisex, furnishing them with a lock, and insuring that they be used only by one employee at a time.[47] More mysterious still is that the final letter that was sent to the complainant on June 12, 1998, also failed to mention at all the third complaint item—that two workers had been disciplined for going to the bathroom. On the contrary, it expressly stated that the complaint encompassed only the two aforementioned items.[48]

In interviews both Jenkins and Gray were just as surprised to learn that OSHA had not cited UPS for having denied Jenkins access to the toilet as they were that the company wound up being cited for an issue that they contended had not been a problem.[49] To be sure, since the violation had occurred five days before the issuance of the Memorandum, which did not state that it had retroactive force, North Carolina OSHA could have declined to enforce the new interpretation. However, since Federal OSHA itself had anticipatorily enforced the interpretation in the Hudson Foods citation[50] almost nine months before issuing the Memorandum—which, moreover, asserted that the interpretation had always been implicit in the standard—it would in any event have been open to North Carolina

---

[47]United Parcel Service, Inc., Insp. No. 301931424 (issued June 12, 1998). UPS refused to provide any information concerning the matter. Telephone message left by Terry Thomas, UPS safety manager for Western North Carolina, in Charlotte, who stated that this refusal was a corporate decision (Dec. 3, 2002). Thomas had been present in 1998 when the OSHA inspector visited the facility.

[48]Letter from Roseanne Morgan, Re OSH Complaint # 200073849, OSH Inspection No. 301931424, to [ ] (June 12, 1998) (redacted and faxed to author by NC OSHA)..

[49]Telephone interview with Jenkins; Telephone interview with Gray (Dec. 4, 2002). Although Jenkins stated that OSHA had never interviewed him, it appears that it did. Telephone interview with Edwards. The shop steward named in the Inspection Narrative as present at inspection could not explain why the OSHA record did not mention Jenkins' suspension. Telephone interview with Chris Plemmons, Clemmons, NC (Jan. 2, 2003). Ironically, believing that OSHA had actually cited UPS for having denied Jenkins access, Gray, the business agent who had processed Jenkins' grievance and is now a vice president of Local 391, which had been aware of the Memorandum, was of the opinion that OSHA's enforcement has been effective in the sense that UPS has not denied toilet access to workers at Winston-Salem since. Telephone interview with Gray. Donny Brown, the current business agent, confirmed this statement. Telephone interview with Brown (Oct. 30, 2002).

[50]See above ch. 2.

OSHA to apply it as well.[51]

In the midst of processing the UPS complaint, North Carolina OSHA issued an internal Standards Notice, effective May 4, 1998, explaining the Federal OSHA Memorandum. The state agency paraphrased Federal OSHA's reasoning somewhat cryptically: "Along with the requirement to 'provide' toilets came the assumption that employees would be able to access toilets as needed or with reasonable restrictions." Remarkably, however, North Carolina OSHA also went beyond the April 6 Memorandum by extending its applicability to the construction industry.[52] The Notice may have been issued promptly, but its effective dissemination to and actual impact on inspectors were haphazard. As the OSHA official in charge of statewide Education and Training observed, once the Notice got that far, it was in the compliance officers' "lap to familiarize themselves with" such documents.[53] Some did while others did not.

Even four and a half years later, one compliance officer (who as District Health Compliance Supervisor had assigned the UPS complaint to an inspector and signed the citation) with almost two decades of OSHA experience not only was not familiar with the Memorandum, but stated affirmatively that what the agency had to follow in this matter was North Carolina labor law and specifically wage and hour law.[54] Indeed, several days later, even after having discussed the matter with colleagues who were familiar with the Memorandum, this same official insisted, in response to a hypothetical question, that if a worker, instead of obeying an order not to go to the bathroom and urinating in her pants, were fired for disobeying an order not to go to the bathroom, there would be nothing for OSHA to do because the agency cannot get her her job back and her only recourse would be filing a discrimination complaint with another state agency.[55]

---

[51]North Carolina OSHA might also have availed itself of another basis for inaction—that the denial had merely been a one-time incident attributable to a low-level supervisor—but its legitimacy would have been undermined by the fact that the incident had had such drastic consequences (suspension) for the worker.

[52]North Carolina Department of Labor, Division of Occupational Safety and Health, Standards Notice 58A (May 4, 1998) (fax furnished by N.C. DOL).

[53]Telephone interview with J. Edgar Geddie, Health Standards Officer, North Carolina OSHA, Raleigh (Dec. 9, 2002).

[54]Telephone interview with Roseanne Morgan, compliance officer and former supervisor, North Carolina OSHA, Winston-Salem (Dec. 9, 2002).

[55]Telephone interview with Morgan (Dec. 13, 2002). When the author, in an effort to draw out the logic of this approach and its disastrous consequences for deterrence, asked whether OSHA would stand idly by while an employer discharged all its employees for disobeying orders forbidding them to go to the bathroom, Morgan accused him of putting words in her mouth, but she never retracted her original claim. In fact, OSHA is empowered to seek reinstatement and back wages for workers who have been fired for

In contrast, another official stated that he would have cited an employer for not letting workers go to the bathroom even before April 6, 1998, although he had not issued any such citations and was not aware of any that his agency had issued.[56] The compliance officer who had conducted the inspection at UPS in Winston-Salem also noted almost five years later that he had never come across a denial of access case and had never heard co-workers mention one in the course of their frequent case discussions.[57]

North Carolina OSHA's timid enforcement activities can perhaps be understood in a related context. Within two weeks of Federal OSHA's announcement of the Memorandum, Representative Cass Ballenger, the North Carolina Republican chair of the House subcommittee that oversees OSHA, stated that he wanted "the new standard tested in three or four workplaces before deciding whether it's an effective way to address the problem. 'We've got to see how they will enforce it. The words "prompt," "restrictive" and "reasonable" could very well cause an OSHA inspector to go off the deep end.'" Ballenger's solicitousness on behalf of employers was not merely representational: he himself owned a factory in North Carolina that manufactured—irony of ironies—plastic bags for adult diapers. He preferred the free-market alternative to mandatory labor norms: "Ballenger says the current economy may correct the problem without government interference. 'With such low unemployment, you better take care of your employees in every possible way.... You can't get away with not letting your employees go to the bathroom. People will quit and go somewhere else.'"[58] To be sure, Ballenger failed to explain the free-market solution in periods of high unemployment or how he personally was able to "get away with" this regime at his factory: "'What we do in my company, is people go when there's a set-up on the machines; sometimes you can go five or six hours without a set-up,' the congressman said."[59]

Editorial reaction to OSHA's action was largely positive, waxing sarcastic over the discovery of the infantilizing treatment to which employers subject

---

exercising their rights under the Act. 29 USC sect. 660(c)(1) and (2) (2000).

[56]Telephone interview with Edwards.

[57]Telephone interview with Pat O'Brien, compliance officer, North Carolina OSHA, Winston-Salem (Dec. 10, 2002).

[58]Shannon Buggs, "Your Innate Right: You Can't Be Stopped from Going," (Raleigh) *News & Observer*, Apr. 19, 1998, at E1 (Westlaw). Deborah Berkowitz disagreed with the author's criticism of Ballenger's proposed intervention: "It's almost seems like your [sic] trying to trap the agency into being exactly what the republicans say it is--nitpicking big brother. I know your heart is in the right place--a little unrealistic and overzealous maybe--but in the right place. OSHA cannot do it all--it [sic; should be "if"] it could we wouldn't need you or me or unions." Email from Deborah Berkowitz to Marc Linder (Apr. 20, 1998).

[59]Tumulty, "Bathroom Breaks on OSHA's Agenda.".

workers. The *Chicago Daily Herald* commented: "Who knew when they were interviewing for a job that they should ask about restroom privileges when they discussed the company's benefits? ... Who knew that the practice of raising your hand in the classroom to go to the restroom might be relived later in life?"[60] The Newark *Star-Ledger* sounded a similar theme:

> When you're a kid in school, you have to ask the teacher for permission to go to the bathroom. It is routinely granted. Believe it or not, some grown-up workers have to ask for permission and it is not always granted. ... The Occupational Safety and Health Administration has long required that toilet facilities "be provided in all places of employment" for all workers, but the regulation says nothing about giving workers access to them. Employers looking for an edge noticed.[61]

A few weeks after OSHA had acted, the governor of Vermont on April 27, 1998 signed into law a bill under which: "An employer shall provide an employee with reasonable opportunities during work periods to eat and to use toilet facilities in order to protect the health and hygiene of the employee."[62] In "one of the eternal mysteries of the legislative process,"[63] the predecessor bills had required employers to provide not reasonable opportunities to use the toilet, but "a reasonable break to permit attendance at religious services."[64] The statute, is, in the words of the general counsel of the Vermont Department of Labor and Industry, "not a break law." It "was originally introduced as a law requiring periodic breaks during the work day but was drastically amended to the form that passed."[65] Although "at least some of the sponsors of the early bills were aware that toilet access was a problem at some worksites,...that particular issue did get an increased amount of press attention in 1997. In earlier sessions, sponsors believed that a broader work break bill would also solve toilet access issues. Lobbying by Associated Industries of Vermont, the Chamber of Commerce, and particularly transportation concerns (who argued that any type of fixed break provision would interfere with service) defeated the earlier versions. Those

---

[60]"Workers Have Right to Go," *Chicago Daily Herald*, Apr. 11, 1998, at 6 (editorial) (Lexis).

[61]"The Right to Go," Newark *Star-Ledger*, Apr. 14, 1998, at 12 (editorial) (Westlaw).

[62]21 Vermont Stat. Ann. Sect. 304.

[63]Email from J. Stephen Monahan, General Counsel, Vermont Dept. of Labor and Industry, Montpelier, to Marc Linder (Dec. 10, 2002).

[64]Vermont Legislature, 1995-96 Sess., H.220 , on http://www.leg.state.vt.us/docs/1996/bills/intro/H-220.HTM; Vermont Legislature, 1993-94 Sess., H.44, on http://www.leg.state.vt.us/docs/1994/bills/intro/H-044.HTM.

[65]Email from J. Stephen Monahan, General Counsel, Vermont Dept. of Labor and Industry, Montpelier, to Marc Linder (Dec. 9, 2002).

groups were unable to provide a convincing rationale for opposing the language offered for toilet access, so that is what passed. S.130 language was amended to the version that passed, by the Senate in March of 1997 a year before O.S.H.A. issued its memorandum."[66] As enacted, the law "primarily addresses complaints that certain employers were denying employees access to the bathroom, because they believed that it either slowed down production and/or was being used as a break." In the years since enactment, however, the department has not needed to define what "reasonable opportunities" means because it has not cited any employers for violations and the few complaints it has received were resolved by a call to the employer.[67]

Interviewed more than four years later, the project manager of Vermont OSHA, who had worked for OSHA for more than two decades and whose agency had received no complaints about, and issued no citations for, denial of toilet access, recalled that when the OSHA Memorandum first came out, "we all thought it was a joke": employers were always complaining that OSHA was unnecessary because employers did what was right, and "here OSHA had to regulate something as" basic as urinating.[68]

In September 1998, a research analyst in the Pennsylvania Legislative Research Office wrote a memo requesting that the Legislative Reference Bureau draft legislation on behalf of Representative Sue Laughlin that was to have included the following provision:

> All employees...within the Commonwealth shall be afforded at least one opportunity during every two hour period of employment to use toilet facilities. The time afforded for employee breaks may not be cumulated and expire at the end of each two hour period. The employee shall have sole authority over the timing of toilet breaks while being expected to recognize reasonable accomodation [sic] to workplace needs. Toilet breaks shall not be accounted as concurrent with daily meal breaks.... Employees whose temporary illness requires more frequent toilet breaks shall be accomodated [sic] or provided with a paid sick day. After three days of such illness, or more than two episodes in one month, an employer may require physician confirmation of the illness and a report of its prog-

---

[66]Email from Monahan to Linder (Dec. 10, 2002).

[67]Email from Monahan to Linder (Dec. 9, 2002). According to Joanna Goodrich, the Wage and Hour Program Coordinator, the agency has received no complaints about toilet use. Telephone message from Joanna Goodrich (Dec. 9, 2002).

[68]Telephone interview with Bob McCloud, Project Manager, Vermont OSHA (Sept. 20, 2002). He was far from being alone in voicing that sentiment. Another OSHA official, for example, recalled that when the Memorandum reached her office she thought it odd that employers had to be told something as commonsensical as that they had to let workers use the toilets that firms were required to provide. Telephone interview with Cheryl Gray, Safety and Health Assistant, Omaha Area OSHA Office (Jan. 2, 2003).

nosis. Toilet breaks shall not be subtracted from an employee's time worked when calculating wages. ...
Failure to provide toilet breaks...as required...shall result in a minimum fine of $200 per employee for the first citation, $300 per employee for the second citation, and $1,000 per employee for the third and subsequent citations. ...
Employers found in a court of law to have retaliated against employee filing complaints under this Act shall be liable for treble damages for wages lost....[69]

Although even this carefully crafted proposal fell far short of the medically indicated standard of permitting healthy workers to urinate more often than once every two hours, which was arguably already in force under OSHA, it was nevertheless too radical even to be filed as a bill. Especially unorthodox was its conferral of discretion on workers to determine when to go and its inversion of OSHA's reasonableness criterion, pursuant to which in Pennsylvania workers would have been expected to accommodate the needs of production.

Other high-profile cases of denial of toilet access after the OSHA Memorandum went into effect included the Freshwater Farms catfish processing plant in the very impoverished community of Belzoni, Mississippi. On November 16, 1998, 68 of the exclusively black and largely female workers, whose wages hovered over the minimum wage and whose shifts were not pre-determined but lasted until management said there was no more fish to process that day, protested the company's failure to deal with their demands seriously by standing outside rather than entering the plant and were fired for violating the no-strike clause in the collective bargaining agreement with UFCW Local 1529, which did not support their demands:

A major concern is the company's refusal to allow adequate time for workers to use the bathroom. Their workstations are between 400 and 500 feet from the restrooms. Only seven minutes are allowed for bathroom breaks, three times a day. "By the time you get your gear off and get to the wash room, the seven minutes is practically gone," [worker Joann] Hogan said. "Then you have to get sanitized and redressed up before you can go back to your work station." If they take longer than seven minutes, workers must clock out or they will be written up.[70]

Though the union later stated that it would take their grievance to arbitration, some of the workers instead formed the Catfish Workers of America.[71]

---

[69]Memo from Dave Callen to Rep. Laughlin (Sept. 18, 1998).

[70]Susan Lamont and Ronald Martin, "Catfish Workers Fight for Dignity," *The Militant*, Jan. 18, 1999, on http://www.themilitant.com/1999/632/632_17.html.

[71]Ved Dookhun, "Mississippi Actions Back Catfish Workers' Fight," *The Militant*, Apr. 26, 1999, on http://www.themilitant.com/1999/6316/6316_5.html. See also Howard

Some management publications took the book and problem seriously, calling the former, for example, "a significant contribution to the literature on a little discussed, but important workplace issue."[72] *CIO*, a magazine for corporate chief information officers, went so far as to offer this "solution" to the problem: "Lock stingy employers in a conference room for a few hours with a big pot of coffee. We think the sympathy will flow."[73] By the end of 1998, a monthly publication for small business was warning its readers:

> Have you ever grown weary of watching your employees trotting off to use the facilities every hour on the hour? Ever considered cracking down on what you suspect to be bogus breaks? You'd better proceed with caution, lest you inadvertently drive your health care spending higher.
>
> According to a new book...Void Where Prohibited ..., companies that regulate controls on bathroom visits run the risk of major illness. That could eventually lead to higher health care premiums, to say nothing of the possible exposure to legal liability.[74]

A comprehensive and probing interview with the author also appeared, appropriately enough, in *Corporate Crime Reporter,* a Naderite publication.[75]

Not everyone, however, appreciated an excremental vision of the workplace. The Princeton University Industrial Relations Section may have included *Void Where Prohibited* on its "distinguished list"[76] of 16 noteworthy books in industrial relations and labor economics for 1998,[77] but the discipline's leading journal in the United States, *Industrial and Labor Relations Review*, declined to publish a review of the book on the grounds that the journal's screener had been offended by the title, while six reviewers, possibly offended by the subject as well, had turned it down.[78] These scholar-censors were reminiscent of the city councillors

---

Rambsy, "Fired Catfish Workers Start Their Own Union," *The Progressive*, 5(63):16 (May 1, 1999) (Lexis)..

[72]Mark Lengnick-Hall, Book Review, *Personnel Psychology* 52(10):218-21 at 221 (Spr. 1999). See also Mary-Kathryn Zachary, "OSHA Provides Interpretations of Its Bathroom Break Standards," *Supervision* 59(8):20 (Aug. 1998) (Westlaw).

[73]Christopher Koch, "Give Me a Break," *CIO*, sect. 2, at 16 (Aug. 15, 1998).

[74]John Ettore, "Let My People Go," *Small Business News Cleveland*. Nov. 1998.

[75]"Interview with Marc Linder," *Corporate Crime Reporter* 12(34):11-16 (Sept. 7, 1998).

[76]Letter from Prof. Orley Ashenfelter, Director, Industrial Relations Section, to Marc Linder (Aug. 5, 1999).

[77]Princeton University, Industrial Relations Section, "Selected References," No. 289 (Apr. 1999).

[78]Email from Barbara Lanning, ILRR office manager, to Marc Linder (Mar. 15, 1999); email from Frances Benson, Editor-in-Chief, ILR Press of Cornell University

of Kansas City whom George Bernard Shaw, after they had thwarted performance of his play *Mrs Warren's Profession* by offering the performers "the alternative of leaving the city or being prosecuted" under a local by-law against indecency, characterized as possibly "simply stupid men who thought that indecency consists, not in evil, but in mentioning it."[79]

Ironically, familiarity with the book's critique at times outran knowledge of OSHA's about-face even in left-wing circles with a specific interest in the lack of workplace rights.[80] Newspaper legal advice columnists, too, were eventually enabled to answer questions about workplace toilet access correctly. In late 1999 a reader who asked the *Houston Chronicle* whether it was legal for a supervisor to write up a worker who, unable to hold it in any longer, had gone to the toilet without having found a replacement, was, under the headline, "Law Doesn't Entitle You to a 'Potty Break' at Work," misinformed by Professor Richard Alderman, that, unless a worker is disabled, "your employer can set the conditions of taking a 'potty break.' You agreed to his terms and now you must adhere to them. If you are dissatisfied with the terms of employment you can either try to have them changed or look for another job."[81] After being informed that he

---

Press, to Marc Linder (Mar. 15, 1999). Ultimately the journal did publish a book review, but only because the author shamed the editors into doing so and furnished a list of reviewers who would not be "offended." Email from Marc Linder to Barbara Lanning (Mar. 15, 1999).

[79]Bernard Shaw, "Preface" to *Mrs Warren's Profession*, in idem, *Plays Unpleasant: Widowers' Houses, The Philanderer, Mrs Warren's Profession* 181-212 at 209 (1976 [1894]). Although all of the author's previous books published in the United States had been reviewed in *Choice*, which quickly reviews books for college librarians, this one was not. The editor of the ILR Press found it "a puzzle. ... I do think it's odd they haven't reviewed yours." Email from Frances Benson to Marc Linder (June 1, 1999).

[80]For example as late as June 1999, Barbara Ehrenreich, writing in *In These Times*, was still citing *Void Where Prohibited* as documenting the lack of a "federal guarantee of the right" "to pee." Barbara Ehrenreich, "The Lexus and the Right to Pee," *In These Times*, June 13, 1999, at 9 (Lexis). The record was set straight in Marc Linder, "Bathroom Breakthrough," *In These Times*, Aug. 2, 1999, at 2, col. 1 (letter to editor). But see Barbara Ehrenreich, "Nickel-and-Dimed: On (Not) Getting By in America," *Harper's Magazine*, Jan. 1999, at 37-52 at 48 n.6 ("Until April 1998, there was no federally mandated right to bathroom breaks"), and idem, *Nickel-and-Dimed: On (Not) Getting By in America* 37 (2001).

[81]Richard Alderman, "Law Doesn't Entitle You to a 'Potty Break' at Work," *Houston Chronicle* (undated clipping, ca. early Nov. 1999). Alderman is the incumbent of the Dwight Olds Chair in Law at the University of Houston and two-time recipient of "the highest honor given by the American Bar Association and the State Bar of Texas for his work in educating the public about the law." http://www.law.uh.edu/faculty/.

had disseminated false advice,[82] Alderman published a correction, emphasizing that "employers must make restroom facilities available to all employees, and must allow prompt access to such facilities."[83]

The need for OSHA's new approach to toilet access became all the more manifest when, following the book's publication, the author became aware of a profession many of whose practitioners were urinarily perhaps even more oppressed than chicken processing workers—pharmacists, especially those working long shifts in single-pharmacist pharmacies within supermarkets or department stores of chains such as Wal-Mart.[84] A 1998 study of about a thousand pharmacists in New York State revealed that 44 percent worked in chain stores, which "tend to be where employment opportunities are now available."[85] Overall, 33 percent of pharmacists worked 11-12 hours shifts and 6 percent 13 or more hours; chain-store pharmacists accounted for 70 percent and 68 percent, respectively, of these two groups working long shifts.[86] Respondents reported that 60 percent of all pharmacists, 75 percent of employees of chain employees, and 85 percent of all pharmacists working shifts of 13 or more hours disagreed with the statement: "I am usually able to take a rest or bathroom break when needed."[87] Indeed: "A majority of pharmacists report that they cannot leave their posts, even for a few minutes...." In addition, 76 percent of all pharmacists, 85 percent of chain-store employees, and 95 percent of pharmacists working shifts of 13 or more hours disagreed with the statement: "I am usually able to take meal breaks of sufficient duration."[88]

Wal-Mart in particular has been engaged in such a ruthless struggle to control time at the workplace—two recent lawsuits filed by 1,100 pharmacists against Wal-Mart for failure to pay overtime revealed that some worked 70 hours a week "'without even a bathroom break'"[89]—that it fired a diabetic pharmacist for closing the pharmacy to take an uninterrupted lunch during his 10-hour shifts to

---

[82]Email between Marc Linder and Richard Alderman (Dec. 7, 1999).

[83]Richard Alderman, "Collector Rings Up Debtor—Repeatedly," *Houston Chronicle*, Dec. 15, 1999, at 13, col. 1, 3.

[84]See Diane Lewis, "A 15-Minute History: Book Takes a Look at Troubles of Many Workers, Firms Over Bathroom Breaks," *Boston Globe*, Nov. 29, 1998, at F4

[85]Pharmacists Society of the State of New York, *1998 Workplace Survey* 3 (June 1998).

[86]Pharmacists Society of the State of New York, *1998 Workplace Survey* at 9.

[87]Pharmacists Society of the State of New York, *1998 Workplace Survey* at 17.

[88]Pharmacists Society of the State of New York, *1998 Workplace Survey* at 16.

[89]John Accola, "Judge Rules Against Wal-Mart: Pharmacists Shorted on Overtime Pay," *Rocky Mountain News*, Sept. 24, 1998, at 1B (Lexis). See In re Wal-Mart, Fair Labor Standards Act Litigation, 58 F. Supp.2d 1219 (D. Colo. 1998).

eat a special diet after his noon-time insulin injection.[90] In spite of such practices, Wal-Mart felt no compunctions about asserting: "'We look at them [pharmacists] as professional people who can make responsible decisions, and they can take bathroom breaks or lunch breaks as they deem appropriate.'"[91]

Society may take an interest in pharmacists' problems and override employers' working-time obsessions, but only once it uncovers the externalities of that autocratic system: after evidence surfaced that "overworked" pharmacists were making mistakes in filling prescriptions that proved fatal to patients, pharmacy officials in North Carolina began "trying to force drug stores to cut back on pharmacists' hours and to require lunch and bathroom breaks in an effort to improve safety."[92] When that state's Board of Pharmacy proposed a rule prohibiting employers from requiring pharmacists to work more than 12 hours a day and mandating rest breaks, large pharmacy chains, in the face of pharmacists' complaints that "they cannot even take bathroom breaks"[93] and of a study showing that 91 percent of pharmacists in North Carolina favored breaks,[94] opposed the intervention on the grounds that their employees "'don't want rules mandated to them. That would change the flexibility of the profession.'"[95] Fortunately for employers alleging that pharmacists "'set their own hours,'"[96] the Rules Review Commission ruled that Board's proposal was not reasonably necessary. At the end of 2002 the Board's judicial appeal was still pending.[97]

Other, less systematic reports revealed theretofore unpublicized extreme abuse of other white-collar workers, which had become blatantly unlawful under the OSHA Memorandum. For example, train dispatchers worked an eight-hour shift without toilet or meal breaks; though they could and did eat at their work

---

[90]Orr v. Wal-Mart Stores, Inc., 2002 U.S. App. Lexis 14659 (8th Cir. 2002).

[91]RuthAnn Hogue, "No Relief in Sight," *The Arizona Daily Star*, Apr. 11, 1999, at 1D, col. 2, at 4D, col. 3-4 (quoting Jessica Moser, Wal-Mart company headquarters spokeswoman).

[92]Sheryl Stolberg, "The Boom in Medications Brings Rise in Fatal Risks," *N.Y. Times*, June 3, 1999, at A1, col. 1 (Lexis).

[93]Aissatou Sidime, "Pharmacists Cite Workload in Suit," *Tampa Tribune*, June 19, 1999, at 1 (Lexis).

[94]Carol Ukens, "Patrons and Pharmacists Like Breaks, Study Finds," *Drug Topics* 144(16):32 (Aug. 21, 2000) (Lexis).

[95]Don Patterson, "Time for a Break: Retail Pharmacists Who Work Long, Busy Days May Put Patients at Risk," *News & Record* (Greensboro), Feb. 9, 1999, at D1 (Lexis) (quoting CVS spokesman).

[96]Sabrina Jones, "Pharmacists Could Use a Break a Day," *News and Observer* (Raleigh), Mar. 28, 1999, at E1 (Lexis) (quoting CVS spokesman).

[97]Carol Ukens, "North Carolina Board Sues to Keep 12-Hour Rule on Work Shifts," *Drug Topics*, July 19, 1999, at 20 (Lexis); http://www.ncbop.org/news.htm.

stations, they had to run to the toilet and void "as quickly as" possible because they "remain responsible for all actions on their territory."[98] A woman who worked at a Space Operations Center Computer Operations room in New Mexico which under company rules could not be left unattended even for a moment, was often not relieved for eight hours in order to be able to go to the bathroom. Moreover, the employer told her that "male employees did not have that problem and that she should relieve herself in a trash can or she could pull up a floor tile." As a result she developed severe septicemia and had to be taken by ambulance to the hospital.[99] And OSHA's new approach also gave hope to transgendered workers, who are sometimes barred from using any bathroom at work.[100]

Reports of additional blue-collar occupations facing toilet access problems also surfaced after the Memorandum was issued. Crane operators in aluminum smelters may work up in their cabs for as long as four consecutive hours; climbing down and up, doffing and donning protective equipment, and walking to and from the bathroom alone may take as much as 10 minutes; and although they are technically free to go after they have disposed of a load, they are under pressure from workflow, management, and their own co-workers not to leave.[101]

Another group of workers especially cut off from toilet access are employed by utility companies to install or repair electrical lines. For example, in California, according to a professor who was a consultant to a firm marketing urination/defecation bags to utility companies, linemen have such heavy work loads that they do not have the time to come down from their cherry pickers to go urinate in the woods. Instead, many urinate into the cherry picker basket, and in at least one instance a worker was reported to have been electrocuted while standing in his own urine.[102] One week before Federal OSHA issued its Memo-

---

[98]Email from Steve Popkin to Marc Linder (Jan. 11, 1999). Popkin, who worked for the engineering-consulting firm Foster-Miller, was working on a project studying the workload, stress, and fatigue of train dispatchers.

[99]Sanchez v. PRC, Inc., First Amended Complaint for Damages (D. N.M., CV 96 1093, Feb. 5, 1997).

[100]Email from Prof. Phyllis Randolph Frye, Thurgood Marshall School of Law, to Marc Linder (Jan. 29, 2000); Phyllis Randolph Frye, "The International Bill of Gender Rights vs. The Cider House Rules: Transgenders Struggle with the Courts Over What Clothing They Are Allowed to Wear on the Job, Which Restroom They Are Allowed to Use on the Job, Their Right to Marry, and the Very Definition of Their Sex," *William and Mary Journal of Women and the Law*" 7:133-216 at 182-88 (Fall 2000).

[101]Telephone interview with Donny Lawrence, president, United Steelworkers Local No. 4895, Alcoa plant, Rockdale, TX (Oct. 10, 2002).

[102]Telephone interview with Paul Holt, Carlsbad, CA (Oct. 21, 2002); *Corporate Crime Reporter*, Sept. 7, 1998, at 13. Holt had heard this account from many in the utility industry, but could not confirm it independently. American Innotek, for which he had

randum, Cal/OSHA informed the firm that the toilet requirements of the state's construction safety order "are not specific enough to preclude the use of the 'Brief Relief/Disposable John & Portable Tent System,'" but that the device "is not a viable alternative in lieu of the facilities required" by the General Industry Safety Orders." Consequently, the agency concluded that the product "would be a valuable asset for those employees, employers and persons engaged in certain activities where standard facilities are not readily available. Mobile crews...and construction cites...are two examples."[103]

The unusual problems of this occupational group also shed light on the special problems of women workers, which OSHA could deal with after April 6, 1998. When the Central Illinois Light Company's sole female lineman sued her employer for sexual harassment on the grounds that it had failed to provide her with restroom facilities, Richard Posner, the most intellectually diversified judge in the United States, who prides himself on being a tough realist, had no compunctions about depicting the male workers as unconcerned about their working conditions:

> Linemen work where the lines are, and that is often far from any public restroom; nor do the linemen's trucks have bathroom facilities. Male linemen have never felt any inhibitions about urinating in the open, as it were. They do not interrupt their work to go in search of a public restroom. Women are more reticent about urinating in public than men. So while the defendant's male linemen were untroubled by the absence of bathroom facilities at the job site, the plaintiff was very troubled and repeatedly but unsuccessfully sought corrective action, for example the installation of some sort of toilet facilities in the linemen's trucks.[104]

Speaking on behalf of the Seventh Circuit Court of Appeals in 2000, Posner agreed that a reasonable person would think an absence of bathroom facilities in

---

been a consultant, had heard the same accounts, but could also not confirm them. Telephone interview with Bruce Salerno (Oct. 23, 2002); fax from Terry Cassidy to Marc Linder (Oct. 23, 2002).

[103]Letter from Frank Ciofalo, Deputy Chief, California Div. of Occupational Safety and Health, to Bob Locher, American Innotek, Inc. (Mar. 30, 1998). The company markets its product to employers on these grounds: "Corporate liability and image, work site health and safety risks, job site down time and the risk of having your employees caught on film are all issues that need to be addressed when determining the personal sanitation needs of your employees. Is 'holding it' part of your corporation's safety policy? The risks associated with 'delayed voiding' or 'holding it' are not worth the reward. Just take a look at some of the many diseases associated with unsanitary lavatory facilities and poor sanitation practices." http://www.briefrelief.com.

[104]DeClue v. Central Illinois Light Co., 223 F.3d 434, 436 (7th Cir. 2000).

most workplaces an "intolerable working condition"; he also agreed that if such an absence deterred women, but not men, from seeking or holding a type of job, that absence may be a form of actionable sex-based discrimination if "those facilities can be made available to the employees without undue burden to the employer." He also emphasized that "women are not 'unreasonable' to be more sensitive about urinating in public than men; it is as neutral a fact about American women, even though it is a social or psychological rather than physical fact, as the fact that women's upper-body strength is on average less than that of men, which has been held in disparate-impact litigation to require changes in job requirements in certain traditionally male job categories." But the federal appeals court rejected the plaintiff's claim because she brought it based not on disparate impact, but on hostile work environment or sexual harassment, which requires a showing that co-workers or supervisors sought to make the workplace intolerable or "severely and discriminatorily uncongenial" to women. Posner conceded that since the chief defense to such a charge is that the employer had done all that he could to prevent the harassment, "as a purely semantic matter it might be possible to argue that an employer who fails to correct a work condition that he knows or should know has a disparate impact...is perpetuating a working environment that is hostile to that class." But the court rejected that argument on the grounds that it would make the two types of discrimination one.[105]

The dissenting judge, Ilana Rovner, took a step toward adopting for the workplace the substantive rather than merely formal equality for women that legislators had created outside the workplace.[106] Legislatures have been much more solicitous of the bladders of the (female) public outside of the workplace than of workers' toilet access. For example, Wisconsin's contribution to the potty-parity movement explicitly stresses "speed of access":

> The owner of a facility where the public congregates shall equip and maintain the restrooms in the facility where the public congregates with a sufficient number of permanent or temporary toilets to ensure that women have a speed of access to toilets in the facility where the public congregates that equals the speed of access that men have to toilets and urinals in that facility where the public congregates when the facility where the public congregates is used to its maximum capacity.[107]

Rovner went beyond Posner in rooting the heavier burden women bear in voiding outdoors not simply in reticence: "The fact is, biology has given men less to do in the restroom and made it much easier for them to do it. If men are less

---

[105]DeClue v. Central Illinois Light Co., 223 F.3d at 436-37.

[106]See Marc Linder and Ingrid Nygaard, *Void Where Prohibited: Rest Breaks and the Right to Urinate on Company Time* 154-56 (1998).

[107]Wisc. Stat. Ann. sect. 101.128(2)(a) (1997).

reluctant to urinate outdoors, it is in significant part because they need only unzip and take aim," whereas women face "a more cumbersome, awkward, and time-consuming proposition." Rovner was also willing to find hostile-environment discrimination where an employer fosters it by failing to respond to complaints calling for corrective action of conditions that he knows have a disparate impact in light of prior reported judicial decisions revealing that "some employers not only maintain, but deliberately play up, the lack of restroom facilities...as a way to keep women out of the workplace." Rovner was able to reach this conclusion with even greater cogency because the employer had in fact been able to provide toilet facilities: the company had given the plaintiff use of a port-a-potty for two weeks, and, after she had filed a complaint with the Equal Employment Opportunity Commission, the Central Illinois Light Company (which had otherwise merely offered her the stigmatizing use of a truck to drive 10 to 20 miles to the nearest restroom) "began providing 'Brief Reliefs' (disposable urine bags) and privacy tents for DeClue and the other [male] lineworkers to use at jobsites."[108]

---

[108]DeClue v. Central Illinois Light Co., 223 F.3d at 438-39.

Part IV

An International Comparative Interlude

In the factory, you learn to hold it. You keep the line moving, regardless of what it is you're processing or assembling. It's not a matter of a capitalist jackboot pressing down on the proletariat's throat. It's the simple but all-consuming demands of the job. The line must move. Interruptions cost money.[1]

---

[1]Andrew Dreschel, "A Salute to Those Who Can't Take a Pee Break," *Hamilton Spectator*, Aug. 13, 1999, at A3 (Lexis).

9

Meanwhile Up in Civilized Canada: Pee for a Fee

It is very difficult for most of us to believe that being able to use the bathroom is a serious issue in the American workplace.[1]

The problem of workplace restrictions on toilet use is by no means confined to the United States. In Canada it is not even limited to nonunion workplaces. For example, in 1994, management at Gainers, Inc., an Edmonton, Alberta meatpacking plant organized by the UFCW, asserting that "'systematic abuse of breaks'" had lowered productivity and would cost the company $1 million annually,[2] imposed a "'pee for a fee'" rule on its 890 employees[3]: "If employees need to use the washroom outside of [two 15-minute] scheduled breaks, they must report to supervisors, who record the time of departure and return and dock workers' pay at the end of the week," which worked out to about one dollar per bathroom break. While "angry" workers asked "'[h]ow can they charge you for going to the washroom,'" the Alberta Labour Ministry confirmed that nothing in the provincial employment code "requires a person to be paid when they don't work."[4]

In 1998[5] this innovative approach to suppressing workers' need to void

---

[1]Diversity Training Newsletter, May-June 1998, on http://www.diversitytraining.com/may_june.html.

[2]"Workers Want New Washroom Rule Flushed," *Toronto Star*, Sept. 21, 1994, at A22 (Lexis).

[3]"Workers Are Very Pee-ved," *Hamilton Spectator*, Sept. 28, 1994, at A14 (Lexis).

[4]"Workers Want New Washroom Rule Flushed." According to Howard Levitt, "Taking Frequent Smoke Breaks? You Could Be Stealing Time," *Toronto Star*, Nov. 21, 1994, at D3 (Lexis), workers at Gainers were required to punch out when they took unscheduled bathroom breaks.

[5]In May of that year the author's op-ed piece was reprinted in a Canadian publication; Marc Linder, "Fighting for Restroom Rights," *Workplace News*, May 1998, at 5, col. 1-2, at 6, col. 5.

spread to a Maple Leaf Pork plant in Burlington, Ontario, which was owned by Maple Leaf Foods, Canada's largest food processing company, which in the interim had bought the Gainers plant in Edmonton (and shut it down permanently after the UFCW had struck it as part of a nationwide strike). The employer's demand for restricting bathroom breaks was part and parcel of its plans to compete with U.S. firms in the world market. "Faced with a mature Canadian market," the company decided to orient its growth to Asia, and with 40 percent of the Burlington plant's output being sold in the Far East, Maple Leaf Foods insisted that it had to close the 50-percent gap between Canadian wages and those of its U.S. competitors.[6]

Significantly, Maple Leaf Food's strategy for pushing Canadian workers' wages down to lower, U.S. levels also brought with it the drive to undercut U.S. employers' oppressive toilet-break policies, almost precisely at the time that OSHA was issuing its Memorandum outlawing them. Workers at the Burlington plant struck on November 15, 1997, after the company in contract negotiations had proposed both hourly wage cuts of six to nine dollars and, "'[a]s a deterrent to excess usage' of the washroom," a 20-minute weekly cap on toilet breaks:

> Any employee who goes beyond the limit will see his pay docked at either double or triple time, depending on the severity of the problem. They will continue to be docked for personal breaks until they've gotten their bathroom habits in line for at least a year. Only those with a doctor's note attesting to a medical problem will be excused from the practice.
>
> To put it bluntly, one worker says: "It's crap."
>
> Working in temperatures of 38 to 40 degrees Fahrenheit means "you gotta go," he adds.
>
> Workers maintain the company has lost sight of the fact at least half the time will be gobbled up by disrobing and robing again before they return to work. Most of the workers wear two aprons, arm guards, gloves, smocks and coats.
>
> And they are required to wash their hands for a minimum of one minute after using the washroom as a safeguard against contaminating the raw pork they're working with.[7]

The executive vice-president of Maple Leaf Foods, Pat Jones, conceded that

---

[6]Paul Mitchell, "Boar War: Lower U.S. Labour Costs Are Squeezing Pork Processors and Workers," *Hamilton Spectator*, Oct. 14, 1997, at E1 (Lexis). See also Katy Lerougetel, "Meatpackers in Canada Strike Against Company's Offensive on Wages, Rules," *The Militant* 61(42) (Dec. 1, 1997), on http://www.themilitant.com/1997/6142/6142_24.html; Susan Berman and Robert Simms, "Striking Meat-Packers in Canada Defy Plant Shutdown, Win Support," *The Militant* 61(43) (Dec. 8, 1997), on http://www.themilitant.com/1997/6143/6143_6.html.

[7]Eleanor Tait, "Washroom Trek Limit Last Straw for Union," *Hamilton Spectator*, Nov. 22, 1997, at D9 (Lexis).

all "[t]hat may be true.... But the company has the same contract language at other plants." He then revealed how market disincentives are superior to force as a technique for modifying potentially anti-capitalist working-class behavior: "'Our experience in those factories is once people are docked for going to the washroom or have to pay for excesses, it stops.'" Jones's response to a reporter's question as to whether the proposed regime would also apply to managers—namely, that he was "not aware" of any managers who "'abuse[d] their washroom privileges'"—prompted the journalist to offer this skeptical speculation: "Possibly it's because they are all dedicated employees. More likely it's because nobody is monitoring them the same as the men and women on the line."[8] To justify a measure bound to strike observers as outlandish, Jones offered the same almost reasonable-sounding explanation that would make Jim Beam Brands Company infamous almost five years later:[9]

> Think about it in practical terms, says Jones. You have an opportunity to go to the washroom when you get to work in the morning. About two hours later you have a coffee break and you can go again. A few hours later you can go during your lunch break. Then you have your afternoon break when you can go again, and then two hours later it's quitting time.
>
> But if the washroom breaks are dear to the workers, then let the union come up with another proposal, Jones said. "We're not unreasonable people."[10]

Referring to the hourly wage and benefit amount at the Burlington plant, which the company wanted reduced to the U.S. level of $16.50 (Canadian), Jones admitted: "It's not the washroom breaks that are the problem.... 'The real problem is the $25.08.'"[11]

After the strike had gone on for more than three months, the company threatened to shut down the plant if the union did not accept its final offer, which screwed average hourly wages down from $17.50 to $11.25. While Maple Leaf Foods sought to buy out the workers' contracts with a bonus offer of between $10,000 and $33,000, each worker nevertheless stood to lose about $80,000 over the life of the six-year collective bargaining agreement. In March 1998, 55 percent of those voting accepted the offer.[12] Although UFCW Local 1227 requested

---

[8]Andrew Dreschel, "Maple Leaf Strikers Face New Corporate Toughness," *Hamilton Spectator*, Jan. 21, 1998, at A9 (Lexis).

[9]See below ch. 13.

[10]Dreschel, "Maple Leaf Strikers Face New Corporate Toughness."

[11]Andrew Dreschel, "Maple Leaf Strikers Face New Corporate Toughness," *Hamilton Spectator*, Jan. 21, 1998, at A9 (Lexis).

[12]Craig Sumi, "Unions Vows to Fight On: Maple Leaf Threatens to Shut Down Plant," *Hamilton Spectator*, Feb. 28, 1998, at N1 (Lexis); Craig Sumi, "Maple Leaf: Yes:

that the Ontario Labour Relations Board set aside the election on the grounds that the signing bonuses were bribes to buy votes, the union ultimately signed the contract.[13]

The collective bargaining agreement that the unions and its members were constrained to accept also contained a remarkable provision, which, according to an arbitrator, "initially deducted the time used on each break from an employee's weekly hours worked. The reason for this was that the Company contended there was abuse of the personal break time to the point where there was a significant economic cost."[14] Article 14 of the collective bargaining agreement, labelled, "Personal Breaks," provided:

An employee will not be disciplined or discharged as a result of authorized usage of time granted by the Company for personal breaks. However, as a deterrent to excess usage the following measures apply:

(a) An employee will have the time used on each break deducted from their weekly hours worked. Sections (b), (c), and (d)...will not apply to employees who have a specific medical condition which is supported by medical documentation which is acceptable to the Company which necessitates greater than normal use of the washroom.

(b) An employee who uses more than twenty...minutes total time per week, in any four...weeks in a twelve...month period, shall have all subsequent time used on each personal break deducted from their weekly hours worked at the rate of double the time used.

(c) Any employee who uses more than twenty...minutes total time per week, in any twelve...weeks in a twelve...month period, shall have all subsequent time used on each personal break deducted from their weekly hours worked at the rate of triple the time used.

(d) If an employee reaches a level as outlined in Section (b) or (c)..."they will be held at that level until they have had twelve...clear months without any week being more than twenty...minutes total time used for personal breaks. In such case, the employee will revert to the next lower level of reduction and so on until they are returned to the level of

---

Bitter Burlington Strikers Accept Pay Cuts to Save 900 Jobs," *Hamilton Spectator*, Mar. 7, 1998, at A1 (Lexis).

[13]Craig Sumi, "Labour Deal Hanging as Maple Leaf Workers Return," *Hamilton Spectator*, Mar. 25, 1998, at D6 (Lexis); Ross Longbottom, "Maple Leaf Union Accepts Deal: Workers Pressure Removes Challenge to New Contract," *Hamilton Spectator*, Mar. 28, 1998, at N1 (Lexis).

[14]Maple Leaf Pork and United Food and Commercial Workers International Union Local 1227, 83 L.A.C. (4th Ser.) 78, 79 (Arb. Stanley Beck, 1999). According to the arbitrator, a union spokesman in 1995 agreed with the employer that "'excessive'" time for breaks was a problem, but asserted that it was partly of the firm's own making because, by banning smoking in the bathrooms, it "forced employees to go to the cafeteria." *Id.* at 79.

deduction outlined in Section (a)....[15]

In a memorandum of agreement, signed on March 26, 1998, which appeared as a Letter of Understanding in the contract, a clarification was added: "It is understood that employee will have the first twenty minutes of personal breaks in a calendar week without charge. For more clarity: Twenty minutes personal break time per week will be provided to each employee without deduction from his/her wages." In addition, the memorandum ratcheted down the penalty rate of the sections (b) and (c) from double to single time (which thus became identical with the penalty of section (a)) and from triple to double time.[16]

Jose Cordeiro, who had worked at the plant since 1979 and trimmed jowls on the continuously moving hog cutting line at very cold temperatures, had suffered from benign prostatic hypertrophy (enlarged prostate) since 1982, which caused him to urinate more frequently and urgently than normal men.[17] Ironically, the 58-year-old Cordeiro stated that he had acquired this condition as a result of complying with the earlier plant break rule, which had reduced bathroom breaks on company time from three to two; he refrained from urinating more frequently, developed an infection, and was hospitalized: "'After this I can't hold it anymore like before and I don't want to be sick again because I know what I suffer.... That's why I pay the price.'" Cordeiro meant the literal price of two or three extra daily admissions to the bathroom, which by the time of the arbitration had already cost him about $400 during the previous year (or $10 a week) in addition to his hourly wage reduction from $17 to $10.[18]

Maple Leaf Foods took the position that it was not applying the penalty provisions of sections (b), (c), or (d), but was merely applying section (a), which, unlike those sections, does not exempt workers with a medical condition. In the logic-chopping words of the employer's lawyer, Dan Shields: "'It's not docked (pay). They [sic] didn't work. If you're not working you will not be paid.' But Shields argues the limits are not mean-spirited, but an attempt to save the company money and production. 'There will be evidence of a significant problem with unauthorized use of the washroom which requires the use of spares or

---

[15]Maple Leaf Pork and United Food and Commercial Workers International Union Local 1227, 83 L.A.C. (4th Ser.) at 79-80.

[16]Maple Leaf Pork and United Food and Commercial Workers International Union Local 1227, 83 L.A.C. (4th Ser.) at 80.

[17]Maple Leaf Pork and United Food and Commercial Workers International Union Local 1227, 83 L.A.C. (4th Ser.) at 81.

[18]"Maple Leaf Docks Man's Pay $400 Over Too Many Washroom Breaks," (Kitchener-Waterloo) *The Record*, Apr. 17, 1999, at E2 (Lexis).

shutting down the line,' Shields told the arbitrator."[19] Specifically, Maple Leaf Foods' vice president for manufacturing testified that unscheduled bathroom breaks had cost the company $2 million a year before the new contract had been signed. The company's president Pat Jones also pleaded with those who characterized the policy as "'being mean of us...to understand it's a very substantial impact for us on cost and productivity.'"[20] (In 1997, the company's sales reached $3.7 billion, while its $117 million in earnings from operations generated a 7.7 percent return on net assets.)[21] Jones also sought to legitimize his policy by asserting that other pork processors imposed similar restrictions, though a newspaper reporter could not confirm the existence of this practice elsewhere.[22]

When Cordeiro (or any other worker) needed to go to the toilet, he signaled his need for a relief person. When that person arrived, Cordeiro had to take off his heavy gloves and apron and walk 100 meters to the bathroom and then repeat the process in reverse. Cordeiro estimated that a break took him five minutes.[23] But he did not have to worry about timing it since a supervisor was assigned that task.[24]

The union alleged that the toilet-break provision violated the Ontario Human Rights Code, but only as applied to a worker with a medical condition, and the arbitrator agreed that it did constitute adverse-effect discrimination against Cordeiro, whose handicap made it impossible for him to comply with the rule. The arbitrator also ruled that the employer would suffer no undue hardship in allowing Cordeiro the additional time he needed.[25]

In launching its human rights appeal, the UFCW argued: "'Maple Leaf Meats seems to believe it has the right to punish an employee for having a physical disability. We say that's wrong.'"[26] The union apparently found it entirely acceptable, however, for the employer to punish employees for being human beings, who must void more than four times a week for five minutes outside of breaks

---

[19]"Maple Leaf Docks Man's Pay $400 Over Too Many Washroom Breaks."

[20]Ed Rogers, "Worker Pays a Price for Washroom Visits, Maple Leaf Says They're Too Costly" *Hamilton Spectator*, June 12, 1999, at B7 (Lexis).

[21]http://www.mapleleaf.com/investors/financial_highlights.htm.

[22]Rogers, "Worker Pays a Price for Washroom Visits, Maple Leaf Says They're Too Costly."

[23]Maple Leaf Pork and United Food and Commercial Workers International Union Local 1227, 83 L.A.C. (4th Ser.) at 82.

[24]Ed Rogers, "Toilet Time for Maple Leaf Worker Focus of Arbitration," *Hamilton Spectator*, Apr. 15, 1999, at A1 (Lexis).

[25]Maple Leaf Pork and United Food and Commercial Workers International Union Local 1227, 83 L.A.C. (4th Ser.) at 83-85.

[26]UFCW News Release, "UFCW Local 1227 Launches Human Rights Appeal," Apr. 14, 1999, on http://www.ufcw.ca/news/19990414.htm.

prescribed by their employer. In contrast, a journalist captured the essence of this dispute much more profoundly than the UFCW in commenting that "the only unusual thing about the provision as it applies to [Cordeiro's] brothers and sisters is that the time restriction is formalized by contract language. The fact is, putting up with restrictions on washroom visitations is pretty standard stuff when you're fettered to a factory floor."[27]

Unsurprisingly, neither anyone at the UFCW Local 1227 office nor its president William Foley returned any of the author's many telephone messages. When asked why the union had not contested the validity of the rule per se, the labor lawyer who handled the arbitration for the union, Michael Mitchell, was not forthcoming: "We are not acting for that Local any longer as it has joined the main Ontario Local 175. I am concerned in answering your question about questions of privilege etc. This much is on the public record: The contract containing this clause came after a long and extremely bitter strike in which, inter alia, wages were cut by many dollars an hour and other major concessions were won by the employer. There was a compulsory vote among the members, which the employer can force under Ontario law, and the employer offered many thousands of dollars in one-time payments to employees for in effect surrendering their rights and voting for the new contract. The employer narrowly won the vote and what followed was I believe a memorandum that was actrually [sic] freely agreed to, I believe, (it has been awhile), if memory serves, the memorandum modified the original even more stringent provision regarding washroom breaks. In any event our brief was to handle the matter as we did."[28] More startling is the fact that, despite the large number of articles about Maple Leaf Foods's outrageous toilet tax in the mainstream press, a Canadian UFCW safety and health expert had never heard of the situation.[29]

These "pee-for-a-fee" cases in Canada refute the claim of the lawyer representing the workers in France who in 1995 had sued their employer, yet another slaughterhouse, for instituting a rule that docked their wages by 50 francs for going to the bathroom outside of the three scheduled five-minute breaks at 8:05, 11:20, and 2:05, that "he had searched in France and beyond without finding any other case of a company attempting to restrict employees' rights in this way. 'This was a first, and I think it will be a last.'"[30]

---

[27]Dreschel, "A Salute to Those Who Can't Take a Pee Break."

[28]Email from Michael Mitchell, Sacks, Greenblatt, Mitchell, Toronto, to Marc Linder (Oct. 15, 2002).

[29]"I do not know the answer to why or how this crap got into a collective agreement. Piss on it, I say!" Email from Larry Stoffman, UFCW, to Marc Linder (Sept. 18, 2002).

[30]Andrew Jack, "French Factory Workers Flushed with Victory," *Financial Times*, Mar. 19, 1996, at 3 (Lexis). See below ch. 10.

Four years after OSHA had confessed error and issued its Memorandum, a leading leftist health and safety law professor in Canada (seconded by one of the country's leading union health and safety advocates),[31] when asked about the lawfulness of these restrictions on workplace voiding, still clung to OSHA's discredited interpretation:

> There is nothing specific that I could find in Ontario's OHS [Occupational Health and Safety] Act, although in the Industrial Establishments regulation (Ontario Regulation 851) there is a provision regarding buildings (s. 120) that says that the Ontario Building Code applies to industrial establishments with respect to, inter alia, washrooms. So employers have to provide them, but there is nothing specific about allowing workers free access to them that I can find. The provincial Employment Standards Act only provides for 30 minute eating periods every five hours and says nothing about bathroom breaks. So on the face of it workers might have a hard time claiming a legal entitlement to bathroom breaks either on OHS or employment standards grounds.[32]

The director of Ontario's Occupational Health and Safety Branch of the Operations Division of the Ministry of Labour initially took virtually the identical position as OSHA officials in the United States when the author had questioned them between 1995 and 1998. When asked whether the requirement that a certain number of toilets be available according to the number of workers under the provincial Building Code[33] included the obligation to let workers use them, Ed McCloskey replied: "That's a stretch." When asked about the legality of the so-called fee-to-pee scheme of charging workers (say) one dollar for each minute of bathroom break time beyond the 30-minute meal break guaranteed by the Ontario Employment Standards Act,[34] he replied that he was not clear that the question implicated an occupational health and safety issue as distinguished from a labor standards issue, but that a worker filing a complaint under the Employment Standards Act probably "wouldn't have a leg to stand on." Nevertheless, he emphasized that the problem had never arisen and that neither the ranking officials in

---

[31]While expressing agreement with Eric Tucker "on the legal issues," she added: "In Canada of course we have the right to refuse unsafe work so a worker could simply invoke that if they needed to go to the bathroom." Email from Cathy Walker, Canadian Auto Workers, to Marc Linder (Sept. 6, 2002).

[32]Email from Prof. Eric Tucker, Osgoode Hall Law School, to Marc Linder (Sept. 6, 2002).

[33]Building Code Act Regulation, Ontario Reg. 403/97, sect. 3.7.4.9 (industrial occupancy).

[34]"An employer shall give an employee an eating period of at least 30 minutes at intervals that will result in the employee working no more than five consecutive hours without an eating period." Employment Standards Act, 2000, c. 41, sect 20 (1).

his Branch nor the director of the Employment Standards Branch had ever heard of the problem. He speculated that workplace toilet access was not a significant problem in Canada because employers would not take so rigid a stance as their counterparts in the United States had. However, if it became a problem, McCloskey observed that formulating a response would require a major and lengthy policy discussion.[35]

To be sure, in the version of the aforementioned mass-distributed March 1998 Associated Press article on voiding restrictions[36] that appeared in a Toronto newspaper, Belinda Sutton, a spokesperson of the Ontario Ministry of Labour, had said that the building code requires that facilities be available, but does not address the issue of employees' getting time to use them. However, she went on to say: "'If people are denied access to bathroom breaks, there are general duty requirements of the Occupational Health and Safety Act that could be used to provide such breaks. How much time in a break would have to be decided on a case-by-case basis.'"[37] That general duty clause imposes on every covered employer a duty to "take every precaution reasonable in the circumstances for the protection of a worker."[38] When confronted with Sutton's statement four years later, McCloskey observed:

> Belinda is correct in that the general duty clause...of our act might well be the mechanism we would consider. There are really two issues here, first, the right to leave the workstation to use the bathroom, and second, whether the employer must pay the worker for the time spent away from the workstation. However, no formal policy has been established on either point. Unofficially, I believe the Ministry would find the kind of situations you described to me unacceptable.[39]

In contrast, OSHA administrators in the United States deemed employers' general duty under the OSHA too narrow to support the imposition of a general duty to make toilets available to workers when they need to use them. That clause provides: "Each employer shall furnish to each of his employees em-

---

[35]Telephone interviews with Ed McCloskey (Sept. 9 and 10, 2002).

[36]See above ch. 2.

[37]Maggie Jackson, "On-the-Job Bathroom Breaks Become a Workplace Issue: Some People Risk Being Fired for Using Washroom at Work," *Toronto Star*, Mar. 23, 1998, at D7 (Lexis). In the Yukon Territory, although the law offers no express protection of the right to void at work, if the health and safety authorities received such a complaint, "they would do an employer visit and shame them into ensuring employees were able to use the washroom as required." Email from Lee Ann Campbell, Labour Services Branch, Whitehorse, Yukon, to Marc Linder (Sept. 12, 2002).

[38]Occupational Health and Safety Act, sect. 25(2)(h), Ontario Stat. (2000).

[39]Email from Ed McCloskey to Marc Linder (Sept. 16, 2002).

ployment and a place of employment which are free from recognized hazards that are causing or are likely to cause death or serious physical harm to his employees."[40] Since it would be very difficult to prove that death or serious physical harm would result from a one-time denial of permission to use the toilet, OSHA instead chose to create voiding rights in the promulgated sanitation standard.[41]

The experts' protestations that Canada was not plagued with the problem to the contrary notwithstanding, even a 1996 survey of working conditions at General Motors, Ford, Chrysler, and CAMI plants in Canada by the powerful Canadian Auto Workers concluded: "Across all four companies, workers found it difficult to find a relief worker so that [they] could go to the washroom."[42]

---

[40]29 USC sect. 654(a)(1) (2000).

[41]The promulgation of standards is controlled by 29 USC sect. 655.

[42]CAW-Canada, *The CAW Working Conditions Study: Benchmarking Auto Assembly Plants* 27 (May 1996). Although the text does specify the washroom, in fact the table, on which it is based, deals with control and autonomy and asks whether the worker "Cannot leave work station." CAMI Automotive Inc. is GM's joint venture with Suzuki. For a different account of bathroom breaks under CAW contracts, see below ch. 16.

Chapter 10

Oui Oui to Wee Wee: The Freedom to Go to the Bathroom As a Fundamental Human Right in France

---

> [H]e's the kind of boss...who never left his shop without first going around to see that all the lights were out. ... The same man who knows how many minutes a day his workers spend in the toilet.[1]

It is instructive to contrast the highest-profile dispute over workplace voiding rights in the United States—Jim Beam's offensive strategy, the UFCW's resistance, and OSHA's enforcement[2]—with a similar campaign in France in 1995-96 at the slaughterhouse of Bigard et Compagnie in Quimperle in Brittany, in the northwestern part of the country. Bigard was the second largest meat company in France, employing 3,000 people (1,200 at Quimperle), with sales of six billion francs, and ranked "the most profitable company in its sector. In terms of modern management methods, Bigard is exemplary."[3] On July 28, 1995, management at SA Groupe Bigard—which employed nearly a third of Quimperle's total working population[4]—established new schedules for its deboning operations, which imposed three fixed five-minute toilet breaks; workers who went to the bathroom outside of these three periods were subject to wage docking of 50 francs (about $10) for "'abandoning their post.'" Immediately upon being informed of the new work rule, the communist-led union CGT (Confédération Générale du Travail) protested.[5]

---

[1]Arthur Miller, *All My Sons*, in *Three Plays About Business in America* 187-266 at 240 (Joseph Mersand ed. 1968 [1947]).

[2]See below ch. 13.

[3]France: Quest for Improved Productivity Behind Dispute at Bigard," Reuter Textline: Le Monde, Aug. 13, 1995 (Lexis).

[4]Andrew Jack, "French Factory Workers Flushed with Victory," *Financial Times*, Mar. 19, 1996, at 3 (Lexis).

[5]Lamouroux v. SA Groupe Bigard, No. 9500433-436, slip op. at 2 (Conseil de

The following week witnessed a three-day strike by 250 beef-deboning and hog-cutting workers in protest against having to "faire pipi à heure fixe" such as 8:05 a.m., 11:20 a.m., and 2:05 p.m. Beginning on Friday August 4 and continuing on Monday and Tuesday, the workers marched in the streets of Quimperle slowing lunch-time traffic. The CGT requested that the state factory inspectors investigate the new work rule, which the union believed violated Article L. 122-35 of the Code du Travail, which prohibits work rules that place restrictions on the rights of persons and individual and collective liberties that would not be justified by the nature of the task to be accomplished or proportionate to the goal sought. In view of the determination of both sides and the workers' sense that their dignity had been deeply wounded, the country's leading newspaper, *Le Monde,* predicted that only the state factory inspectors' intervention could resolve the conflict.[6]

According to CGT shop steward Alain Lamouroux (who would soon become the lead plaintiff), the "'dispute sounds like something from the 19th century. We are demanding to go to the toilet according to each person's needs, not at the blow of a whistle. This should be a human right.'" The strike ended when the owner, Lucien Bigard, abandoned the plan of docking 50 francs per day from a worker's holiday bonus for spending too much time in the bathroom or going there at the wrong time; nevertheless, he still insisted on fixed breaks, outside of which "workers would need special permission from bosses." Bigard's justification was that they "were spending too long socialising at the toilets," while the unions argued that he was "trying to boost productivity at an already profitable and expanding business." Thus despite the initial compromise, Lamouroux speculated that they might still "'have to end up in court.'"[7] In general "the CGT feared especially that 'the search for productivity at any price would lead heads of businesses to harden the rules.' The union judges the Quimperle affair as so much the more serious as it is not the deed of a small employer taken by whim, but the decision of a group that employs a total of 2,500 people."[8]

Unlike the non-analytic reporting of the Jim Beam dispute by the press in the United States, *Le Monde* devoted a serious 2000-word article to analyzing the roots of the problem. Basically the newspaper saw Bigard's toilet rule as "the

---

Prud'hommes de Quimper, Mar. 18, 1996); Michele Lecluse, "'Acte unilateral de l'employeur'—Les règlement intérieurs soulèvent peu de contestation," *Les Echos*, Aug. 11, 1995, at 2 (Lexis) (quote).

[6]Malingre Virginie, "À Quimperle, des pauses pipi à heures imposeés," *Le Monde*, Aug. 9, 1995 (Lexis).

[7]"Relief as French 'Pee Break' Strike Ends," Reuters World Service, Aug. 9, 1995 (Lexis).

[8]Lecluse, "'Acte unilateral de l'employeur.'"

outcome of a frantic race for productivity," while social legislation "remains mute on the problem of toilet breaks." The article began dramatically with the symbols of the sea change in French class struggle:

> Each era has the seminal conflict it deserves. In 1973 the workers at Lip, in Besancon, inaugurated twenty years of struggle for employment. In 1995, in Quimperle (Finistère), the Bigard workers defended the right to go to the toilet freely! Don't laugh! This affair of pauses-pipi at fixed times imposed by the management...is not, alas, ridiculous. It is neither a blunder of backward enterprise nor a resurgence of the 19th century. But the logical outcome of a disturbing evolution. A conflict of productivity. A conflict of this end of the century.
>
> Arriving at Quimperle, one expects Zola. One finds a laboratory. ...
>
> According to all the criteria of modern management, [Bigard] is an exemplary success. ...
>
> Bigard can also boast of a sophisticated social management. The wages are above the average for the sector, the restaurant of the enterprise is as modern as the workrooms, where hygiene, cold (7 [degrees centigrade] maximum), and automation prevail. Five years ago the former head of personnel was replaced by a young director of human resources imbued with all the techniques of management. Flexibility, schedules à la carte, skillful use of all the resources of the law and massive recourse to temps—everything has been placed in the service of productivity.
>
> A little too much, precisely. In its desperate race for competitiveness, Lucien Bigard has forgotten that his employees are humans. Modernization at a forced march of the business has been done at their expense. ...
>
> Not that the tasks here are more arduous than elsewhere, but the ambience is execrable, the tension permanent, the rhythms infernal. ...
>
> The new technologies are supposed to make work easier. ... But the modern material increases the rhythms and the stress, hence the accidents, explains a cutter. 240 work accidents in 1994...for 1,200 employees. ...
>
> ...Constraints of delivery compel, the chopped-meat departments felt compelled to work the weekend, at first overtime hours then at normal rates, thanks to the progress of social legislation in the matter of flexibility. ... Summer vacations have been limited to two weeks. ...
>
> ...A system of financial penalties punishes every absence or lateness, even for sickness, at the rate of 50 francs per day....
>
> The introduction, this summer, of toilet breaks at fixed times in certain departments, coupled, here too, with financial penalties in the case of infractions, was, by this logic, only one additional small step. Designed officially to sanction established abuses and, secondarily, to increase productivity, which, according to management, is said to have dropped 10% in six months.
>
> One step more. One step too many. In Quimperle, it is said with humor, the toilet-break affair was the drop of water that made the vase overflow.
>
> ...But the affair has not ended. The toilet breaks at fixed times are going to be experimented with for a month in two workrooms, and one will see where things stand. ...
>
> Management has not even realized that this decision constituted a social regression

and a first in France. ...

Lucien Bigard is not a social employer. Far from it. ... All he is doing is pushing to its conclusion the logic of productivity, forced by international competition and the demands of the market always to do more. An extreme case. Nothing more. ... Secretly, the other employers are applauding.[9]

Lamouroux, the representative of the CGT, the strongest union at Bigard, declared: "'On the eve of the year 2000, it seems aberrant to us to become robots required to go to the toilets at fixed times.'"[10]

On October 5, Lamouroux and four others filed suit in county court (Tribunal de Grande Instance) seeking a temporary injunction requiring Bigard to withdraw its new work rule (under penalty of delay of 50,000 francs per day based on Article 809 of the New Code of Civil Procedure) and to pay each of the petitioners 2,000 francs (based on Article 700). They argued that the rule derogated from the worker's private life and his physical integrity and contravened a simple natural law. They also brought to bear the aforementioned breach of the Labor Code.[11] But the civil chamber of the court, agreeing with the employer's arguments, declared its incompetence to hear the case and referred the plaintiffs to the Conseil des Prud'hommes.[12]

The Conseil, the world's oldest system of labor courts, dating back to 1806, specializes in conciliation of individual labor disputes.[13] Each sitting Conseil

---

[9]Maurus Veronique, "La course à la productivité au coeur du conflit des Bigard," *Le Monde*, Aug. 14, 1995 (Lexis).

[10]Jack, "French Factory Workers Flushed with Victory."

[11]Lamouroux v. Bigard at 3. Article 809 states: "Le président peut toujours, même en présence d'une contestation sérieuse, prescrire en référé les mesures conservatoires ou de remise en état qui s'imposent, soit pour prévenir un dommage imminent, soit pour faire cesser un trouble manifestement illicite. Dans les cas où l'existence de l'obligation n'est pas sérieusement contestable, il peut accorder une provision au créancier, ou ordonner l'exécution de l'obligation même s'il s'agit d'une obligation de faire." Article 700 states: "Comme il est dit au I de l'article 75 de la loi n° 91-647 du 10 juillet 1991, dans toutes les instances, le juge condamne la partie tenue aux dépens ou, à défaut, la partie perdante à payer à l'autre partie la somme qu'il détermine, au titre des frais exposés et non compris dans les dépens. Le juge tient compte de l'équité ou de la situation économique de la partie condamnée. Il peut, même d'office, pour des raisons tirées des mêmes considérations, dire qu'il n'y a pas lieu à cette condamnation."

[12]"Pauses-toilettes: le tribunal se declare incompétent," *Les Echos*, Oct. 26, 1995, at 4 (Lexis).

[13]Code du Travail Art. 511-1; Xavier Blanc-Jouvan, "The Settlement of Labor Disputes in France," in *Labor Courts and Grievance Settlement in Western Europe* 1-80 at 15 (Benjamin Aaron ed. 1971).

consists of four lay judges, two each from the employees' and employers' side.[14] If the Conseil is equally divided in a case, the matter is remanded to the same Conseil, this time presided over by a professional judge from a civil court of first instance.[15]

On October 25, 1995, the petitioners referred the matter to the Conseil des Prud'hommes[16] requesting withdrawal of the employer's measures, and the parties convened a hearing for November 17, at which the plaintiffs, basing themselves on the aforementioned Article 122-35 of the Labor Code, requested both that the principle putting in place the obligatory toilet breaks be declared unlawful and a declaration that, in so far as necessary, the employees' freedom to go to the toilet could not be restricted or subject to the foreman's authorization, except to sanction abuses duly established. For its part, Bigard, calling attention to the new measures taken following the meeting of the works committee on October 31, 1995, argued that the employed had the possibility of leaving after having informed the foreman; consequently it requested that the Conseil declare that the system of toilet breaks at fixed times with deviations did not constitute an attack on the workers' individual liberties. At a proceeding on November 23, the Bureau de Judgment (judgment board), declared itself equally divided, as the two employee members opposed and the two employer members favored Bigard's new rule, and remanded the matter for a *Départage* hearing on January 15, 1996, at which the independent judge would resolve the stalemate between the two groups of lay judges.[17]

At that hearing, the workers, "conscious of the responsibilities weighing on them and of organizing production for the best," made the following proposal to management: 1. the three five-minute breaks would be retained, but they would

---

[14]Code du Travail Art. 512-1.

[15]Code du Travail Art. 515-3.

[16]According to Article 422.1.1: "Si un délégué du personnel constate, notamment par l'intermédiaire d'un salarié, qu'il existe une atteinte aux droits des personnes ou aux libertés individuelles dans l'entreprise qui ne serait pas justifiée par la nature de la tâche à accomplir ni proportionnée au but recherché, il en saisit immédiatement l'employeur.

"L'employeur ou son représentant est tenu de procéder sans délai à une enquête avec le délégué et de prendre les dispositions nécessaires pour remédier à cette situation.

"En cas de carence de l'employeur ou de divergence sur la réalité de cette atteinte et à défaut de solution trouvée avec l'employeur, le salarié, ou le délégué si le salarié concerné averti par écrit ne s'y oppose pas, saisit le bureau de jugement du conseil de prud'hommes qui statue selon les formes applicables au référé.

"Le juge peut ordonner toutes mesures propres à faire cesser cette atteinte et assortir sa décision d'une astreinte qui sera liquidée au profit du Trésor."

[17]Lamouroux v. Bigard slip op. at 3-5; Jack, "French Factory Workers Flushed with Victory."

not be called toilet breaks and could be used for any reason other than going to the toilet; 2. recognition of the employee's absolute freedom to go to the toilet outside of those breaks, in case of necessity; 3. informing the foreman of the fact that the employee has to leave the chain to go to the toilet; and 4. possible sanction of abuses that could be verified in conformity with the work rules. In support of their demands the workers alleged that going to the bathroom did not cause any disruption in the deboning department inasmuch as a co-worker could take over the work. At the hearing the workers recognized that the employer's position had become somewhat more flexible in that they were permitted to go to the bathroom outside of the breaks after having informed their supervisor. However, this permission was subject to the work rules, which provided that in operations requiring a continuous presence, the worker was not permitted to leave his post without making sure that his replacement was present; if such was not the case, he had to alert his supervisor and remain at his post for the time it took the latter to organize the work. Thus the workers argued that according to the work rules, even a momentary absence was subject to the supervisor's authorization and to the condition that a replacement be found. The workers argued that this system was unlawful with respect to the guarantee in Article 9 of the Civil Code that everyone has a right to respect of his private life and to the aforementioned Article L. 120-2 of the Labor Code, inasmuch as subjecting workers' freedom to go to the bathroom to a supervisor's authorization was in principle contrary to the employee's fundamental rights. Indeed, they regarded it as not only a restriction, but "a suppression pure and simple" of that freedom, adding that "the necessities of production cannot in any case permit establishing such a regime that interferes with the employee's physical integrity." Indeed, they went so far as to assert that alone the name "toilet breaks" was an assault on their dignity.[18]

In response, Bigard defended its break system on the grounds that it was situated within the framework of the normal exercise of the employer's power in that it had to rationalize the absences of workers going to the bathroom with respect to the imperatives of production. Bigard also pointed out that it was entitled to organize these breaks since they were part of compensated work time. The employer also stated that it did not understand why the workers violently rejected the term "toilet breaks" in contrast to other terms such as meal and snack breaks. The employer also argued that if there was any affront to rights and individual liberties, it was "justified by the necessities of the work on the chain and completely proportional to the goal pursued." Bigard asserted that by virtue of the interdependence of the various work stations, the sudden absence at one station disorganized the functioning of the chain downstream.[19]

---

[18]Lamouroux v. Bigard, slip op. at 5-7.

[19]Lamouroux v. Bigard, slip op. at 7-8.

At this point, the Conseil, agreeing with the parties that their positions were drawing closer, proposed another hearing for February 19, 1996, but the parties were given until January 25 to negotiate; however, they reached no agreement. On January 24 Bigard made the following proposal: 1. institution of three paid five-minute collective breaks for recuperation and satisfaction of physiological needs; 2. the employees have the freedom to go to the bathroom outside of these breaks if necessary; 3. employees leaving their work station outside of the breaks have to inform their supervisor so that replacement of the worker can be effected beforehand if that turns out to be necessary; and 4. management reserves the possibility of penalizing possible abuses. If the plaintiffs did not accept the proposal, Bigard requested that Madame le Juge Départiteur declare it satisfactory.[20]

Taking exception to point 3, the workers on January 25 counter-proposed the additional clause that "the freedom to go to the bathroom cannot be conditioned on the replacement of the worker leaving his station." To be sure, the workers also proposed another provision requiring Bigard to replace workers immediately. On January 29 Bigard undertook to do what was necessary to make the replacement of workers possible, but the employer insisted that any agreement was applicable only to the beef deboning department. It then made its final offer, which centered on this provision: "In order to permit the replacement of employees before leaving their work station momentarily outside of the breaks," Bigard would put in place a crew of four employees. Bigard also proposed putting this plan into effect for six months during which the parties could return to the Conseil if it proved difficult to apply. On the same day, the workers, insisting that Bigard's plan continued to infringe on their liberty, proposed that workers should be free to go to the bathroom outside of breaks, subject only to informing their supervisor and penalization of possible abuses. In addition, they accepted Bigard's proposal of putting in place a crew of replacements. Importantly, the workers also stressed "the fact that it behooves the employees involved to make proper use of this freedom in connection with a corollary—the sanctioning of possible abuses. The sanctioning of these abuses constitutes a sufficient guarantee for the Bigard Company." Having fallen short of accepting the company's position, the workers then asked the Conseil to resolve the litigation.[21]

In handing down its decision on March 18, 1996, the Conseil observed that agreement between the parties had stumbled over the word "beforehand," which expressed the employer's intent to organize the conditions of the exercise of the worker's freedom to go to the bathroom outside of breaks. The Conseil observed that "the fact of going to the bathroom responds to a physiological need that only the individual is in a position to judge. It is indeed a question of a fundamental

---

[20]Lamouroux v. Bigard, slip op. at 8-9.

[21]Lamouroux v. Bigard, slip op. at 10-12.

freedom of the human being. The principle therefore is that of everyone's freedom to exercise the right to go to the bathroom solely at his convenience. Nevertheless, it is appropriate to admit that certain circumstances (trips, lack of public toilets...) constitute momentary constraints, independent of the will of the affected person but equally independent of the will of a third party." The Conseil however noted that the dispute before it did not fit within that framework because it was the employer, to which the worker is subordinated, that intends to organize the conditions of exercising the right to go to the bathroom. It is based on its power of organization and management.. Even under its revised proposal, Bigard "intends to arrogate to itself the right to decide that the employee should or should not, according to the circumstances (necessity of replacement), defer the exercise of this right...." The Conseil concluded that it could not consider that such an affront to a fundamental right, even if it were justified by the nature of the task, would be proportional to the goal sought. Moreover, "the employer's power of management cannot extend to governing a fundamental right pertaining to the employee's private life. To admit the lawfulness of such an intrusion into private life would certainly open the door to abuses of every sort and more particularly of the 'little bosses' with regard to employees vis-à-vis whom they could be opposed with respect to a disputed matter of whatever nature." And in this regard the fact that Bigard took every measure for providing a rapid replacement of the employee could not justify validating its proposal.[22]

Thus the Conseil, agreeing with the workers, declared that the obligatory toilet breaks were unlawful, that "the right to go to the bathroom cannot be subject to authorization by a third party or prior replacement of the parties concerned," and that the workers' final proposal was satisfactory and in conformity with the law.[23] Two days later *Le Monde* observed that the tribunal's ruling had adjudged the employer's effort at "taylorisation de la pause-pipi" as based on an unlawful practice, which, the newspaper added, was "above all profoundly humiliating and stupid."[24]

The French resolution of the conflict was thus both more principled and more radical than OSHA's method. As is only appropriate for an agency charged with protecting workplace safety and health, OSHA felt that it was constrained to root its policy of insuring urinary well-being in clinical urological findings concerning urinary tract infections and other adverse health effects associated with excessive urine retention. And even within this strictly medical framework, OSHA shunned an absolutist approach, grafting, instead, on to at-will bathroom breaks a reasonableness criterion that conferred some discretion on employers to give limited

---

[22]Lamouroux v. Bigard, slip op. at 12-14.

[23]Lamouroux v. Bigard, slip op. at 15.

[24]Georges Pierre, "Règlement intérieur," *Le Monde*, Mar. 20, 1996, at last p. (Lexis).

weight to their own economic interests by interposing reasonable delays before workers could stop work to go to the bathroom.[25] In contrast, the Conseil des Prud'hommes deprived the employer of any right to determine when and how often workers go to the bathroom; instead, it focused the resolution of the dispute on the workers' absolute and fundamental moral and human rights of autonomy in their private lives, even when they were living those lives at the workplace and on the employer's clock. The Conseil merely accorded the employer, as the workers themselves had already done, a subsidiary, after-the-fact, role in punishing abuses.

To be sure, no U.S. court has issued a decision that even remotely combines this expansive practical and moral sweep. Nevertheless, at least one federal judge in 1997 made a start with regard to a state tort claim for intentional infliction of emotional distress brought by a cashier against an electric utility company. In spite of a letter from her physician to the employer stating that her medication for Major Affective Disorder might make her need to go to the bathroom more often than usual, the company systematically denied her requests for an accommodation; instead, it permitted her to go the bathroom only twice a day. In denying the employer's motion for summary judgment, the court declared that it "realize[d] that using the restroom is a basic human function and to limit it in an unreasonable way would be to deny an individual's 'humanness,' and is an outrageous restriction on human rights thus raising it to a level of being 'utterly intolerable in a civilized community.'"[26]

---

[25]See above ch. 6 and below Appendix II.

[26]Barton v. Tampa Electric Co., 1997 U.S. Dist. Lexis 3168, at *12 (M.D. Fla. 1997).

Part V

Post-April 6, 1998 Enforcement

OSHA's focus on this issue has left some people shaking their heads in disbelief.[1]

---

[1]Haynsworth, Baldwin, Johnson and Greaves LLC, "OSHA Clarifies Requirements on Bathrooms," *Florida Employment Law Letter* 10(4) (June 1998) (Lexis).

11

The Few, The Proud, The Citations

> One OSHA watcher said this issue was really an eye opener for him. "We talk about implementation of new standards and how important they are for worker safety and pat ourselves on the back for making workplaces safer. But the reality for many people is that it is still the Dark Ages. They work for employers who don't allow them to use the bathroom. They are not accorded even that basic human necessity."[1]

Which Is Worse—No Access or No Toilets?

Since OSHA officials themselves freely admit that they do not know whether the Memorandum has had any impact on compliance,[2] it is necessary to try to piece together disparate indicators in order to gain some empirical sense of the overall national enforcement effort and compliance response. This chapter is devoted to constructing a nationwide statistical overview of the citations that OSHA has issued to employers for restriction of employees' access to the toilet since April 6, 1998 and trying to explain why the agency has not cited more employers.

Because the OSHA Memorandum securing workers the right to go to the bathroom when they need to is an interpretation of 29 CFR section 1910.141 (c)(1)(i), which on its face requires employers to provide their employees with a certain number of toilets, any citations that the agency issues for violations of this standard are statistically ambiguous: computerized data bases that capture only the subsection of the standard (in this case, 1910.141(c)(1)(i)) that has been violated do not and cannot distinguish between the two distinct concrete violative circumstances—provision of too few toilets and restriction of access to (what may or may

[1]"OSHA Clarifies Bathroom Policy," Newsline (Apr. 3, 1998), on http://www.penton.com/oh/member/newsline/4-3.html. For a slightly different version, see Lisa Finnegan, "OSHA Directive Clarifies Bathroom Use Policy," *Occupational Hazards* 60(5):13 (May 1, 1998) (Westlaw).

[2]E.g., telephone interview with John Hermanson, Assistant Regional Administrator for Enforcement Programs, Chicago Region (Nov. 27, 2002).

not be the proper number of) toilets. So long as these two different kinds of violations share the same standard subsection and summary statistical space, they cannot be sorted out. The only way to segregate one from the other is by examining the actual citations, which do specify the factual details of the violations.[3] The results of that examination constitute the core of this chapter, but an initial review of the larger universe of toilet standard violations will place restrictions of toilet access in a broader context of degraded and degrading working conditions.

Over the three decades of its existence OSHA has issued many thousands of citations for violation of its toilet standard. From July 1972 through the end of 2002, Federal and state OSHA cited employers more than 2,893 times for violating 29 CFR section 1910.141(c)(1)(i)—an imprecision resulting from the fact that only Federal OSHA's computerized records go back to 1972, while those of the state agencies begin in various years from 1979 to 1990.[4] Inclusion of the citations issued by state programs that use different designations in their code of regulations increases the total by at least 1,418.[5]

The fact that the hundreds of citations that OSHA has issued since April 6, 1998 to employers for failure to provide (or even to have) the proper number of toilets are not the focus of this study does not mean that this particular violation is unimportant. Indeed, where, as in many cases, employers failed to provide any toilet at all, the end result for the workers may have been exactly the same as in those workplaces where employers denied their employees access to toilets that did exist. Even where there were toilets in a workplace, but not enough (or properly functioning ones), several OSHA inspectors in Indiana, for example, expressly stated on citations that the possible injuries had been a "urinary tract infection"[6] or "Bladder damage caused by employee not using bathroom facility as needed."[7]

Nevertheless, there are differences between these types of violations. Most strikingly, the failure to provide any toilet at all is not only a blatantly undeniable violation of an unambiguous OSHA standard (and public health laws and ordi-

---

[3]On the details of the citations and the statistics, see below Appendix V.

[4]Search ("19100141 c01 i") conducted in Lexis-Nexis OSHAIR file (Dec. 31, 2002).

[5]According to searches performed on Lexis-Nexis, from July 1990 on, Cal/OSHA issued 586 citations under sect. 3364(a) (number of toilets required) and 697 under 3364(b) (clean and accessible toilets); from Jan. 1990 Washington issued 81 under section 24-12007(1)(a); and from Oct. 1988 Michigan issued 28 under R4201(3)(a)(i). Lexis-Nexis OSHAIR file (Dec. 31, 2002). A search conducted by Washington OSHA in its own computerized data base going back to 1986 identified 107 citations. Print-out from Barbara Harris-Williams, WISHA, to Marc Linder (Oct. 31, 2002).

[6]Steve's Flower Shop, Insp. No. 125417683, May 25, 1999, Indiana; Worksheet provided by Indiana OSHA.

[7]Avon Community School Corp., Insp. No. 125358358, Dec. 15, 1998, Indiana; Worksheet provided by Indiana OSHA.

nances), it is also such an outrageous affront to universally accepted standards of hygiene that offending employers would, if their identities were publicly disclosed, be regarded as contemptible pariahs. Their tone was captured by an OSHA inspector, responding to a complaint from workers at a nonunion freight terminal in Secaucus, New Jersey, where "[t]oilets and urinals had been inoperable for several months." The terminal manager, who admitted to the inspector that "the toilet room had not been in use for about 6 months," in the presence of other managers responded to the inspector's "query as to why the toilet room had not been repaired...They (dock workers) are animals."[8]

Surprising as it may seem in the twenty-first century, OSHA has issued numerous citations to employers for failing to provide any toilet at all. For example, in Kentucky in 2001, a nonunion sawmill manufacturing pallet boards had no restrooms: "The employer said that there was an outhouse out back, but this inspector did not follow-up on this. There is one female employee on site, and she is allowed to go to the McDonald's restaurant. The employer said that he would get a portable toilet for the employees."[9] Scarcely better was a nonunion sawmill in El Dorado Springs, Missouri, which was fined $800 (reduced to $120 following a formal settlement), after a Federal OSHA inspector in 2000 found that male and female employees

> were provided with one 'outhouse' for use as a toilet. The facility lacked any type of plumbing or sanitary waste disposal system and consisted of a roofed three-wall building (the door, which would have made up the fourth wall, was missing), an interior wooden floor and a wooden frame constructed over a hole in the ground. The toilet 'seat' consisted of two loose boards laying across the otherwise open-top wooden frame. No toilet paper was provided.[10]

Since OSHA had revoked the toilet paper requirement 22 years earlier,[11] the nod to long-gone regulatory sensibilities was, to be sure, a nice inspectorial touch.

A small wood products company in North Carolina that provided no toilets in 2001 told its "'employees to go to the woods.'"[12] A Kentucky OSHA compliance officer perfected the circumlocution when, responding to a complaint, he cited (but

---

[8]Arrowpac International Inc., Insp. No. 303263867, Worksheet (2000).

[9]C & F Lumber Co. Inc., Insp. No. 304699440 (Oct. 3, 2001); Narrative furnished by Kentucky OSHA.

[10]Citation and Notification of Penalty, Superior Lumber Co., Insp. No. 303207062 (issued Dec. 14, 2000).

[11]See above ch. 4.

[12]Vickery Farms, Inc., Insp. No. 304278831, Narrative OSHA -1A (Feb. 27, 2001). The employer was fined $150 for a serious violation. http://www.osha.gov/cgi-bin/est/est1vd?30427883101001A

imposed no monetary penalty on) a six-employee, nonunion sawmill, where "no restroom is provided to employees. Employees use the restroom in the woods...."[13]

The absurdly lenient attitude that some inspectors have adopted toward employers committing such violations can be gauged by the following compliance officer observations in a case, triggered by a complaint, involving numerous safety and sanitation violations by a nonunion metalworking firm with seven employees in a small town in Kentucky:

> The building has no running water at all, hence no drinking water, no hand-washing facilities, and no toilet. ... At the present time, the employer allows the employees (usually only two) to go to a filling station approximately one-quarter mile away to use their restroom facilities. On hot days, he supplies gatorade to his employees. While the employer is clearly not in compliance with OSHA Standards, he is not totally unreasonable in his treatment of his employees.[14]

Nor was the inspector "totally unreasonable in his treatment" of the employer, proposing a zero-dollar penalty.[15]

On the other hand, even in such workplaces workers may actually face few or no restrictions on their ability to stop work to go find a toilet elsewhere or to void in the open. The cheapness displayed by employers who refuse to pay for the purchase and installation of toilet facilities may be irrationally confined to those sunk costs and not extend to the additional time lost to production while their employees are searching for or walking to remote toilets. ("'[D]isgusting'" rather than irrational cheapness characterized the practice enforced by the Fairfax County Maryland public works department of requiring the employees who repair its sewer pipes to urinate into buckets, open a manhole cover, and dump the urine into the sewer system "because it would delay work if they went looking for a public restroom.")[16] In fact, according to the executive director of the Portable Sanitation Association International, which is charged by the American National Standards Institute with developing its Sanitation in Places of Employment standard, such irrational cheapness is not at all uncommon among smaller nonunion construction employers, who refuse to spend the money to rent portable toilets in spite of being

---

[13]Cross Chase Manufacturing, Inc., Insp. No. 302079942, CSHO Observations (June 10, 1998).

[14]CMC Metalworking, Insp. No. 303125017 (Nov. 24, 1999); Narrative furnished by Kentucky OSHA.

[15]This information, which is lacking on the OSHA website, stems from a computer print-out of citations for violations of 29 CFR 1910.141, which Kentucky OSHA prepared for the author (Sept. 23, 2002).

[16]Michael Shear, "When Nature Calls, Fairfax Workers Visit Bucket," *Washington Post*, June 17, 1998, at B1 (Lexis).

presented with actual studies proving that the savings in wages that do not have to be paid to workers for driving to and from distant toilet facilities far exceed the rental cost.[17]

There Really Are Violations and Even Some Citations

When, in the course of interviewing Sheldon Samuels, who for many years had been the energetic and capable director of Occupational Safety, Health, and Environmental Affairs of the Industrial Union Department of the AFL-CIO, the author mentioned that part of the research for this book involved identifying all the OSHA citations that had been issued since 1998 for restricting workers' access to toilets, Samuels's immediate reaction was: "You're not going to find any citations." The basis for his prediction of the lack of any evidence of enforcement was twofold: employers would be too embarrassed to deny access and getting to go to the bathroom was simply not that big a problem.[18] And yet there are many denials of access by employers and some citations. One of the chief reasons for presenting thickly described narrative accounts of the circumstances surrounding the cited denials of access is to persuade skeptics, including some who should know better, that such implausibly oppressive workplace practices really do exist.

The exposition of the citation statistics begins with the Federal OSHA system, followed by an overview of the state-plan OSHA programs. In the two following chapters, the two state OSHA programs that have most vigilantly enforced the new interpretation embodied in the Memorandum are examined in detail: Iowa's citation of several large animal slaughter plants and Kentucky's citation of Jim Beam Brands, which the media turned into a high-profile case. Special attention is then focused on two additional states, Washington and California, because the former had taken the position before 1998 that denial of access was unlawful and actually

---

[17]Telephone interview with William Carroll, executive director, Portable Sanitation Association International, Bloomington, MN (Jan. 17, 2003); Scott Bruce, "Provisions for Sanitation Facilities at Construction Sites" (Aug. 15, 1988) (the studies); Portable Sanitation Association International, "Putting One on the Right Spot" (ca. 2001) (flyer stating that four dollars could be saved for every dollar spent).

[18]Telephone interview with Sheldon Samuels (Nov. 23, 2002). In 1972, at the principal OSHA public hearings on the future of the toilet standard, Samuels had testified on behalf of the Industrial Union Department of the AFL-CIO: "I have no objection to the proposal on the numbers of toilets.... I'm sorry we have to be so concerned about toilets when we are worried about carcinogens, but we are not really terribly excited about that issue." United States Department of Labor, Occupational Safety and Health Administration, "Hearing on Proposed Revision of Sanitation Standards" at 108 (Nov. 9, 1972, Washington, D.C.).

issued a citation on that basis, whereas the latter is the only state-plan agency that not only has never issued such a citation, but programmatically refuses to do so on the grounds that it is not obligated to enforce the standard as effectively as Federal OSHA does.

The citations that OSHA has issued to employers for having restricted employees' access to the bathroom since April 6, 1998 are summarized in Tables 1, 2, and 3. Tables 1 and 2 present summary information about the nine citations issued by Federal OSHA and 11 state OSHA citations for denial of access, respectively. Since the citation is linked to a violation of standard 1910.141(c)(1)(i) (or the equivalent state OSHA standard), which also encompasses the obligation to provide a certain number of toilets, Table 3 shows both the total number of citations issued by state-plan agencies under this standard and the subset dealing with denial of access. This very small number of access citations over a period of almost five years should be contrasted with the erroneous figure published by the *Wall Street Journal* in its account of the Jim Beam case (which was then repeated by others): "OSHA cites 30 to 50 employers a year for violations."[19]

Federal OSHA

Half a year after OSHA had promulgated its Memorandum, the author requested information from the agency on the citations it had issued pursuant to the new interpretation; in response, OSHA disclosed that it had issued two such citations for "serious" violations pursuant to complaints made by employees at union-

---

[19]Carlos Tejada, "Work Week," *Wall Street Journal*, Aug. 28, 2002, at B3, col. 1. For a similar claim (presumably taken from the *Wall Street Journal*), see *Employment Law Report*, 6(14):2 (Oct. 2002) (Lexis) ("OSHA cites 30 to 40 employers a year for violations of the ['timely access'] rule"). Tejada's source, Bill Wright at OSHA's Office of Public Affairs, emailed him data for violations of 29 CFR 1910.141(c)(1)(i) for fiscal years 1995 through 2002; they totaled 29, 21, 42, 39, 40, 46, 31, and 27 (through Aug. 23, 2002), respectively. Email from Bill Wright, OSHA Office of Public Affairs, to Carlos Tejada, Wall Street Journal, forwarded to Marc Linder (Aug. 26, 2002). Wright later stated that he knew that the data he had given Tejada did not distinguish between failure to provide the proper number of toilets and failure to give prompt access, and that he did not know that, let alone why, Tejada had gotten them wrong. Wright did not himself generate the data—he has access only to OSHA's public website, which can generate data only for 1910.141 (and not its subsections) and only for the immediately preceding fiscal year—which he obtained from someone at OSHA he did not identify. Telephone interview with Bill Wright (Dec. 23, 2002). The reason for the discrepancy is not clear, but a search of the Lexis-Nexis OSHAIR file found two to three times more citations for these years than those mentioned by Wright, who expressed little interest in the discrepancy.

Table 1: Federal OSHA Citations under 1910.141(c)(1)(i) for Denial of Toilet Access, 1998-2002

Employer name	Location	Business or occupation	Union status	Inspection type	Violation type	Penalty ($)	Year
Ree's Contract Service	Kansas City MO	Guard	Union	Complaint	Serious	1,125/ 562.50	1998
Steel of W. Virginia	Huntington WV	Steel plant	Union	Complaint	Serious	1,500/750	1998
SIM & S	St. Louis	Dispatcher	Nonunion	Complaint	Other	0	1998
Spectaguard	Jersey City NJ	Guard	Nonunion	Complaint	Other	0	1999
Raven Management	Kearneysville WV	Guard	Union	Complaint	Serious	1,250/100	1999
Summit Security Services	New York City	Guard	Union	Complaint	Other	2,000/800	2000
Liz Claiborne	Montgomery AL	Clothing dis-tribution ctr.	Union	Complaint	Other	1000/900	2001
Convergys	Killeen TX	Call center	Nonunion	Complaint	Serious/ Other	1,125/0	2002
Securiguard	Kingsland GA	Guard	Union	Complaint	Serious	975/600	2002

Table 2: State-Plan OSHA Citations under 1910.141(c)(1)(i) for Denial of Toilet Access, 1998-2002

Employer name	Location	Business or occupation	Union status	Inspection type	Violation type	Penalty ($)	Year
City of S. Tucson Communication Center	Tucson AZ	Police call center	Nonunion	Complaint	Other	0	1998
Initial Security	Indianapolis	Guard	Nonunion	Complaint	Serious	0	2002
Excel	Ottumwa IA	Animal slaughter	Union	Complaint	Willful/ serious	36,000	1999
John Morrell	Sioux City IA	Animal slaughter	Union	Complaint	Serious/ other	2,000	2001
Swift	Marshall-town IA	Animal slaughter	Union	Referral	Serious/ other	1,875/ 1,000	2000
Swift	Marshall-town IA	Animal slaughter	Union	Complaint	Repeat	5,000/ 2,500	2002
Ardco	Elkton KY	Glass mfr.	Union	Complaint	Other	0	1998
CR/PL	Somerset KY	Bathroom partitions	Union	Complaint	Serious	975	1999
Jim Beam Brands	Clermont KY	Bourbon distillery	Union	Complaint	Other	0	2001
S. Carolina Dept. of Juvenile Justice	Columbia SC	Correction officer	Nonunion	Complaint	Other	0	2000
Custom Apple Packers	Quincy WA	Apple packing	Nonunion	Complaint	Other	0	1999

Table 3: Citations of Employers for Violating OSHA Toilet Standard 1910.141(c)(1)(i) (or State Equivalent) and Denying Toilet Access in State-Plan States from April 6, 1998 to Fall 2002		
State	**Total Toilet Standard Violations**	**Of which Access Denials**
Alaska	2	0
Arizona	6	1
California	181	0
Connecticut	0	0
Hawaii	0	0
Indiana	12	1
Iowa	6	4
Kentucky	14	3
Maryland	13	0
Michigan	7	0
Minnesota	1	0
Nevada	4	0
New Jersey	0	0
New Mexico	0	0
New York	9	0
North Carolina	14	0
Oregon	11	0
South Carolina	10	1
Tennessee	5	0
Utah	2	0
Vermont	0	0
Virginia	11	0
Washington	21	1
Wyoming	0	0
Total	**329**	**11**

ized workplaces.[20] The first citation was issued in July 1998 to Ree's Contract Service, Inc., which "provides security services for federal government properties in several states under contracts with the Federal Protective Service of the General Services Administration."[21] In the course of providing security services at the Bannister Federal Complex in Kansas City, Missouri in June, according to OSHA, "employees were denied necessary use of bathroom facilities. Employees are not able to leave their posts for bathroom access during a four to five-and-half hour shift."[22] The $1,125 proposed penalty was halved by means of an informal settlement agreement.[23]

The second citation was issued in September 1998 to Steel of West Virginia, a steel minimill, in Huntington, West Virginia, which produces specialty steel products such as I-beams,[24] and in 1997, the year before it was acquired by a larger steel company, had sales of $113 million.[25] Its approximately 500 production workers are represented by Local 37 of the United Steelworkers (USW), which organized the plant in 1937.[26] According to the "instance description" on the inspector's worksheet: "Company supervisor told an employee who had been having trouble with intestinal problems that he, the employee, could not go to the restroom until he had relief. The employee had to wait at least ½ hour and was not provided with relief. He defecated in his pants." At the closing conference with OSHA: "The employer representative admitted that he told the affected employee that he could not go to the restroom until he had relief. He gave the employee a directive not to shut the Torrit [waxing] Line down again. This was after the employee had previously shut the line down because he did not have relief to go to the restroom. The torrit line takes 9 employees to operate. Before the incident

---

[20]Fax from Helen Rogers, OSHA, Office of General Industry Compliance Assistance (Oct. 13, 1998).

[21]Reed v. Kelly and Ree's Contract Service, Inc. 37 S.W.2d 274, 276 (Mo. App. 2000).

[22]Ree's Contract Service, Insp. No. 302241682, OSHA-2 (July 20, 1998). OSHA also conducted a follow-up inspection; Insp. No. 302241773 (Oct. 22, 1998). The employer at that location is organized by Local 20 of the United Government Security Officers of America. http://ugsoa.com/happeningGSA.html. In fiscal year 2001 Rees was the eighth largest federal civilian agency contractor in Missouri. Federal Procurement Data System, Federal Procurement Report 47 (2002), on fpdc.gov/fpdc/FPR2001a.pdf.

[23]"Informal Settlement Agreement" (Aug. 11, 1998); http://www.osha.gov/cgi-bin/est/est1vd?30224168201001.

[24]The company was bought up in 1998 by Roanoke Electric Steel Corp. See http://www.swvainc.com.

[25]http://www.roanokesteel.com.

[26]Telephone interview with Scott Ramey, President, USW Local 37, Huntington, WV (Dec. 16, 2002).

the employee [ ] had shut the line down to go to the restroom. Mr. Dave McMillon, supervisor of the area came by and noted that the line was shut down and not in operation. Several employees told Mr. McMillon that the other employee [ ] was in the restroom and was having intestinal problcms. Mr. McMillon started the line up (he was the 9th employee that it takes) and worked it. When [ ] came back from the restroom, Mr. McMillon acknowledged that he told [ ] not to shut the line down again." The inspector then added this "NOTE: There are two relief people for the line but they provide relief for employees taking [rest] breaks. If the two relievers are relieving and the two employees who were being relieved leave the area, then there is no relief. This is why the employee shut down the line prior to Mr. McMillon coming to the area a[nd] re-starting it, and then giving [ ] directive not to shut the line down again unless he had relief." Finally, the inspector made the following "POINTS- 1. There is not relief for this bathroom emergency. As stated the relievers relieve workers for their breaks. 2. The manager told [ ] he could not leave unless he had relief. 3. Employee stated that because the way he was directed by the supervisor not to shut down the line he feared being charged with insubordination."[27]

Background information provided by the president of Local 37, who had been head of the grievance committee at the time, sheds considerably more light on the labor-management dynamics prevailing in the plant. According to Scott Ramey, the worker in question is black and the supervisor, who was in fact superintendent of fabrication, "does not like black people," of whom there are very few in the plant and several of whom have been fired in recent years. Although the president noted that bathroom access had never been a problem before or after this incident—workers also get 15-minute breaks every 90 minutes plus a meal break—he stressed that it fit well within the management's recent pattern and practice of taking "sadistic glee in dehumanizing people," which he dated back to the advent of a new vice president for human resources at the beginning of 1998, who boasted of his union-busting career. Unlike reports from the USW in several other steel mills, which stressed that workers' freedom of access to the bathroom was a function of the fact that they were little supervised and basically ran the operation,[28] Ramey's account noted that workers at Steel of West Virginia were very closely supervised and frequently harassed by an inefficient overabundance of foremen. He had personally heard them declare their management policy that "you treat a

---

[27]Steel of West Virginia, Insp. No. 300460292, Worksheet, OSHA-1B (Sept. 1, 1998). The square brackets [ ] indicate that OSHA deleted the name of the worker. In fact, in two places the FOIA officer neglected to redact the name, but there is no need to divulge it. According to the company's human resources director, the wax line is really called "torrid" and not "torrit." Telephone interview with Bruce Groff, Huntington (Dec. 17, 2002).

[28]See below ch. 16.

guy like crap, he's gonna work harder." The president also contested the OSHA finding that the complaining worker had shut down the line: he contended that it was not necessary because other workers could take over the work for a colleague who left the line for a brief toilet break. The worker who had been made to defecate in his pants was a member of the grievance committee, but still chose not to grieve his treatment; instead he filed a discrimination claim with the Equal Employment Opportunity Committee in addition to the OSHA claim.[29]

The vice president for human resources, Bruce Groff, offered an entirely different view of events. Of central importance was his denial that the worker had ever defecated in his pants: although neither side had proof that he had or had not done so, Groff believed that the employee, whom he described as a disciplinary problem and "disenfranchised guy" who was as "unfortunate" for the union as the company, had simply lied as part of a pattern of causing problems for the company, including the filing of a race discrimination suit. Despite his belief that there was no basis to the citation, the human resources director decided not to contest the citation because the "adverse publicity" and being "dragged through the mud" were not worth the fight. While emphasizing that the company was not at all shy about litigating—in fact, Steel of West Virginia had a "huge" litigation bill, which he attributed to the "Appalachian factor" and an "I'm entitled, I'm entitled" attitude—Groff, when reminded of and furnished additional information about the Jim Beam dispute, expressed both incredulity at that firm's decision to expose itself to such public ridicule and criticism of a management style that would have prompted such a confrontation in the first place.[30]

This set of events underscores how even at a unionized place of employment, protected by a relatively strong union, individual workers may be subject to unhygienic forms of harassment by supervisors. Ultimately OSHA cited the company on the grounds that: "Employee was not permitted to use the restroom during a period when he was having intestinal problems." But at the informal conference, at which the $1,500 penalty was halved, the violation language was softened to read: "'releif [sic] persons were not readily available.'"[31]

In the beginning of 2000, an attorney interested in the problems of toilet access

---

[29]Ramey did not know what had come of the EEOC complaint. Telephone interview with Ramey.

[30]Telephone interview with Groff. Since it is very unlikely that OSHA would ever appear on the scene quickly enough to be in a position to inspect fresh physical evidence of such alleged defecation, the evidentiary superiority of flinging the freshly sodden pants at a supervisor, whose instinctual imposition of instant discipline would then preclude him from denying the truth of the allegation, is easy to appreciate. For a perhaps unique case, see Marc Linder and Ingrid Nygaard, *Void Where Prohibited: Rest Breaks and the Right to Urinate on Company Time* 166-67 (1998).

[31]Steel of West Virginia, Insp. No. 300460292 (Sept. 30 and Oct. 13, 1998).

for the transgendered made a Freedom of Information Act request to OSHA for all citations and settlements based on the April 6, 1998 Memorandum. OSHA replied that a search of its citation data base had identified 66 citations issued between April 6, 1998 and December 31, 1999 under 29 CFR section 1910.141(c)(1)(i), but that according to reports from the field offices: "The vast majority of the...citations were issued for lack of any toilet facilities, not having a sufficient number of toilets, being located too far away from the work area, or for not having a door that could be locked when the facilities were shared by males and females. Only...five were found to be responsive to your request."[32] Since two of these five had already been disclosed to the author, only three additional citations had been issued during the last quarter of 1998 and all of 1999 (within the Federal OSHA system).

The first of these three involved a non-union employer in the service industry (not otherwise classified) headquartered in "Tullahomo" [sic; should be Tullahoma], Tennessee, but operating in St. Louis, whose "[t]oilet facilities were not readily available for employee(s) who are required to work eight to ten hour shifts and are not allowed to leave their work stations unattended." For this "other than serious" violation the citation, triggered by a complaint, included no monetary penalty.[33] The citation did not disclose that the employer, SIM & S, Inc. (Systems Integration/Modeling, and Simulation) is "a provider of advanced systems engineering and other information technology solutions.... Although classified as a small disadvantaged 8(a) Business [for Small Business Act purposes], SIM&S believes in open competition." The inspection took place at 1520 Market Street in St. Louis, the location of the Federal Building and the General Services Administration, for whose satellite communications center SIM&S "provides trained and certified dispatchers and operates an Alarm Monitors Center 24 hours a day, seven days a week...."[34]

The second citation, also triggered by a complaint from a nonunion workplace, was issued to Spectaguard, Inc., another guard service, in New Jersey, which "did not provide toilet facilities that were accessible without restrictions for employees required to guard outside parking lot stations/posts." The violation was classified as other than serious and OSHA imposed no monetary penalty. The employer abated the problem—which had arisen from the fact that an employee needing to void had contacted the security station at the hospital where a shuttle driver was dispatched to relieve the officer—by adding a parking lot supervisor as a relief person, thus reducing response time to one or two minutes.[35]

---

[32]Letter from Richard Fairfax, Director, Directorate of Compliance Programs, OSHA, to Phyllis Randolph Frye (Mar. 9, 2000).

[33]S.I.M. & S., Inc., Insp. No. 302253919 (Dec. 2, 1998). The employer was located in Tennessee, but the inspection site was located in St. Louis.

[34]http://www.sim-s.com.

[35]Spectaguard, Inc., Insp. No. 302131917 (Mar. 9, 1999).

The third and final citation involved Raven Management Corporation, a firm with an office in Dunn, North Carolina, which had been providing security services for entry control and monitoring monitors and alarm systems at an Internal Revenue Service facility in Kearneysville, West Virginia, where security was so tight that it took two hours for the OSHA inspector to obtain the IRS's okay to enter the building: the IRS "was at a level 5 alert status (because of the conflict overseas and a concern for terrorism)."[36] The workers were represented by Local 170 of the United Plant Guard Workers of America, which is not affiliated with the AFL-CIO. The "primary concern" of all the employees interviewed by the inspector was that "since day one" they had "not been allowed to leave their work stations to use the rest room when they need to do so." They went on to explain that "it's common knowledge that they know that if they leave their command post without properly being relieve[d], then the IRS security could pull their security clearance, and they would lose their job. Each employee stated that they had request[ed] to go to the restroom several times, but were never sent any relief by their company, so they could go. ... Several employees stated that their [sic] had been times that they wanted to go but they had to wait until after their shift...." Raven's supervisor informed the inspector that after employees had on numerous occasions addressed this concern to her, she informed the company president, who told her that "he wasn't going to pay for an extra person. She also stated that she had taken the issue to the union, which said it could not do anything about it and "that they should notified [sic] OSHA." The supervisor also contacted the General Services Administration (the government agency that administers such government contracts), which stated that it was Raven's "responsibility...to provide a way for its employees to go to the restroom when they need to, without any unreasonable restrictions on the employees to use the facilities." The IRS's concern was that it gave Raven "a large sum of money for providing service, and that they the subcontractor should be able to pay for extra employees."[37]

The company president's recalcitrance was poignantly on display during the telephonic closing conference when he expressed the view that "OSHA was siding more with the employee, and was not taking into consideration that the reason why this complaint was filed was because the employees did not want to take a lunch break during their shift. The employer felt that if the employee told [sic; must be took] a lunch break, then they would be able to use the restroom then, and there would not be a problem with someone having to relieve the guards. The employer also stated that he had corrected the problem when he let a guard come in and work the day shift (on their day off) [to give relief]. CSHO [Compliance Safety and Health Officer] explained to the employer that this was not a solution, and the

---

[36]Raven Management, Insp. No. 116520800, Safety Narrative, OSHA-1A (May 13, 1999).

[37]Raven Management, Insp. No. 116520800, Worksheet, OSHA-1B (May 13, 1999).

employer stated that this is the way it was going to be handled."[38]

In the end, OSHA cited the company for failing to "provide an effective relief worker system that allowed employees prompt access to the toilet facilities, without imposing unreasonable restrictions on the employee use of the facilities...." OSHA imposed a $1,250 penalty for this "serious" violation, but reduced it to a mere $100 through an informal settlement agreement, which also stipulated that if the employer, which was no longer at "this IRS facility," returned there, it agreed to insure effective relief facilities.[39]

Thus all five citations issued during the first 21 months the Memorandum was in effect were generated by complaints at relatively small and little known employers (except Steel of West Virginia), four of which were in service industries, and three of which were unionized. Two of the citations were issued in Missouri, where OSHA had also issued its first (pre-Memorandum) citation for denial of access to Hudson Foods, and two in West Virginia. Nevertheless, the Assistant Regional Administrator for Federal and State Operations in the Kansas City office confirmed that those were the only two such citations and that inspectors in that region "very, very, very infrequently" received complaints about the issue.[40] Overall, then, OSHA's enforcement effort was purely reactive and geographically narrowly based. In contrast, in fiscal year 2001, only 26 percent of all 35,897 Federal OSHA inspections were generated by complaints (or accidents), while 50 percent were high-hazard targeted inspections; the corresponding figures for the 55,948 state-plan state inspections were 27 percent and 59 percent respectively.[41]

Pursuant to another FOIA request submitted by the author on September 9, 2002, Federal OSHA identified 74 citations that it had issued from January 1, 2000 until October 24, 2002 for violations of section 1910.141(c)(1)(i) (one of which was misidentified, leaving a total of 73).[42] The author then requested and collected all the citations issued by Federal OSHA from January 1, 2000 to the fall of 2002, and also identified an additional 11 cases, which the OSHA national office had

---

[38]Raven Management, Insp. No. 116520800, Closing Conference Notes, OSHA-1A (May 13, 1999).

[39]Raven Management, Insp. No. 116520800 (June 4 and 29, 1999). The company apparently no longer exists, but according to the person who answered the phone at the building where the firm used to be located, it had provided contract security services to the government, including the IRS in Kearnesyville; a partner in Raven Management had also been a partner in the CPA firm now located there. Telephone interview with Rivenbark CPA firm, 1207 W. Cumberland, Dunn, NC (Dec. 2, 2002).

[40]Telephone interview with Steve Carmichael (Nov. 27, 2002).

[41]http://www.osha.gov/as/opa/oshafacts.html.

[42]One citation was for violation of 1910.141(c)(1)(iii) involving sewage disposal endangering employees' health. Mansion Memorial Park, Insp. No. 303155824 (2000).

omitted because the cases had not yet been closed, bringing the total to 84. Only four of the 84 citations can arguably be regarded as having been issued for restriction of access; two of these employers were private guard companies.

Perhaps the most blatant violation was committed by one of the largest corporations to which OSHA issued a citation under the toilet standard. Liz Claiborne, Inc., a clothing and accessories firm with more than $3 billion in sales in 2001,[43] stood accused in the 1990s of producing its commodities in sweatshops in the Northern Mariana Islands, Honduras, and El Salvador.[44] Stung by the adverse publicity, the firm purported to adopt a "workplace code of conduct," which constituted "a comprehensive human rights and workplace safety policy":

> No Harassment or Abuse. All employees shall be treated with respect and dignity. No employee shall be subject to any physical, sexual, psychological or verbal harassment or abuse. ...
>
> Health and Safety. Employers shall provide a safe and healthy working environment to prevent accidents and injury to health arising out of, linked with, or occurring in the course of work or as a result of the operation of employer facilities.[45]

Whether or not Claiborne has complied with this code in the Third World, its disrespect for its employees' bladders and dignity at its Montgomery, Alabama distribution center for women's clothing would have seemed quite familiar to workers in Third World sweatshops.[46] It may have "[h]ired a Salvadoran personnel manager with an OSHA background"[47] in its sweatshop in El Salvador, but it was not enamored of OSHA personnel in Alabama, preferring instead off-duty police. According to the then president of Local 310 of the Union of Needle and Industrial Trades Employees who had worked there for about eight years, in 2001 the company, which previously had not prevented workers from using the toilets outside of fixed breaks, began locking the bathroom doors outside of the scheduled breaks (15 minutes in the morning and afternoon and 30 minutes for lunch) on the grounds that employees were going too often and staying too long. After about a month of this regime, the union filed a complaint with OSHA,[48] prompting an

---

[43]http://www.lizclaiborne.com.

[44]See, e.g., Philip Shenon, "Made in the U.S.A.?—Hard Labor on a Pacific Island," *N.Y. Times*, July 18, 1993, sect. 1, at 1, col. 2; Anne Quindlen, "Public and Private: Out of the Hands of Babes," *N.Y. Times*, Nov. 23, 1994, at A23, col. 1; Bob Herbert, "In America: Not a Living Wage," *N.Y. Times*, Oct. 9, 1995, at A17, col. 5 (Lexis).

[45]http://www.lizclaiborne.com/lizinc/lizworks/workers/conduct.asp

[46]Linder and Nygaard, *Void Where Prohibited* at 3-4.

[47]http://www.lizclaiborne.com/lizinc/lizworks/workers/faqs.asp

[48]Telephone interview with Bernard Burton, former president of UNITE Local 310, Montgomery, AL (Nov. 22, 2002).

inspection by the Mobile Area Office in August. The problem, in the words of the OSHA citation, was: "Throughout the facility: The restroom facilities were locked and were not available for immediate use by employees, except during breaks and lunch periods."[49] Managers kept the bathrooms locked at all other times because "they had a serious problem with theft and that was the only way they could eliminate some of the theft." Claiborne, according to the inspector's opening conference notes, "used off-duty Montgomery Police Dept. personnel to monitor the area to reduce the problem the company had with theft. The MPD personnel were required to unlock the restrooms during the morning, afternoon breaks and lunch."[50] (The union president denied the allegation of theft, expressing wonderment at how workers would have been able to steal clothing in the bathroom.)[51] The workers, according to the inspector's closing conference comments, "would have to go to their supervisors to get a written permission slip when they had to use the restrooms. The slip was given to the security personnel at the main entrance as that was the only restroom which could be used when all other restrooms were locked." Although the company official at the closing conference "was not pleased that the restroom would have to be left open for the employees," she stated that "they would have to make arrangements to keep the restroom doors unlocked." The inspector found that the workers were "exposed to long periods of time without being using [sic] the restrooms located adjacent to the work area," thus creating the possibility of bladder infections. The inspector imposed a monetary penalty because the lack of other opportunities to use the toilet

> created a problem when approximately 300 employees had to stand in line during the entire break and lunch period to go to the restrooms. Several of the female employees were pregnant and required frequent use of the restrooms but had to get written permission from the supervisors in order to do so, which caused a potential health problem. On one occasion, an employee was dismissed from her job because management determined she had gotten too many restroom passes, which took time away from her job.[52]

---

[49]Liz Claiborne, Inc., Inspection No. 304455629 (issued Nov. 21, 2001).

[50]Liz Claiborne, Inc., Inspection No. 304455629, Safety Narrative, OSHA-1A (Nov. 20, 2001)

[51]Telephone interview with Burton. The state director of UNITE for Alabama confirmed that the company had at first made that allegation, but she added that it was untrue and that later the company had shifted its position, claiming that workers had been spending too much time in the bathroom. Telephone interview with Willie Jones, Montgomery, AL (Nov. 25, 2002).

[52]Liz Claiborne, Inc., Inspection No. 304455629, Worksheet, OSHA-1B (Nov. 20, 2001). There is an error in the inspector's written comments as to the identity of the company official, but, based on information from the union president, it appears to have

Management, according to the union president, complied with the OSHA abatement order during the year following the issuance of the citation.[53] Whether the $900 penalty (reduced from the proposed $1,000)[54] would have continued to deter a $3 billion corporation from reverting to the deployment of police to interdict its workers' responses to the call of nature is speculative: Claiborne permanently closed the Montgomery distribution center on October 6, 2002.[55]

OSHA also found a violation at a call center operated by Convergys Corporation in Killeen, Texas. Convergys, which declares that "[w]e're the global leader in integrated billing and customer care services, and employee care services," "employs more than 44,000 people in our contact centers, data centers and offices in the United States, Canada, Latin America, Europe, the Middle East, and Asia. ... Revenues in 2001 topped $2.32 billion."[56] Although it advertises to potential customers that "Convergys can improve your employee satisfaction,"[57] not all of its own employees were satisfied with the company's bathroom access policy. In April 2002 the Austin Area Office of OSHA issued a citation proposing a penalty of $1,125 for a "serious" violation of the toilet standard on the grounds that: "Mens [sic] and womens [sic] restrooms are not effective in that employees are not allowed to use the facilities when needed."[58]

Convergys, none of whose call centers is unionized,[59] employed 442 workers at its Killeen location, one of whom called the Austin OSHA Office to file a complaint about not being permitted to use the bathroom. The Assistant Area Director, Elizabeth Slatten, offers what may well be the best hope for rigorous enforcement that can be expected from this agency. This 14-year OSHA veteran and former compliance officer is knowledgeable, competent, energetic, frank, and forceful. Yet, echoing the it-all-comes-out-in-the-wash self-doubts that have troubled labor-standards enforcers throughout the ages, even she wondered aloud whether vigorous enforcement of employers' obligation to let workers go to the bathroom might prompt some firms to eliminate some or all scheduled breaks to the extent that they are not mandated by state or federal labor standards laws. She underscored that when the April 6, 1998 Memorandum reached her office, it was read attentively and immediately recognized as marking a change in di-

---

been the human resources director.

[53]Telephone interview with Burton.

[54]Liz Claiborne, Inc., Inspection No. 304455629.

[55]Telephone interviews with Burton and Jones.

[56]http://www.convergys.com/company_overview.html.

[57]http://www.convergys.com/employeecare.html

[58]Convergys Corp., Insp. No. 304389711 (Apr. 17, 2002).

[59]Email from Dave LeGrande, director of safety and health, CWA, to Marc Linder (Nov. 7, 2002).

rection for enforcement activity; for that reason it was discussed at a staff meeting. Whereas prior to April 6, 1998, she and others gave callers who complained about being denied breaks the "pat" answer that breaks of any type were not part of OSHA law, now the office was ready to investigate complaints about denial of bathroom access. Apart from the U.S. Postal Service, a few of whose mail sorters have called OSHA, the focus of such complaints in the Austin Area was call centers,[60] which have been a frequent source of violations and complaints elsewhere in the United States as well.[61]

As Slatten explained the pattern at Convergys (which was not unique), in the abstract it might be difficult to convince an administrative law judge that an employer that offered one hour and 11 minutes of breaks—30 minutes for lunch, two 15-minute fixed breaks, and 11 minutes of discretionary break—was denying workers access. However, since, in her view, OSHA cannot directly cite employers for the number of minutes workers have to wait between bathroom visits—and, again, she fears that if OSHA did prescribe a fixed ceiling, the result might be that some employers would eliminate some of the rest breaks they voluntarily afford employees—the real defect in Convergys's policy lay in imposing bans on breaks during certain times. Slatten reported that the Austin Office had investigated other complaints from call centers, which, however, were resolved through the phone-fax procedure without issuing citations.[62]

As the inspector's internal inspection narrative noted:

> Complaint is e[mploye]es not allowed to use restrooms—the mens and women are more than adequate for the number of employees and are very clean and had paper towels and waste cans and hot and cold water.
>
> The problem is that the employees have quota of phone calls and minutes on the phone in their 8 and one-half hour shift. They get two fifteen minutes break[s] and a 30 minute lunch with so called 11 minutes of extra personal minutes. But according to the employees interviewed different team leaders schedule and allow these breaks. Often the first break i[s] one hour after the e[mploye]es['] shift begins. So then they have to work 3 straight hours and get placed on a written development plan if they exceed their allowed minutes off/away.[63]

---

[60]Telephone interview with Elizabeth Slatten, Assistant Area Director, OSHA Austin Area Office (Nov. 7 and Dec. 24, 2002). For a more detailed account of the complaints, see below this chapter.

[61]See above ch. 1 and below ch. 16. For a report on restricted toilet access at a predecessor of Convergys in Tucson, see RuthAnn Hogue, "No Relief in Sight," *The Arizona Daily Star*, Apr. 11, 1999, at 1D, col. 2, 3.

[62]Telephone interview with Slatten. For further discussion of these other call center complaints, see below this chapter.

[63]Convergys Corp., Insp. No. 304389711, Inspection Narrative, OSHA-1A (Apr. 15, 2002) (copy furnished by Austin OSHA).

The inspector went on to note: "Team Leaders on Saturday and Monday, which are the busiest days will not allow employees the time to use restroom when the urge strikes and many employees have been put on disciplinary development plans for using restroom." Pointing to the realm of the possible, the inspector added that "urinary and fecal matter in and on employees and chairs and clothing could cause HEP B."[64] Although a penalty of $1,125 was proposed for a serious violation, at an informal settlement conference OSHA changed the citation to an other-than-serious violation and eliminated the monetary penalty altogether in exchange for fast abatement. The agreement required Convergys to retrain its team leaders on its policy of allowing employees to use the bathrooms as well as on its enforcement of their implementation of that policy, of which the company was required to give OSHA a copy.[65] In its abatement letter the employer summarized the policy for the agency as containing these points: "Employees are not to be restricted from restroom use and team leaders should not state that an employee is not free to use the restroom when needed. Team leaders should notify Human Resources in any situation where an employee suggests that a more frequent use of the restroom is causing the employee not to meet weekly performance variance requirements. Further, team leaders will discontinue performance discussions on variance in that situation. ... Human Resources and Operations will continue to work with employees to insure that they feel free to use the restroom while assisting employees in meeting their performance goals."[66] In particular this final aspect of the policy, while vague and ambiguous, is nevertheless inconsistent with the rigid rule governing the proportion of the workday that workers have to spend on the telephone that, according to the Senior Human Resources Manager, was in effect half a year later.[67] Having filed away this policy statement and informed the

---

[64]Convergys Corp., Insp. No. 304389711, Worksheet, OSHA-1B (Apr. 15, 2002) (copy furnished by Austin OSHA). Slatten explained that the reference to possible hazards did not mean that fecal material had been observed on chairs. Email from Elizabeth Slatten to Marc Linder (Dec. 23, 2002).

[65]Convergys Corp.,. Insp. No. 304389711, Informal Settlement Agreement (May 9, 2002).

[66]Convergys Corp.,. Insp. No. 304389711, Abatement Letter from Convergys Corp. to OSHA (May 31, 2002)

[67]The employer's senior human resources manager, though eager, after gaining approval from the corporate legal department, to offer the company's version of events, failed to present a coherent counter-account. Apart from denying that managers—at least to her knowledge or with her approval—instructed employees not to leave the phones to go to the bathroom, she did not mention the 11 minutes of extra personal time (in addition to the 60 minutes of lunch and rest breaks). Instead, she spoke of 55 (which she thought might also be only 45) minutes of time the workers were permitted away from the phones, which were calculated on the basis of workers' being required to spend 88 percent (of which she was also not absolutely certain) of their working hours on the phone (12 percent

complainant of the outcome, OSHA, pursuant to its normal protocol, conducted no follow-up inspection to determine whether the problem still existed—although the informal settlement agreement obligated the employer to give OSHA access to the workplace to confirm compliance—trusting, instead, that the original complainant would be intrepid enough to call back if Convergys denied access yet again.[68]

Three citations that Federal OSHA issued for violation of section 1910.141 (c)(1)(i) copies of which could not be acquired were issued by its Manhattan Area Office, located at 6 World Trade Center, which was totally destroyed on September 11, 2001.[69] One of them was a denial of access case.[70] This citation was, coincidentally, issued for violations committed at 2 World Trade Center. According to the summary inspection report available on the OSHA website, a complaint involving unionized guards[71] prompted OSHA to conduct an inspection in June

---

of 450 minutes, which equal an eight-hour day minus the two 15-minute rest breaks, equals 54 minutes). Management monitored this 88-percent work effort on a daily basis, but workers who failed to sustain it were given the opportunity to achieve it over a week; she believed but was, again, not sure, that a computer program now made it possible for workers to track the percentage themselves. She did not clarify the exact scope of purposes to which these minutes could be put: they could, for example, be spent on such work-related activities as walking across the room to ask a question concerning a customer, but they could also be used to go to the bathroom; although she did not make clear how much of this time could be used for voiding, she did state expressly that workers were not free to use the full 55 minutes and that they were not permitted to use the time for breaks, and if supervisors "catch someone on a break," disciplinary consequences would follow. Finally, despite the fact that OSHA had not divulged the name of the employee who filed the complaint, she believed that she knew who that person was. Telephone interview with Terry Hall Senior Human Resources Manager, Convergys, Killeen, TX (Dec. 3, 2002).

[68]Telephone interview with Slatten.

[69]Letter from Richard Mendelson, Area Director, OSHA Manhattan Area Office, to Marc Linder (Nov. 18, 2002).

[70]Of the other two, one employer, after denying that there had been any problem or that OSHA had issued a citation, explained that the violation had been failure to have separate toilets for men and women. Telephone interview with unidentified person at Clarion Design Jewelry, 15 W. 46th Street, New York, NY (Nov. 15, 2002); Inspection No. 302944921 (2001). The proposed penalty was reduced from $450 to $250. The other employer's telephone number had been disconnected or was no longer in service without any indication of a new number. 99 Excel Fashion Inc., 658 62nd Street, Brooklyn, NY, (718) 567-8266 (Nov. 15, 2002); Inspection No. 302941661 (2000) ($0 penalty). The inspector confirmed that it was not a denial of access violation; to the best of her recollection the employer had not provided the proper number of toilets. Telephone interview with Laverne Ogiest, OSHA Manhattan Area Office (Nov. 25 and Dec. 2, 2002).

[71]According to the inspector's Narrative, the union was the Allied International

2000, to find a violation of section 1910.141(c)(1)(i), and to impose a monetary penalty of $2,000 against Summit Security Services, Inc., which contested the citation; a formal settlement agreement then reduced the penalty to $800.[72]

Private security guards, who have figured so prominently in OSHA's marginal enforcement of its Memorandum, are by no means a quantitatively small category of the labor force; they constitute a large and low-paid group of workers: in 2001 the median hourly wage among 995,560 such guards was only $8.94.[73] At 83.4 percent, the industry also recorded the highest proportion of firms with violations of the premium overtime requirement of the Fair Labor Standards Act

---

Union, headquartered in Mineola, NY. Summit Security Services, Inc. Insp. No. 302941315, Narrative, OSHA-1A (Aug. 1, 2000). According to Summit Security Services, Inc.'s website, "all full-time (working in excess of 32 hours per week) security personnel are affiliated with the Allied International Union." http://www.summitsecurity.com/faq/main.htm. The person who answered the phone at the union office refused even to confirm or deny that the union had been the collective bargaining representative at the World Trade Center, and told the author to put the question in writing to the union president. Telephone interview with unidentified person (Dec. 20, 2002). On the undemocratic practices of the union, see Metzler v. Allied International Union, 1998 U.S. Dist. Lexis 2882 (S.D.N.Y. 1998). On the alleged gangsterism of its leadership, see James Cook, "Brother Cunningham and the Guards," *Forbes*, Feb. 14, 1983, at 107 (Lexis); Roy Rowan, "The 50 Biggest Mafia Bosses," *Fortune*, Nov. 10, 1986, at 24 (Lexis). SEIU Local 32-BJ was at the time of the inspection engaged in a legal dispute over representation and its general counsel, who had first told the author that he would tell an SEIU official with knowledge of the World Trade Center collective bargaining agreement to speak to the author and then instructed that official not to speak to him, refused even to state whether the union had had anything to do with the OSHA complaint. Telephone interview with Larry Engelstein, New York City (Nov. 27 and Dec.20, 2002).

[72]Summit Security Services, Inc., Insp. No. 302941315 (Lexis and http://www.osha.gov/cgi-bin/est/est1xp?i=302941315). According to the inspection report, the complaint was filed on Jan. 5, 2000, but it was five months before an inspection took place. It is impossible to explain this extraordinary delay from the surviving report: OSHA's "customer service goal is to inspect an establishment within 5 days of receiving a formal written complaint from a current employee." Email from Elizabeth Slatten, Assistant Area Director, OSHA Austin Office, to Marc Linder (Nov. 28, 2002).

[73]http://www.bls.gov/oes/2001/oes_33Pr.htm; http://www.bls.gov/oes/2001/oes339032.htm. When Summit Security Services was hired by the Port Authority to provide security service at the World Trade Center in 1996, its $7 an hour wages were about one-half that of the workers they replaced who had been earning $13. Juan Gonzalez, "PA's New Trade Center Bomb," *Daily News*, Feb. 13, 1996, at 11 (Lexis). According to the general counsel of SEIU Local 32-BJ, which had organized the guards who had worked for Burns, the previous employer, a prolonged and complex legal dispute with the Port Authority arose over the successorship contract, which was not resolved until 2000. Telephone interview with Larry Engelstein (Nov. 27, 2002).

uncovered by U.S. Department of Labor investigations in 1997 among seven "problem industries."[74] The crux of their bathroom access problem is that some employers fail to establish the relief systems needed to enable guards who are forbidden to leave their posts to take the time to go to the bathroom.

In spite of the destruction of OSHA's Manhattan Area Office, because Federal OSHA had required Area Offices to secure the Regional Office's approval to issue a citation for restriction of access,[75] an official in the New York Regional Office, which was not located at the World Trade Center, happened to have kept a partial case file, including the inspector's Narrative and Worksheet, which have more detail than the citation. According to this work product, which was based on interviews with 41 employees of the 383 employed by Summit Security Services at the World Trade Center, the guards, whose collective bargaining agreement afforded them a 30-minute lunch break but no other breaks, had filed a complaint with OSHA stating that they were not being provided bathroom breaks. The inspection determined not only that the guards had been "advised by management that they are not to abandon their post under any circumstance,"[76] but that even if they had been free to go whenever the pleased, physical access would still have been a problem:

> The CSHOs [Compliance Safety and Health Officers] observed that there were also no bathroom facilities readily available/provided to the employees. During the inspections the CSHOs observed the employer unlock the bathroom facilities and indicated to CSHOs that employees could use the public toilet facilities in the WTC. Employees working inside the WTC informed the CSHOs that they would ask the nearby stores in the WTC mall to use their toilet facilities. These stores would sometimes deny outside employees the usage of their toilet facilities and the WTC mall public toilet facilities would have long waiting lines. ... Employees in the WTC and on street locations informed the CSHOs that the work place toilet facilities were kept locked and therefore were not accessible to them.[77]

When the inspectors asked the resident manager of the security firm where the toilets used by employees were located, he told them that the guards could use

---

[74]Brian Tumulty, "Many Low-Wage Workers Getting Shortchanged on the Job," Gannett News Service, Dec. 20, 1997 (Lexis).

[75]See above ch. 6.

[76]Summit Security Services, Inc., Insp. No. 302941315, Narrative, OSHA-1A (Aug. 1, 2000).

[77]Summit Security Services, Inc., Insp. No. 302941315, Worksheet, OSHA-1B (June 28, 2000). The second inspector had coincidentally accompanied the inspector to the inspection site, which was very close to the OSHA office. Telephone interview with OSHA employee who did not identify herself (Nov. 14, 2002).

the public bathrooms in the World Trade Center mall; the manager also told the inspectors that although the toilets had been locked because employees had vandalized the walls, the keys were kept with the dispatcher or receptionist, and that there was also a sign-out log. However, when they asked her for the keys, the dispatcher/receptionist neither had nor knew anything about the keys; they were also unable to find a log. The manager's statement to the OSHA officials that the employees had access to toilets on level B-1 also foundered when they found those facilities locked.[78]

By the same token, even if the toilets had not been locked, the guards

> were not given prompt access to toilet facilities. Employees radioed the dispatcher for a Post Relief and would be told to 'standby' for hours before receiving relief or they would not be relieved at all. As a result, some employees have urinated and defecated on themselves. [T]he employer used Post Reliefs to work various post [sic] when there is a shortage in staff, therefore eliminating the post relief person(s).[79]

A somewhat different account emerged from an interview with an official of the security company who expressed understanding for the guards' voiding needs and laughed in disbelief on hearing that prior to 1998 OSHA had taken the position that employers were not obligated to let workers use the toilets they were required to provide. For years, according to him, the firm had had an elaborate relief worker system in place, under which 10 to 15 relief guards relieved the up to 113 guards on duty in 2 World Trade Center so that they could take breaks including toilet breaks. Despite this relief arrangement, it still sometimes took too long for a relief guard to reach guards who needed to go to the bathroom, as the company official himself admitted, although he insisted that real security issues were the primary factor in the delay since guard posts could not be left unattended. The problem in his view was that the toilets available to the guards were located in the company's offices in the sub-basement of the tower, to which they had to take an elevator no matter how high up in the building they were patroling, because there were no toilets on those floors to which they had access: either they

---

[78]Summit Security Services, Inc., Insp. No. 302941315, Worksheet, OSHA- 1B (June 28, 2000). The manager was killed trying to save others on Sept. 11, 2001. http://www.cnn.com/interactive/us/0109/missing/files/hoerner.ronald.html; Collins Conner, "Memories of a Vanished Son," *St. Petersburg Sun*, Sept. 11, 2002, at 1 (Lexis).

[79]Summit Security Services, Inc., Insp. No. 302941315, Worksheet, OSHA-1B (June. 2, 2000). Warren Simpson, an industrial hygienist in the New York Regional Office, who had read the worksheet before the citation had been issued, kept a copy of the worksheet, of which he gave a copy to the inspector, Laverne Ogiest. Simpson laughed briefly when he read to the author on the telephone the sentence about guards' voiding on themselves. Telephone interviews with Simpson and Ogiest (Nov. 25 and Dec. 2, 2002).

were located in tenants' offices or required keys (in part as a result of heightened security after the bombing in 1993). After the inspection, OSHA reached an agreement with the employer that the latter would in the future guarantee that it would provide relief within 15 minutes of a guard's calling in.[80]

Interestingly, Summit Security did not even allude to what the Area Director of the OSHA Manhattan Area Office, based on his involvement with the case two years earlier, called the employer's main defense—namely, that the Port of New York Authority had refused in its contract with Summit to pay for the additional relief guards necessary to make it possible for guards to get to go to the bathroom promptly. To be sure, because OSHA did not deem the Port Authority to be a joint employer and it is the employer's obligation to shoulder the expense of complying with the standard, the agency considered the defense invalid. Even if OSHA had considered the Port Authority a joint employer, as a governmental agency it is not covered by Federal OSHA; and although it is covered by the New York State OSHA program and Federal OSHA has given thought to doing enforcement actions on a multiple-employer cross-agency basis, the Manhattan Area Director was quite certain that OSHA would not use the toilet standard as a test case.[81]

The last of the four citations for restriction of access was issued in 2002 to Securiguard Inc., which provides security guards at the Kings Bay Naval Submarine Base in Kingsland, Georgia. One employee told the OSHA inspector that the guards sometimes had to wait 45 minutes for relief and that requesting relief was "'an aggravating experience'"; another employee reported that men "'take pee cups with them in case relief doesn't arrive in time.'" Facing what he perceived as a lack of "hard evidence," the inspector could not verify the complainant's account that he had had to wait two hours, but "even the employer's initial investigation determined that relief did not reach the guard for 50 minutes."[82] In the Notice of Alleged Safety or Health Hazards, the inspector informed the employer: "On 3/26/2002, between 11:00 a.m. and 1:00 p.m., one of your employees requested to be relieved to go to the bathroom. Relief came after two hours, and at that time the employee had already defecated on himself/herself. After two hours, the affected employee went to the bathroom and discarded her/his underwear. On a prior occasion, another employee had to urinate

---

[80]Telephone interview with an official of Summit Security Services, Inc. (Nov. 15 and 19, 2002). The interviewee's statement that no monetary penalty had been imposed is contradicted by the inspection report and other OSHA-system data.

[81]Telephone interview with Richard Mendelson, Area Director OSHA Manhattan Area Office (Dec. 4, 2002).

[82]Securiguard, Inc., Insp. No. 303771547, Worksheet, OSHA-1B (May 17, 2002).

in a trash can, because timely relief was not provided."[83] In the sanitized final report, the employer was ultimately cited because the unionized guards "have not been relieved in a timely manner hindering access to sanitary facilities." For this serious violation OSHA fined the employer $975, which was reduced to $600 following an informal settlement.[84] Offered an opportunity to explain the employer's version of the incident to the author, the firm's director of human resources refused to discuss the facts of the case.[85]

In addition to the aforementioned citations, the atypical facts of two other cases make them difficult to classify, but suggest quasi-restriction of access. Skill Staff is a private profit-making employment agency in Kansas City, Missouri that recruits day laborers for employers. The OSHA inspector's Worksheet described the hazard this way: "Between 50 to [sic] 80 employees are present at the site between 05:00 to [sic] 08:00 and at other times during the day to accept temporary work assignments. Employees are not allowed reasonable access to a toilet facility. ... Employees (and others) are exposed to health risks from having to urinate and/or defecate in a public alley while awaiting a work assignment." The inspector also noted that the "[e]mployer had been advised of violative condition when referral-based non-formal investigation was initiated" four months earlier.[86] The inspector's Health Narrative observed that "it became apparent that the respondent provided false information in responding to the investigation. Mr. Thornton [the regional vice president] alleged 'verbal permission' from another employer in the neighborhood that would allow Mr. Thornton's employees access to their restrooms. However, CSHO called the other employer who denied ever having had a conversation with anyone regarding the issue. Also, other business owners in the proximity to this establishment continued to express concern about the situation. Decision to inspect was based on the apparent factual misrepresentation coupled with the continued concern of other business owners in the neighborhood."[87] OSHA cited the employer because "[m]ale and female employees required to be at employer's facility to accept a temporary work were not provided with reasonable access to toilet facilities," and fined it $1,800 for this "serious" violation, which it reduced to $900 for an other

---

[83]Securiguard, Inc., Insp. No. 303771547, Notice of Alleged Safety or Health Hazards, OSHA-7 (Apr. 29, 2002).

[84]Securiguard, Inc., Insp. No. 303771547 (May 22, 2002).

[85]Telephone interview with Ms. Howard, Securiguard, Inc., McLean, VA (Dec. 2, 2002). The CEO of the company, Patrician DeL. Marvil, is a member of the executive committee and treasurer of the Hispanic Business Roundtable. http://hbrt.org/exec.htm.

[86]Skill Staff, Insp. No. 303307708, Worksheet, OSHA-1B (Aug. 1, 2000).

[87]Skill Staff, Insp. No. 303307708, Health Narrative, OSHA-1A (Aug. 1, 2000).

than serious violation at an informal conference.[88]

The problem, according to the Skill Staff officer manager, was that she did not want the day laborers wandering around the building lest they impair the staff's safety. Therefore, any day laborer who wanted to use the bathroom had to be accompanied by a staff member who would make sure that he or she went to the bathroom and not elsewhere; the manager characterized this system as providing "limited access." Since the staff also had their regular clerical work to do, the staff member often told a day laborer that he or she would have to wait until that work had been completed. Since the bathroom was only for one person at a time, several day laborers might be waiting. Although the manager alleged that the day laborers, who had "gotten smart to systems that interrupt the business," had complained to OSHA out of spite over a personal dispute with the company's driver, in fact the inspection was prompted not by a complaint, but a referral. Since she admitted that the health department had also inspected the office (and among other things told her she would have to get an industrial-type toilet paper holder), it is possible that the health department had referred the case to OSHA. Since one toilet would clearly be inadequate for the number of day laborers waiting in the office at peak times in the morning, the author asked whether OSHA had discussed this matter and/or the broader issue of whether the day laborer job applicants, whose time at the office while they were waiting to see whether they would be hired each day was presumably not compensable under the Fair Labor Standards Act (FLSA),[89] were employees or customers and thus whether OSHA or the public health department had jurisdiction over the number and condition of the toilets, but she admitted that her memory had become blurry as to which agency had said and done what.[90] Since the definition of "employee" under the FLSA is broader than under OSHA, it is unclear how the workers could have been covered by OSHA and thus entitled to bathroom access when they needed to go if they were not, at the time, (some employer's) employees for purposes of the FLSA and thus not entitled to the minimum wage for their waiting time.[91]

---

[88]Skill Staff, Insp. No. 303307708 (Aug. 2, 2000); http://www.osha.gov/cgi-bin/est/est1vd?30320770801001A.

[89]See Sakas v. Settle Down Enterprises, 90 F. Supp2d 1267, 1279 (N.D. Ga. 2000): "Defendants contend Plaintiffs are not entitled to compensation for the time they were on the premises prior to being notified that work was available. On this much the parties can agree...." See also GAO, *Worker Protection: Labor's Efforts to Enforce Protection of Day Laborers Could Benefit from Better Data and Guidance* 24 (GAO-02-925, 2002).

[90]Telephone interview with Heather Moyer, Office Manager, Skill Staff, Kansas City (Dec. 2, 2002).

[91]See Marc Linder, "Dependent and Independent Contractors in Recent U.S. Labor Law: An Ambiguous Dichotomy Rooted in Simulated Statutory Purposelessness," *Com-*

The alleged violation description on another citation (imposing no monetary penalty) involving a Springfield, Missouri, unit of Git N Go, a chain of convenience stores/gas stations, also gave rise to some ambiguity: "The employer did not effectively communicate to each employee its policies and provisions for ensuring employees, including those working alone, had prompt access to toilet facilities in accordance with the standard."[92] The employee's complaint to OSHA had read: "Personnel not allowed to use restrooms." The inspector's findings stated: "Employer states they have a policy that employees working alone can lock the door when no customer's [sic] are in the store if they need to use the restroom. There is no written policy. Policy not effectively communicated to each employee and no provisions for employees to be relieved during times of steady customer traffic in the store."[93] According to the company's human resources director, a "disgruntled" female employee had made the complaint to OSHA. The firm's policy was to tell employees working alone to lock the door and put up a sign saying that he or she would be back in five minutes before using the toilet, which was located 15 feet from the counter. The director added that, for the sake of safety and protection of corporate assets, this policy did not apply so long as the employee was waiting on a customer. When asked what would happen if there was a line of customers, the director, in effect confirming OSHA's finding, replied that the employee would have to wait until all the customers had been served and left, although he contended that employees know when the busy times are and can use "common sense" to plan accordingly.[94] Thus, like the OSHA citation, this interview left unanswered the question as to whether any employee had in fact been denied access. The night employee at the Git N Go store to which the citation had been issued stated that she had no problem going to the bathroom, did hang up such a sign, and had never heard of the OSHA inspection, although she had been working there at the time.[95]

The group of 84 employers cited for having violated some provision of the toilet standard in 2000-2002 can be statistically compared with that of the nine employers cited for restricting workers' access from 1998 to 2002 (four of which are included in the 84). Of the 84 citations that Federal OSHA issued under section

---

parative Labor Law & Policy Journal 21(1):187-230 at 196 (Fall 1999); IBP, Inc. v. Herman, 144 F.3d 861, 865 (D.C. Cir. 1998).

[92]Git N Go, Insp. No. 304304900 (Nov. 14, 2001).

[93]Git N Go, Insp. No. 304304900, Safety Narrative, OSHA-1A (Nov. 8, 2001).

[94]Telephone interview with Russell Estes, Corporate Human Resources Director, Git N Go, Tulsa, OK (Dec. 3, 2002).

[95]Telephone interview employee at Git N Go, 2953 N. National, Springfield, MO (11:40 p.m., Dec. 2, 2002).

1910.141 (c)(1)(i), 67 (80 percent) involved unionized employers and only 17 (20 percent) nonunion employers. In contrast, two-thirds of the employers restricting access were unionized and only one-third nonunion. In terms of the type of inspection, 58 (69 percent) of the 84 were triggered by complaints from workers; the other inspections were programmed/planned (7), referrals (7), related to other inspections (4), done in connection with fatalities (4) or accidents (2), unprogrammed-related (1), or planned (1). In contrast, 100 percent of the inspections leading to citations for restricting access were triggered by complaints. Geographically, 26 (31 percent) of the toilet standard citations were issued to workplaces located in the South, that is, the seven states of the Confederacy covered by Federal OSHA (Arkansas, Alabama, Florida, Georgia, Louisiana, Mississippi, and Texas, plus one federal citation in the state-plan Tennessee program), with Texas alone accounting for 15 of them. Outside the South, the states with the largest number of citations were New York (12), New Jersey (8), Ohio (7), Pennsylvania (6), and West Virginia (5). (In addition, three citations were issued to clothing sweatshops in the Northern Mariana Islands.) In contrast, one-third of the workplaces where employers restricted access were located in the South.

With regard to monetary penalties, 53 (63 percent) of all toilet standard citations imposed none, while 31 (37 percent) did, whereby two of these were later reduced to zero. In contrast, 67 percent of the restriction-of-access citations imposed monetary penalties (excluding one reduced to zero). The initial penalties in the larger group ranged from a high of $2,500 to a low of $150, while the highest final penalty was $1,250. The mean initial penalty of these 31 was $919, while the mean final amount for the 29 employers with penalties was $575. Averaged over all 84 cited employers, these total penalty amounts were $339 and $199, respectively. Among the employers cited for denial of access, initial penalties ranged between $2,000 and $975, while the highest final penalty was $900; the mean initial penalty was $975 and the mean final amount averaged among all nine employers was $412. Similarly, only 16 (19 percent) of all citations were "serious," while five (6 percent) fell into the "repeat" category. In contrast, 56 percent of the denial-of-access citations were classified as "serious" (but one was changed to "other than serious" by an informal agreement, lowering the overall proportion to 44 percent). The specific nature of employers' violation of the toilet standard can be categorized as follows: 27 (32 percent) provided no toilets; 24 (29 percent) failed to provide the required number of toilets; eight provided toilets that were broken or not working; seven did not provide or make toilets available; five provided toilets that could not be locked; four denied employees access to toilets; two locked the toilets; two failed to provide the required number of toilets for each sex; two provided a toilet with a broken seat or no seat; one provided a port-a-potty; and one provided an outhouse. Finally, in terms of industrial or occupational focus: 15 cited firms were in manufacturing; five citations were issued to the U.S. Government (including the U.S. Postal Service); four involved security guards:

four were issued to sawmills; four involved a single taxi company at four locations; and three were issued to call centers or telemarketers. In contrast, among the nine employers that restricted access, five employed guards, one a dispatcher, and one operated a call center. Interestingly, four of the guard companies performed services for the Federal Government, while one did so for the Port of New York Authority.

The proliferation of government contracts with private firms to provide security guards for Federal building and facilities, including a submarine base of the United States Navy, which presumably does not suffer from a lack of trained armed personnel, is said by insiders to date back to efforts by the Clinton administration to be able to claim that it had shrunk government, if only by virtue of having effectively transferred its employees onto private firms' payrolls.[96] Under these circumstances it should, as one OSHA official suggested, at the very least, be incumbent on the Federal Government to include the need for extra relief guards in the contracts for which it solicits bids from private employers.[97] That bathroom access for guards is a widespread problem was confirmed by a public-interest labor lawyer in San Francisco, who reported that virtually all of the low-wage workers on whose behalf he had intervened with employers in this matter were guards.[98]

State-Plan States

In the year 2000, when more than 100 million workers employed by 6.5 million employers were covered by federal and state OSHA programs,[99] the federally approved OSHA programs in the 21 state-plan states (plus three states—Connecticut, New Jersey, and New York—with approved programs for public employees only) covered a total of 53,211,536 employees employed by 3,288,028 employers; of these, 44,742,525 employees were employed by 3,183,572 private-

---

[96]Telephone interview with Jerry Peyton, Branch Manager, Initial Security, Indianapolis (Dec. 26, 2002).

[97]Telephone interview with Cheryl Gray, Safety and health Assistant, OSHA Omaha Area Office (Jan. 2, 2003). Efforts to find the Federal Government officials responsible for contracts with private security guard companies proved inconclusive. The official at the General Services Administration in Fort Worth who was in charge of the master contract took no part in the actual bidding process, which was the responsibility of the individual departments and agencies for whose buildings the firms provided guards, and did not know the names of any of the departmental officials who were involved in the process.

[98]Telephone interview with Mike Gaitley, Employment Law Center, San Francisco (Dec. 2, 2002).

[99]http://www.osha.gov/as/opa/oshafacts.html (data for fiscal year 2001).

sector employers, while 8,469,011 employees were employed by 104,456 public-sector employers.[100] In the state-plan states, the number of citations issued for violations of section 1910.141(c)(1)(i) (or the corresponding state version) and, as a subset thereof, the number of citations for restriction of toilet access in cases opened after April 6, 1998 are presented in Table 3.[101]

Of the 329 citations issued for violations of the toilet standard by the state OSHA agencies—of which California alone accounted for 55 percent—only 11 dealt with restriction of access. These 11, however, were not evenly distributed: Iowa and Kentucky, with four and three respectively, accounted for 58 percent of them, while Arizona, Indiana, South Carolina, and Washington issued one each. None of the other state programs (including those approved only with respect to the protection of public employees) issued any.

These 11 are summarized in Table 2. All 11 citations were triggered by complaints (except for one referral), of which seven came from unionized places of employment. In six cases the agency imposed no monetary penalty; the principal exception was Iowa OSHA, which initially classified all four violations as serious, repeat, or willful and fined all four cited employers amounts ranging from $1,000 to the national record of $36,000 (Excel Corporation). In terms of industry composition, all four Iowa cases involved animal slaughter plants, while all three Kentucky employers were manufacturers. Of the four remaining cases, three involved guards or police.

The combined Federal and state OSHA data for all 20 citations for restriction of access reveal that all were triggered by complaints (except for one referral), 65 percent of which were filed at unionized workplaces. Forty-five percent of cited employers were ultimately not assessed a monetary penalty, which averaged $2,309 for all 20 employers or, excluding the outlier Excel, $536.

Narratives of the cases are presented below except: the four Iowa cases, discussion of which occupies chapter 12, the Kentucky Jim Beam case, which forms the subject of chapter 13, and the Washington case, which is dealt with in chapter 14.

In what may have been the first case to have arisen in the United States after

---

[100]Occupational Safety and Health State Plan Association, *OSHAPA Report: Grassroots Safety and Health in the Workplace: 1999-2000 State Plan Activities Report* 35 (n.d.), on http://www.dir.ca.gov/dosh/dosh_publications/oshsparpt.pdf.

[101]In a few instances it was not possible to verify the precise basis for issuance of a citation for violation of the toilet standard because the paper file had been lost or purged after expiration of the retention period. Virginia OSHA lost the file of Tarmac America, Inc., Insp. No. 301817383 (Oct. 30, 1998). Telephone interview with Jay Withrow, Director, Office of Legal Support, Virginia OSHA (Oct. 31, 2002). Cal/OSHA lost 20 case files. For the methodology and sources used in compiling the data in this table, see below Appendix V.

April 6, 1998, Kentucky OSHA on April 25, 1998 received a complaint from workers, members of the Teamsters Local 236, at Ardco, Inc., a plant with 290 employees manufacturing industrial refrigerator glass doors, in Elkton, a small industrial town. The workers' allegation read: "Employees must sign a sheet to get to the bathroom or use the phone, etc. You may be placed on a waiting list with 20-225 [sic; must be 25] people before you are issued a bathroom pass, sometimes may take an hour or more for your turn. ... Sixty-five...people share 2-one stall bathrooms." Based on a remarkably thorough and insightful (though syntactically shaky) inspection report, the inspector cited (but did not impose a monetary penalty on) the employer despite the workers' having had four daily bathroom breaks in addition to three scheduled breaks:

> The company has stated that in the contract with the union there are two ten minute breaks and a half hour lunch. ... The employer also stated that the employee [sic] are given four extra additional breaks in the day to go to the bathroom and use the phone. ...
>
> There can be as many as seventy employees in any one area, if the two water closets in the area are occupied then the employees may have to walk up to three hundred feet to the next restroom. The company allows so many bathroom passes per five percent...of the employees in the area. For example if there are 50 employees in the area, the employer will have two to three passes for that group. Employees place there [sic] name on a list to receive the next pass, once receiving the pass they must place the time they leave and the return time from the restroom. If the time exceeds 9 minutes then the employees get a verbal or written warning. If there [are] several write ups then the employee is suspended.
>
> The Federal Occupational Safety and Health Administration issued an interpretation concerning 29 CFR 1910.141(c)(1)(i) dated April 6, 1998. In that interpretation letter it basically states that the employer can not place unreasonable restrictions on employees who need to use the restroom. Each individual is different in how often we expel waste from our bodies, medications and medical conditions can alter that mechanism and increase the frequency and during [sic; must be duration] of ridding waste from our bodies. The employer was also informed in times of heat extremes that increase of liquids is vital to prevent heat related illnesses. By the addition of liquids, increased rest breaks and restroom breaks can be encountered.
>
> The policy of the company that places a time limited [sic] and frequency [sic] of restroom breaks on the employees, without the consideration that employees who are on medications that require them to urinate or defecate more frequently, prevent[s] the employee from using the restroom until their turn is available. This policy is therefore considered unreasonable and thus is considered to be a violation of 29 CFR 1910.141 (c)(1)(i). The issue concerning telephone usage is not an occupational issue.[102]

---

[102]Ardco In., Insp. No. 302081336 (July 21, 1998); Inspection Outline furnished by Kentucky OSHA.

This zero-dollar admonition was presumably not the reason that Ardco, Todd County's largest factory, shut the plant down in January 2001.[103]

In 1999, following a complaint, this time by Teamsters Local 651, Kentucky OSHA cited for a "serious" violation and imposed a $975 penalty on CR/PL, a manufacturer, ironically enough, of bathroom partitions and toilet bowls, in Somerset. The company, which owned Crane Plumbing, was one of the largest manufacturers and distributors of plumbing fixtures and specialty plumbing products in the United States, and employed 281 employees in this plant.[104] OSHA found that employees "were denied the ability to use the rest room until other employees returned with the pass." After the inspection the employer revised the policy so that "employees will have the amount of time away from their work station tracked, but they will not be denied the availability of the rest room."[105]

Arizona OSHA has issued only one citation to an employer for restricting access. Although it classified the violation as "nonserious" and imposed no monetary penalty, in 1998 the agency, responding to a complaint, cited the City of South Tucson's nonunion Communications Center (Police) because: "Communication employees were not afforded unrestricted use of toilet facilities in that relief for their position was often not made available causing them to wait for extended periods of time, thereby subjecting the employees to adverse health effects and serious infections."[106] Despite the lack of a monetary penalty, the city contested the citation, which was upheld by an administrative law judge.[107]

In the one Indiana case, in which the injury/illness was specified as "Bladder damage," the lack of a toilet merged with denial of access ("No toilet available"). The employer, Initial Security, a security guard company, is owned by the ominously named Rentokil Initial, a British-headquartered international business services firm with 94,000 employees. Ironically, the parent company's hygiene division provides industrial bathroom supplies and specialist cleaning of washrooms.[108] For the nonunion employees posted at a guardhouse of a UPS facility

---

[103]http://www.owensborodio.org/archives/newspaper/2002/5help.html.

[104]"Plumbing Supplier Tries to Plug Loss," Crain's Chicago Business, Oct.. 10, 1988, at 9 (Lexis); http://www.somersetpd. com/webdoc5.htm. Although the OSHA report mentioned only partitions, the person who answered the phone at the company (on Dec. 23, 2002), now called Crane Plumbing, stated that the factory also manufactured bowls and sinks.

[105]CR/PL, LLC, Insp. No. 302404157 (Nov. 8, 1999); Inspection Outline furnished by Kentucky OSHA.

[106]City of South Tucson - Communications Center, Insp. No. 127001626 (Aug 27, 1998).

[107]City of South Tucson - Communications Center, Insp. No. 127001626 (Feb 25, 1999); http://www.osha.gov/cgi-bin/est/est1vd?12700162602004.

[108]http://www.rentokilinitial.com; http://www.initialsecurity.com.

in Indianapolis in 2002, there was no restroom to use "unless they call for someone to relieve them." On the first shift "there is no extra person to relieve and employees sometimes have to wait several hours for someone to get there to relieve them. Employees have to call dispatch and then wait for someone to show up. One employee had used the bushes in the past." Even when a relief worker did show up, since there was no toilet on site, the employees had to walk over 150 yards to use another firm's (UPS's) toilet. Although Indiana OSHA deemed this violation of section 1910.141(c)(1)(i) "serious," it assessed no monetary penalty.[109]

South Carolina, too, has issued only one citation for restricting access. In 2000, responding to a complaint at a nonunion workplace, state OSHA issued a citation, without a monetary penalty, to the South Carolina Department of Juvenile Justice because "Juvenile correction officers were not provided immediate access to restroom" at two units of a correctional facility.[110] The basis for the citation was ambiguous because it was unclear whether the employer refused to permit the workers to go or whether the toilets were physically inaccessible or both. However, since the inspector's own Worksheet notes referred to the OSHA Memorandum (a copy of which was also appended to the report) in connection with the amount of time it took to give the employees relief, it may well qualify as a restriction of access citation. The employer was cited because "Juvenile correction officers not provided immediate access to restroom" in the building unit to which they were assigned. However, a more detailed Hazard Description stated: "Employees have limited access to restroom facilities located in the units due to them being locked when social worker is away from office and on weekends. Employees have been informed to use Omega Dorm however they are not allowed to leave co-workers alone in the units. Management has been informed to no avail." Employees stated that they could use the restroom when the social workers were in the office, but after that office was locked, they had to call the supervisor to come open it or relieve them so they could go to a bathroom in another building. "Some employees are on medication which causes them to use the restroom more often. Takes supervisor 40-5?? min[ute]s to come and open restroom door. See interpretation letter dated 4-6-98." Management stated that

---

[109]Initial Security, Insp. No. 305200115, Worksheet, OSHA-1B (July 19, 2002); provided by Indiana OSHA. At the same time, however, the firm was fined $3,150 (reduced by informal agreement to $1,170) for lack of potable water. http://www.osha.gov/cgi-bin/est/est1xp?i=305200115. The branch manager declined to be interviewed about the case until he received corporate approval, which he purported to seek but apparently did not obtain. Telephone interview with Jerry Peyton, Indianapolis (Dec. 26 and 31, 2002 and Jan. 10, 2003).

[110]Insp. No. 3020442108 (Sept. 18, 2000); Citation and Notification of Penalty furnished by South Carolina OSHA.

it had "had to lock the restroom doors located in the social workers area and take keys away from the officers because the doors were being left unlocked and juveniles were entering the area. They can call their supervisor and they will come and allow them to go and use the restroom."[111]

Despite having issued this one citation, South Carolina offers some sense of the lax enforcement that certain state OSHA's have engaged in with regard to toilet access. An OSHA standards officer and former inspector there knew, based on her own experience, that most citations under the toilet standard were for lack of a sufficient number of toilets. She was aware of one case, in 1999-2000, when she was monitoring, in which an employer refused to let workers go to the bathroom except at fixed times. She merely suggested that the employer, a processing business, provide relief workers, but did not issue a citation; the employer's "compliance" took the form of merely letting supervisors give relief.[112] Yet ironically, the very same day this officer was interviewed, the receptionist who answered the phone in the office, priding herself on hailing from a more civilized state (New Jersey), revealed that just the previous day at intake a case had been discussed concerning a factory worker who had been denied access, which prompted the receptionist to wonder out loud why things were so backward in South Carolina.[113]

Possible Explanations of the Small Number of Complaints and Citations

The number of citations issued by OSHA to employers for restricting workers' access to the bathroom may not be indicative of the extent of the violations because many (especially nonunion) workers are too intimidated or inhibited to complain about anything, let alone about a matter as personal and intimate as eliminating waste from their bodies. For that reason all enforcement and compliance officials interviewed agreed that no conclusions about the prevalence of toilet access could be drawn from the number of complaints filed.[114] One long-time OSHA official who had been an iron foundry manager before working for the state of Tennessee observed that unskilled workers do not complain about anything, including safety

---

[111]South Carolina Dept. of Juvenile Justice, Insp. No. 302442108, OSHA Worksheet, DOSH-C 1A (Oct. 12, 2000).

[112]Telephone interview with Gwen Thomas, South Carolina OSHA (Sept. 20, 2002).

[113]Telephone conversation with unidentified receptionist, OSHA office, Columbia, South Carolina (Sept. 20, 2002).

[114]E.g., Telephone interview with George Vigil, Program Manager for Statistics Section, New Mexico OSHA, Albuquerque (Sept. 20, 2002); email from Lee Ann Campbell, Labour Services Branch, Whitehorse, Yukon, to Marc Linder (Sept. 12, 2002).

and health violations.[115] Enforcement must remain inadequate as long as inspections are triggered chiefly by (unionized) employees' complaints. That the vigor with which OSHA may be prosecuting even union complaints may leave much to be desired was suggested by one OSHA official's question as to why unions were not able to deal with bathroom breaks on their own, and why, if they were not, it was any of OSHA's business.[116] Even OSHA officials who took the problem of denial of bathroom access seriously and acknowledged the injuries that it inflicts on workers' health and dignity tempered that sympathy with the admonition that the higher priority of preventing workers from being electrocuted or losing limbs crowds out enforcement activity to compel compliance with the sanitation standard in a chronically underfunded and understaffed agency.[117]

Interviews with officials in all ten Federal OSHA Regional Offices, many Federal OSHA Area Offices, and all 21 state-plan programs leave no doubt that they regard both employers' violations of their obligation to make toilet facilities available to employees when they need to use them as a low-priority issue and the perceived small volume of worker complaints as therefore appropriate. For example, the director of the Federal OSHA Denver Area Office, who has worked for the agency for 25 years, stated that toilet access complaints were very rare.[118] The Regional Administrator for the Denver Region, encompassing Colorado, Montana, North Dakota, South Dakota, Utah, and Wyoming, reported that there had been very little activity, even of an informal nature, averaging perhaps one or two complaints per area office annually, all of which were resolved by the phone-fax procedure with the employers, dispensing with the need for on-site inspections.[119] Neither the Acting Regional Administrator for the Boston Re-

---

[115]Telephone interview with Mike Maenza, Manager for Standards and Procedures, Tennessee OSHA (Sept. 23, 2002).

[116]Telephone interview with Maenza.

[117]Telephone interviews with Maenza and Mary Bryant, Administrator, Iowa OSHA, Des Moines (Sept. 25, 2002). Maenza's good faith was underscored by an anecdote he related about why he had decided years earlier to leave his job as a foundry supervisor: One day while he was standing in the cafeteria line he distinctly smelled human feces; seeing that the seat of the pants of the worker in front of him was soiled, Maenza tapped him on the shoulder and asked him to step outside to talk. When Maenza asked him whether he realized that he had defecated in his pants, the man replied that he had, but that his job did not permit him to prevent it or do anything about it. Maenza told him to go home, wash, change his pants, and return to the foundry, and that he would be paid for the time.

[118]Telephone interview with Herb Gibson, director, Denver Area OSHA Office (Oct. 31, 2002).

[119]Telephone message from Adam Finkel, OSHA Regional Administrator, Denver Region (Dec. 9, 2002).

gion[120] nor the Deputy Regional Administrator for the Philadelphia Region was aware of any change in the volume of complaints since the Memorandum had been issued. Moreover, the Philadelphia Regional official asserted apodictically that she did not think that the small and unchanged volume of complaints could be explained by workers' being unaware of their right to go to the bathroom, at least not in her region. Rather, she was "willing to bet" that the problem was centered in the southern states, though, revealingly, she immediately slipped back into speaking of the problem as a lack of bathrooms altogether (and not denial of access).[121]

Of greatest interest is the response of OSHA's Assistant Regional Administrator for Compliance Programs for the Atlanta Region, who reported that his area offices in Alabama and Georgia used to receive a considerable volume of complaints from poultry processing workers, but no longer did. He and two assistants even jumped from the perceived absence of complaints to the conclusion that the problem itself had "gone away." Asked to speculate on the possible reasons, he suggested, in addition to the Memorandum (of which poultry management had been well informed and to which it had not objected), the impact of the UFCW's organizing activities directly on the plants that it has organized and indirectly on others that may need to offer equivalent working conditions to fend off unionization.[122]

To be sure, representatives of the UFCW itself did not seem quite so sanguine about the alleged disappearance of the problem, at least not in nonunion plants, let alone its cause,[123] but the Area Director for Local 1996, which has organized a dozen poultry plants, reported significant progress in several unionized plants.[124] Rick Brown, the executive assistant to the president of UFCW Local 1996, emphatically denied that bathroom access was no longer a problem in nonunion poultry plants; on the contrary, the prospect of being "able to take a pee can make a difference" during an organizing drive. Brown also dismissed the notion that an absence of complaints could be sensibly interpreted as an absence of violations. A huge proportion of the poultry plant labor force consisting of recent

---

[120]Telephone interview with K. Frank Gravitt (Nov. 27, 2002).

[121]Telephone interview with Marie Cassady (Nov. 27, 2002).

[122]Telephone interview with Benjamin Ross (Dec. 18, 2002). Two of Ross's subordinates, team leader Bill Fulcher and senior industrial hygienist Jim Drake, who also participated in the interview by way of speaker-phone, agreed with Ross's account.

[123]Telephone interview with Robyn Robbins, Assistant Director, Occupational Safety and Health Dept., UFCW, Washington, D.C. (Dec. 18, 2002); email from Jackie Nowell, Director, Occupational Safety and Health Dept., UFCW, to Marc Linder (Dec. 19, 2002).

[124]Telephone interview with Curtis Williams, Sewanee, GA (Dec. 26, 2002). For further details, see below ch. 16.

immigrants, many of whom lack proper documents, would, in his experience, "certainly not call a government agency of any sort to report that they didn't get a bathroom break."[125] Edgar Fields, the International representative of the Retail, Wholesale, and Department Store Union, which is affiliated with the UFCW and has organized three poultry plants in Georgia, pointing out that workers at the unionized plants still had to deal with unlawful restrictions, scoffed at the suggestion that bathroom access was no longer a problem in the 80 percent of plants there without a union. As an example, Fields mentioned the Cagle's plant in Perry, Georgia—at which the union had recently lost an election—where denial of access had gone to the point at which one worker urinated in his pants.[126]

Despite these qualitative assessments by management-level OSHA officials who may be too far removed from the area offices' daily intake routine to be intimately familiar with the volume and composition of telephone complaints, it is possible that the actual universe of toilet-access complaints may be considerably larger than they imagine. Indeed, there is good reason to believe that no single person in an OSHA office would be in a position to furnish an accurate answer without doing a tedious hands-on review of the complaint records.[127]

OSHA's recordkeeping and retention procedures generally make it impossible to determine how many toilet-access-restriction complaints OSHA has received nationally that were resolved short of the stage of on-site inspection and citation.[128] As one state-plan official observed: "When we receive complaints that are handled via telephone, facsimile or letter and resolution of alleged conditions occurs, those activities are not tracked as inspections. Therefore, this issue may be the subject of more complaints that find their resolution without initiation of an inspection. An exact number of such complaints is not available nor is it easily derived."[129]

Although Federal OSHA offices retain such complaint records for three years, they are generally filed under the name of the employer and cannot be accessed by standard number, even in their abbreviated computerized form.[130]

---

[125]Telephone interview with Rick Brown, Executive Asst. to the Pres., UFCW Local 1996 (Jan. 8, 2003).

[126]Telephone interview with Edgar Fields, International Representative, RWDSU, Atlanta (Jan. 6, 2003).

[127]Email from Elizabeth Slatten to Marc Linder (Dec. 30, 2002).

[128]Telephone interview with Byron Orton, Iowa Labor Commissioner, Des Moines (Oct. 18, 2002).

[129]Email from Chris Ottoson, Health Analyst, Enforcement Policy Section, Oregon OSHA, to Marc Linder (Sept. 24, 2002).

[130]Telephone interview with Elizabeth Slatten (Nov. 7, 2002); email from Slatten to Marc Linder (Dec. 17, 2002).

However, at least one Federal OSHA Area Office has created a computer program that does make it possible to access informally resolved phone-fax complaints and to identify the corresponding paper files by standard (though not by subsection). By examining all the section 1910.141 complaint files for fiscal year 2002 (running from October 1, 2001 to September 30, 2002), the Omaha Area Office identified seven (out of a total of 358) telephone complaints dealing with restriction of toilet access.[131] Despite the additional work caused by purely hand retrieval, a review of the paper files was also performed in the Federal OSHA Austin Area Office, which determined that in fiscal year 2002, of a total of 644 complaints that it received, including those addressed through an on-site inspection, those investigated by phoning and faxing a letter to the employer (which does not usually progress to an inspection),[132] and those referred to another agency or dismissed for lack of jurisdiction, 11 dealt with restriction of toilet access. Of a total of 134 complaints that the Austin office received during the first quarter of fiscal year 2003, three involved bathroom access, making a total of 14 during those 15 months. In addition, the office in fiscal year 2002 conducted one inspection based on a restriction of access complaint that did not result in a citation.[133]

If the Austin and Omaha Area Offices are typical—and there is no obvious reason, in terms of the demographic base and employment composition of the geographic area they cover, to think otherwise—the 85 Federal OSHA Area Offices could be receiving almost 800 restriction of toilet access complaints annually. If the parallel state-plan OSHA system, which has jurisdiction over an approximately equal number of workers, receives an equivalent volume of complaints, the annual nationwide total might be more than 1,500. This number would by no means be trivial, especially when it is considered that for every complaint filed an indeterminate number of workers may be suffering a denial of access in silence.

Even if this quantitative extrapolation turned out to be excessive, an examina-

---

[131]Telephone interview with Bonita Winningham, Assistant Area Director and Duty Officer, Omaha Area Office (Dec. 17, 2002); telephone interview with Cheryl Gray, Safety and Health Asst. and Duty Officer, Omaha Area Office (Jan. 2, 2003); letter from Bonita Winningham to Marc Linder (Jan. 3, 2003).

[132]For a description of the phone-fax system, see http://www.osha.gov/as/opa/worker/complain.html; http://www.osha.gov/as/opa/worker/handling.html.

[133]The initial hand count, which retrieved all the bathroom complaints, was later supplemented by a report of the total number of complaints. Email from Elizabeth Slatten to Marc Linder (Jan. 31, 2003). One of the complaints in fiscal year 2003 was filed against a construction company regarding work at a construction site and was presumably subject to the construction industry sanitation standard, which is not subject to the April 6, 1998 Memorandum. Complaint, Wilson Construction, Austin (2002).

tion of these complaints that OSHA declined to process to the point of on-site inspections can shed light on the question of why OSHA has issued so few citations. In Nebraska, six of the seven employers were well-known national or regional companies, including the U.S. Postal Service,[134] two home improvement and/or lumberyard chains (Lowe's[135] and Menard's[136]), APAC Customer Services,[137] Aquila (electric and gas utility),[138] and Carneco Foods, an animal slaughter plant, which is a joint venture between Tyson and Lopez Foods Inc., one of the country's largest Hispanic-owned businesses.[139] Two of the complaints concerned guards (Menard's and Visinet, Inc., a business that provides monitors who accompany parents during court-supervised child visitations),[140] whose complaints echoed those already discussed—namely, lack of a relief guard enabling them to leave their post). Two (APAC and Aquila) involved the by now familiar complaints of call-center workers that they are not allowed to stop answering the telephone to go to the bathroom outside of fixed breaks.[141]

The information that several of these employers furnished OSHA in response to the complaints should at the very least have raised some suspicions at the agency, prompting it to explain employers' obligations under the Memorandum to the firms in greater detail. That it did not, may in part have resulted from the fact that the information received, though revealing of potentially deep-seated problems with the employer's policy, was merely incidental to resolving the specific allegations of the complaint, on which OSHA was exclusively focused. For example, Carneco Foods' human resources director wrote:

> Going back a few months, it was common practice for a group of employees to leave the production floor for the restroom whenever their production line went down. Numerous times the line would come back up while the employees were still away from their work stations. When supervision would go looking for the employees, they would be found visiting in the hallway as opposed to returning to the work station. Therefore, a

---

[134]OSHA, U.S. Postal Service, Complaint No. 203760798 (Apr. 26, 2002).

[135]OSHA, Lowe's, Complaint No. 203760392 (Apr. 1, 2002).

[136]OSHA, Menards, Complaint No. 203762422 (Sept. 18, 2002).

[137]OSHA, APAC Customer Service, Complaint No. 203761291 (May 31, 2002).

[138]OSHA, Aquila, Complaint No. 203761739 (July 17, 2002).

[139]http://www.estradusa.com/mr/camr/21500.htm; http://www.hispanicbusiness.com/research/500/view.asp?companyid=849

[140]OSHA, Visinet, Inc., Complaint No. 203761036 (May.15, 2002).

[141]Information furnished by Cheryl Gray (Jan. 2 and 3, 2003). Gray noted that guards at Menard's had complained several times before; although their pay checks come from a private security company, OSHA regards them as employed by Menard's because Menard's controls all their conditions of employment, monitors them, and tells them what to do.

"pass system" was implemented. A meeting was held with employees...at which...employees were advised of the 'pass system'. When an employee requested to use the restroom facility, they would be given a pass to leave the production floor and use the restroom. Upon their return, someone else may take the pass and use the restroom. This "pass system" was not very well accepted by the employees. They were told restroom privileges were being abused and until they improved, we would use the "pass system".

During our investigation today, I was advised by the supervisors, the "pass system" is no longer being used. Employees have been very cooperative and not abusing the use of restrooms.[142]

Here the employer revealed above all that it had still not grasped that, pursuant to the OSHA Memorandum, workers have a right to void when they need to, not merely "restroom privileges," to be revoked at the employer's discretion. Exacerbating this unlawful approach was the practice of punishing all employees for what the company claimed was "abusing" behavior by some employees (although even according to the firm's own account it is not clear that it had informed those employees how long the downtime would be). And finally, the pervasive paternalism and correlative infantilization that culminated in the introduction of the "'pass system'"—the human resources director's use of distancing quotation marks hinting that even the company was somewhat embarrassed by its adoption—mimicking the regime imposed on schoolchildren should have given OSHA pause about the firm's limited attitudinal capacity for voluntary unassisted compliance.

In the Austin area, more than half of the employers about whom workers made toilet-access complaints were large national or Texas companies, including the U.S. Postal Service (two complaints), Southwest Airlines (reservations call center), Nationwide Insurance (claims call center), MCI Worldcom, Best Buy, Corrections Corporation of America, and the H.E. Butt Grocery Company (H-E-B supermarket chain).

Remarkably, under OSHA procedures, employers do not have a heavy burden to carry in providing the type of "adequate response" that shifts the burden to the worker to dispute it.[143] Although some of the employer responses were detailed, others merely denied the truth of the complainant's account; in the most extreme example, it sufficed that the employer merely conclusorily summarized its alleged investigation by declaring the worker's complaint "unsubstantiated"

---

[142]Letter from Carneco Foods to U.S. Department of Labor, Omaha (May 15, 2002), in OSHA, Complaint No. 203760996, Carneco Foods (May 14, 2002).

[143]OSHA, CPL 2.115: Complaint Policies and Procedures, sect. L.6 (June 14, 1996), on http://www.osha.gov/pls/oshaweb/owadisp.show_document?p_table=DIRECTIVES &p_id=1.

(underlined three times).[144] Even this minimalist response prevailed because the worker failed to respond, let alone to dispute it; in fact, non-response was the norm. In Texas, in 12 of the 14 cases OSHA never heard from the worker again; and even in the other two cases, the additional OSHA involvement did not proceed to the point of inspection. In Nebraska, none of the complainants responded to the employer's response.[145]

In fact, OSHA officials agree that silence is the overwhelmingly typical response of complainants. The Assistant Area Director in the Omaha Area Federal OSHA Office estimated that 99 percent of telephone complaints come from ex-employees and that 99 percent of complainants drop the matter at the time that OSHA requests a response from them to the employer's denial of their allegations. This pattern of abandonment prevails despite (or perhaps in part because of) the fact that some OSHA offices inform ex-employee complainants at the outset that if the employer denies their allegations and refuses OSHA permission to enter the premises to perform an inspection, it is very unlikely that OSHA would be able to obtain a judicial warrant to enter based solely on an ex-employee's statement.[146] In this sense, it may hardly matter at all whether the ex-employee responds to an employer's denial or not. At the same time, the official conjectured that the vast preponderance of telephone complainants are ex-employees because current employees may fear for the security of their jobs.[147]

---

[144]The employer was Bryan Centex Countertops.

[145]Telephone interviews with Slatten and Cheryl Gray.

[146]Telephone interview with Bonita Winningham, Asst. Area Director, Omaha (Dec. 30, 2002). In fact, there is no principled impediment to obtaining a warrant based on an ex-employee's statement: "Central Mine argues that, under the Act, whenever probable cause is asserted on a basis other than a general administrative plan, it must be founded upon information supplied by someone currently employed by the employer. Section 8(a) of the Act contains no such restriction upon OSHA's authority to inspect. Section 8(f)(1) merely describes the procedures to be followed when an employee does complain of a safety or health standard violation. Probable cause to issue an administrative search warrant under the Fourth Amendment may be established upon information from any reasonably reliable and credible source." In the Matter of the Inspection of Central Mine Equipment Co., 1978 WL 18636 at 8-9 (E.D. Mo. 1979). The issue is more that of the staleness or recency of the ex-employee's information. Telephone interview with Stephen Reynolds, attorney, Solicitor's Office, OSHA Kansas City Regional Office (Jan. 27, 2003). The Assistant Regional Administrator for Federal and State Operations, OSHA Kansas City Regional Office, which oversees the Omaha Area Office, agreed with Reynolds and was unaware of any practice of informing ex-employee complainants that ultimately there was little OSHA could do if the employer denied the agency access. Telephone with Steve Carmichael (Jan. 27, 2003).

[147]Telephone interview with Winningham. Elizabeth Slatten, the Asst. Area Director of the Austin Area Office, agreed that the vast majority of complainants are ex-employees,

Although another OSHA official in Omaha, who as duty officer took complaint calls two weeks of every month, estimated the proportion of telephone complaints coming from ex-employees at 20-25 percent, she believed current employees accounted for one-third, and others, including customers, spouses, and mothers, for the largest share. In addition, she estimated that one-third of all calls were made by people who declined to identify themselves at all, some of whom explained their caution by expressly stating that if they were to take the process to the next higher level, "they're going to know it's me."[148]

One actual example of a worker's abandonment of a complaint may illustrate the typical process. In 1999 a legal services organization complained to OSHA on behalf of a former employee that workers at a frozen foods plant in South Texas worked 12-hour shifts with only one 10-minute break in the morning and one in the afternoon and no meal period: "Employees are not allowed any other restroom break throughout the shift. There is no relief personnel for employees taking breaks to the restroom. If they need to use the restroom more than twice per day they are assumed to be sick and must have a doctors [sic] excuse." Instead of conducting an on-site inspection, OSHA relayed this complaint to the employer, requesting a reply within four days.[149] The employer then denied all accusations, asserting in particular that: "There is no limit on the number of restroom breaks taken as long as it is not abused or excessive." Based on this reply, OSHA informed the complainant that "OSHA feels the case can be closed on the grounds that the hazardous conditions have been corrected (or no longer exist)," and gave the complainant six days to inform OSHA of any disagreement. Because the worker no longer worked there and would not have been able to offer additional information or informants, OSHA, in the absence of a reply disputing the employer's claim, took no further action.[150]

These patterns suggest that even workers who have had the fortitude to file a complaint may lack the wherewithal for further confrontation with what they perceive to be a hostile and determined employer that might, after the next round of exchanges, be able to identify the anonymous complainant (and/or blacklist an ex-employee) if it has not already done so. A major research project in its own right would be needed to determine the range of reasons causing so many com-

---

but did not think they comprised 99 percent. Telephone interview with Slatten (Dec. 30, 2002).

[148]Telephone interview with Cheryl Gray.

[149]Letter from OSHA, Corpus Christi Area Office, to Southern Frozen Foods, Alamo, TX (Mar. 22, 1999).

[150]Letter from R.L. Mike Hunter, OSHA Area Director, Corpus Christi, to Jose Torres, Texas Rural Legal Aid (Mar. 30, 1999); email from Larry Norton, Texas Rural Legal Aid, to Marc Linder (Oct. 23, 2002).

plainants to abandon their claims. The point here, however, is merely that, whether for a good or bad reason, the enforcement process has been terminated without OSHA's having acquired sufficient information to be able to make an independent judgment as to whether the employer violated a standard—and denial of toilet access is no exception—because the agency either lacks the resources for, or has decided to allocate its resources in such a way as to preclude, following up complaints that it has categorized as unpromising.

Finally, even when the rare employer concedes that there was some validity to the complaint and promises to undertake what OSHA calls "corrective action," the agency will never know with certainty whether the employer actually implemented such action if the complainant never calls back.[151] Thus, for example, in response to a complaint that employees at the gas station at H.E.B. in Pflugerville were not given restroom, water, or lunch breaks in a timely manner, the employer declared that all of management agreed that a better effort was required to meet the needs of "our partners" including giving them relief when they have to go to the bathroom.[152]

The formal criteria that OSHA has developed for closing complaints may have warranted its inaction in some of these cases, but the peculiar facts in a number of these cases, which differ from those discussed elsewhere in the book, may have raised difficult questions concerning the intersection of OSHA law and wage and hour and/or labor-management relations law that agency officials may have wished to avoid having to resolve. For example, in the Southwest Airlines reservations call center case, the complainant stated that, because of bathroom renovations, workers having to wait to use the toilet had been penalized for taking too few calls.[153] A complaint directed against an Austin post office stated that management had refused to authorize time for letter carriers to take bathroom breaks while on a delivery route and had consequently written them up for using unauthorized overtime. In response, the employer countered: "Management does not and will not refuse an employee the right to take a bathroom break while on a delivery route. However, if a carrier's authorized overtime is extended, management has the responsibility of ascertaining the reasons for the overtime. If the supervisor has reason to know that an employee while 'on the clock performed

---

[151]Although none of the seven employers whom complainants accused of restricting bathroom access conceded that there was truth to the complaints, the Omaha Area Office Duty Officer estimated that, in her experience, overall about half of employers admit that there is something to the charges and state that they have taken corrective action. Telephone interview with Cheryl Gray.

[152]Complaint, H.E.B. Pflugerville (Apr. 19, 2002).

[153]Complaint, Southwest Airlines, San Antonio (2002).

no work,' the supervisor may take appropriate action."[154]

Similarly, at the Nationwide Insurance Claims Call Center in San Antonio, which employed 400 workers, the complainant stated that the employer was restricting the employees' use of the bathroom to 15 minutes every four hours, and that employees with medical conditions requiring frequent access to the bathroom had been disciplined.[155] The employer, agreeing that it gave workers two 15-minute breaks, responded that employees could divide up these breaks into whatever segments they wished. In addition, the company declared to OSHA that it had rescinded two disciplinary actions that it had taken against employees with medical conditions and had initiated a process for obtaining in the future medical considerations under the Americans with Disabilities Act.[156] OSHA felt warranted in closing this complaint because the employer had taken corrective action dealing with the complainant's specific complaint.[157] Nevertheless, had OSHA been in the habit of vigorously enforcing workers' right to go to the bathroom when they need to, it would have noticed that the employer's response was, on its face, a clear admission of an unlawful policy: even if the employer did permit employees to divide up their 15-minute break during each four-hour period so that they could go to the bathroom several times, the employer was conceding that once workers had used up their 15 minutes, they would, unless they had some type of medical excuse, presumptively be subject to discipline for going to the bathroom when they needed to do so. This particular OSHA office was especially remiss because its intake officers already knew that call centers were the focal point of complaints about restrictions on bathroom use. Here OSHA's acquiescence in the employer's unlawful policy failed to enforce the letter or spirit of the Memorandum and therefore violated the agency's own directive that "[t]he complaint shall not be closed until OSHA is certain that the hazard has been eliminated/abated."[158]

To be sure, OSHA does require the employer to post a copy of OSHA's letter detailing the nature of the alleged hazard mentioned in the complaint and to certify to OSHA that it has posted it.[159] But if current employees are too intimi-

---

[154]Complaint, U.S. Postal Service, Austin. OSHA did not even investigate or send the U.S.P.S. a letter concerning another complaint because it had investigated the complaint two months earlier and, based on the employer's response that there was no problem, closed it. Complaint, U.S.P.S., Austin, Cross Park (July 2002); Complaint, U.S.Postal Service, Austin (Sept. 2002).

[155]Nationwide Insurance, Complaint No.204010011, Notice of Alleged Safety or Health Hazards, OSHA-7 (June 14, 2002).

[156]Letter from Nationwide Insurance to Austin Area OSHA Office (June 24, 2002).

[157]Letter from Austin Area Office (June 27, 2002).

[158]OSHA, CPL 2.115: Complaint Policies and Procedures, sect. L.8

[159]OSHA, CPL 2.115: Complaint Policies and Procedures, Appendix C: Sample

dated to file complaints, there is little reason to assume that workers' continued silence following this posting signals their agreement with the employer that no hazard exists or that it has been abated.

Even if workers have filed more complaints about toilet access than some OSHA officials realize, one possible reason for the meager enforcement of OSHA's Memorandum of April 6, 1998 was suggested by a state OSHA official: Perhaps no one at OSHA read it. Pointing to the hundreds of memoranda that state-plan officials receive every year, he asserted that if they read all of them, they would have no time to do anything else all day. Moreover, he noted, for really important directives (of which he said there had been ten in 1998), OSHA had established a two-way fax procedure, which serves to downgrade the significance of the memoranda not subject to the procedure.[160]

To be sure, OSHA's website lists 137 standard interpretations that were issued in 1998, but only the toilet standard memorandum of April 6 was an interpretation issued not as an opinion letter in response to a question from an employer, employee, union, legislator, or State OSHA official (although it was in reality a response to a request from the UFCW for clarification), but as a "memorandum," directed to the Federal OSHA regional offices and state-plan states, and admonishing the latter that their interpretations had to remain "'at least as effective as'" the Federal OSHA standard.[161] Nevertheless, the manager of one Cal/OSHA district office ventured the opinion that Federal OSHA had not been very serious about the Memorandum if it did not devise some more effective method of bringing it to the attention of the state-plan states.[162]

In contrast, John Miles, who as Director of Compliance Programs had issued the Memorandum in 1998, noting that the fax-fax procedure is used only for new standards, stressed that since the Memorandum had been a "hot potato" accompanied by a great deal of media attention, there was little likelihood that OSHA officials around the country had been unaware of it, especially since OSHA had discussed it repeatedly at various meetings. To be sure, when informed that some state-plan state officials were still unaware of it in 2002, he acknowledged that

---

Letters. The Austin Area Office also requires employers to post their response to OSHA. Telephone interview with Slatten (Dec. 30, 2002).

[160]Telephone interview with Maenza.

[161]http://www.osha.gov/pls/oshaweb/owasrch.search_form?p_doc_type=INTERPRETATIONS&p_toc_level=2&p_keyvalue=DATE_1998&p_text_version=FALSE&p_status=CURRENT. Two other general memoranda that year issued amendments to a standard and corrected typographical errors in a standard.

[162]Telephone interview with Luis Mireles, Manager, San Diego District Office of Cal/OSHA (Nov. 11, 2002).

attrition and the advent of new personnel could explain such ignorance.[163]

A revealing statement by Federal OSHA's Manhattan Area Director—echoed by other OSHA officials[164]—provides yet another possible reason for lack of effective enforcement: he doubted very much whether any inspector in the course of interviews conducted during a programmed inspection would ask employees whether their employer was not letting them go to the bathroom.[165] Indeed, one official observed that since many employees were reluctant to answer any questions, in part because they were not sure whether OSHA inspectors were associated with the employer, asking about something as intimate as voiding would, to say the least, not be productive.[166] Similarly, while not at all pooh-poohing the hazard associated with involuntary suppression of the bladder, one state-plan inspector observed that during the limited time available to inspectors for interviews with employees during planned inspections, they would simply be forced to focus on other more immediately life-threatening hazards.[167] The Manhattan Area Director's observation that the agency has also not advertised to employers the existence of their new obligation[168] finds confirmation in some large firms' ignorance. For example, in responding to a complaint of denial of restroom access, a store manager of Lowe's—a Fortune 100 company with 110,000 employees and $22 billion in revenue operating 800 stores in 43 states[169]—faxed OSHA three pages on breaks from the corporate Human Resources Management Guide, which includes in tabular form information on state statutory requirements. Though revised more than a year after the issuance of the Memorandum, the Guide expressly states that "the Company is not required by Federal law to provide breaks...."[170]

Overall, then, OSHA's timid and purely reactive approach to enforcement of and compliance with voiding rights is not geared toward changing the status quo.

---

[163]Telephone interview with John Miles, OSHA Dallas Regional Administrator (Nov. 12, 2002).

[164]E,g., telephone interview with Mike Rohde, Research Analyst, WISHA, Olympia (Sept. 25, 2002).

[165]Telephone interview with Richard Mendelson.

[166]Telephone interview with Mike Maenza, Manager for Standards and Procedures, Tennessee OSHA, Nashville (Sept. 23, 2002).

[167]Telephone interview with Pat O'Brien, compliance officer, North Carolina OSHA, Winston-Salem (Dec. 10, 2002).

[168]Telephone interview with Mendelson.

[169]http://www.lowes.com

[170]Lowe's Human Resources Management Guide, Policy No. 312 (5/15/99), in OSHA, Lowes [sic], Complaint No. 203760392 (Apr. 1, 2002). The firm nevertheless "supports work breaks to maintain alertness and energy resulting in better customer treatment, satisfaction and employee safety."

12

Relatively Vigorous Complaint-Driven Enforcement: UFCW-Organized Animal Slaughter Plants in Iowa

[G]oing to bathroom at work is just a bad habit.[1]

Iowa OSHA was, as already noted, the first to act expressly to require employers to let workers void when needed,[2] in large part because the author and Mark Smith, the president of the state Labor Federation, pressed for the change, which was implemented by an openly labor-friendly commissioner, Byron Orton, who has a national reputation for administering a vigorous state-plan OSHA program.[3] Iowa OSHA, therefore, is important to examine as a case study of the kind of results that this relatively favorable constellation of forces can generate.

Enforcement in Iowa differs from that of other jurisdictions in that four of the six toilet-standard citations issued after the new interpretation had gone into effect (in the case of Iowa, on January 21, 1998) were for failure to provide prompt access, while the other two were issued to employers that had failed to provide any toilets at all.[4] All four citations for failure to provide access were imposed against large firms in a major Iowa industry: animal slaughterhouses. One each was issued to John Morrell and Excel and two to Swift. All of these plants operate under collective bargaining agreements with the UFCW and all of the inspections were triggered by complaints (or, in one case, referral).

These unlawful conditions at animal disassembly plants are not confined to Iowa. The same incessant pressure to maximize throughput and a frenzied pace

[1]Iowa OSHA, Excel. Corp., Insp. No. 300375060, Inspector's Notes at 200 (1999) (statement to OSHA inspector by worker at Excel quoting a supervisor).

[2]See above ch. 3.

[3]Telephone interview with Mike Wright, Director of Occupational Safety, Health, and Environment, United Steelworkers, Pittsburgh (Oct. 10, 2002).

[4]Area Residential Care, Dubuque, IA, Inspection No. 115090235 (July 15, 1998); Overton's Disposal Company, Inc., Davenport, IA, Inspection No. 115098535 (Feb. 25, 2002).

on slaughterhouse production lines that prevents workers from taking ergonomically recommended short breaks to avoid repetitive stress injuries also prevents them from going to the toilet throughout the United States.[5] For example, some workers at the IBP plant in Wallula, Washington, which permits only a 30-minute lunch break and one 15-minute rest break during a seven-hour and 56-minute shift, have voided in their pants.[6]

Iowa OSHA's administrator, Mary Bryant, indicated that all of its inspections originated in complaints because it was too short-staffed to perform programmed inspections. Moreover, even if it were in a position to carry out such inspections, they would focus on violations suggestive of high injury rates; since lack of toilet access is not ordinarily associated with loss of work time or a reportable injury, such violations would not figure as high-priority items. Thus even in the case of an industry such as slaughterhouses with a documented record of repeated violations, Iowa OSHA continues to rely almost exclusively on complaints.[7]

The first of the two violations that Swift and Company (which at the time was a subsidiary of ConAgra, the second largest food company in the United States) committed at its Marshalltown plant was revealed by a referral from the Division of Latino Affairs of the Iowa Department of Human Rights to OSHA in June 2000: "Employees were not provided access to toilet facilities within a reasonable time frame: (a) Throughout establishment - Employees requesting to use toilet facilities were repeatedly told to wait. During the waiting period employees urinated and menstruated on themselves." OSHA proposed a penalty of $1,875, which was reduced by settlement to $1,000.[8] A little more than a year later, responding to a complaint, Iowa OSHA cited Swift for a repeated violation: "Toilet facilities were not provided in accordance with TABLE J-1 of this Section: (a) On the Loin Boning Line - On or about 10-25-01, at approximately 11:00 pm, an employee was not able to use toilet facilities and subsequently urinated in his clothing. The availability of utility relief workers and employee information were contributing factors." Iowa OSHA proposed a penalty of $5,000, which was reduced through settlement to $2,500[9]—not exactly a

---

[5]Marc Linder, "I Gave My Employer a Chicken That Had No Bone: Joint Firm-State Responsibility for Line-Speed-Related Occupational Injuries," *Case Western Reserve Law Review* 46:33-143 (1995).

[6]Nancy Cleeland, "For Meatpackers, Walkout Was Step Forward and Back," *L.A. Times*, July 9, 1999, at A1 (Lexis).

[7]Telephone interview with Mary Bryant, Des Moines (Sept. 25, 2002).

[8]Iowa Occupational Safety and Health Administration, In the Matter of Swift and Company, IOSH No. 300378031 (Oct. 13, 2000). The reduction is shown in a computer print out of cases that Iowa OSHA made available and is also accessible in the Lexis OSHAIR file.

[9]Iowa Occupational Safety and Health Administration, In the Matter of Swift and

powerful financial deterrent vis-à-vis a very wealthy recidivist.

Ironically, just a few months before the first of these OSHA inspections in Iowa, ConAgra's manager of corporate relations, in response to complaints by Hispanic workers in slaughter plants in Nebraska that they were being denied permission to go to the bathroom and some were urinating on themselves, had declared that "she was appalled when she read reports that workers at some plants don't get bathroom breaks. She said that doesn't happen at ConAgra plants. 'I can assure you we allow our employees to go to the bathroom.... They can go when they need to go.'"[10]

In 2001 Iowa OSHA imposed an (uncontested) penalty of $2,000 on John Morrell & Company of Sioux City, which is organized by UFCW Local 1142, for a whole series of violations. Morrell is a subsidiary of Smithfield Farms, "the largest vertically integrated producer of processed meat and fresh pork in the United States."[11] Generally, the employer restricted access "through extended delays, limits on the number of times employees were allowed to use the toilet per day, and limits on the time allowed to be away from the work area." OSHA then adduced the following specific instances, which in their variety and totality offer some insight into the lack of freedom to void even in unionized slaughterhouses:

An employee was required to wait 1 and ½ hours to use the toilet facilities. The employee asked his/her supervisor for a restroom break or spell out and was not allowed to use the restroom until their scheduled break time. The employee felt he/she would be fired for leaving the line without a utility person replacing them. The employee was delayed to the point that one or more of the following 3 events occurred—urinated, defecated, and/or heavy menstruation in their clothing. The employee has reduced the intake of liquids to avoid needing restroom breaks or spell outs during his/her shift.

An employee was required to wait 45 minutes to use the restroom. The employee asked his/her supervisor for a restroom break or spell out and was told "no". The employee left the work area to use the restroom and was verbally reprimanded by his/her supervisor.

An employee was afraid to ask for a restroom break. The employee felt if he/she requested too many breaks or spell outs they would be fired. The employee waits to use the facilities for a scheduled breaktime, the delay created [sic] one or more of the following to occur—urination, defecation, and/or a heavy menstruation in their clothing.

Company, IOSH No. 304790132 (Jan. 24, 2002).

[10]Mike Sherry, Cindy Gonzalez, and Leslie Reed, "Meatpacking Inquiry Opened," *Omaha World-Herald*, Sept. 11, 1999, at 59 (Lexis) (quoting Joan Lukas). On Lukas's position, see http://www.conagra.com/media/news.jsp?ID=19990712.

[11]http://www.johnmorrell.com/

An employee was required to wait approximately 50 minutes to use the restroom. The employee asked his/her supervisor twice and another supervisor twice before he/she was replaced to use the facilities. The employee felt that they would be written up if they left without a utility worker present.

An employee was denied a restroom break or spell out because the supervisor believed the employee had received a restroom spell out break already that day. The employee had not received a spell out or restroom break. The employee pleaded with the supervisor until the restroom break was granted. The employee had to wait 55 minutes for a restroom break. The employee felt they would be written up if they left the line without a utility worker or supervisors [sic] approval.

An employee was verbally reprimanded for using the facilities on their way to be interviewed by the inspector. The employee was told to report to the conference room and was caught using the facilities prior to reporting to the conference room.

An employee was required to wait 45 minutes to 1 hour to use the restroom. The utility worker was busy working on the line to replace an absent employee. The utility worker has been doing that job for several weeks. The line was short workers. The employee was reprimanded for leaving the line to use the restroom.

An employee was required to wait 45 minutes to use the restroom. The employee asked his/her supervisor three times. The employee shut down the line to use the restroom. The employee was sent to the personnel office.

An employee, with a doctor's excuse slip to use the toilet facilities because of a medical problem, was required to wait because he/she had already went [sic] once that day. Supervisor and other employees gave the employee a "hard time" about the frequent visits to the restroom. The employee did not believe they could leave the line without a utility person replacing them. The employee was delayed to the point that one or more of the following events occurred—urinated, defecated, and/or a heavy menstruation in their clothing.[12]

Against the background of such a broad array of blatant transgressions, the question, once again, arises as to whether the meager financial penalty can be expected to deter a wealthy corporation that may have calculated that the monetary value of the additional production it can squeeze out of a workforce deprived of toilet breaks (until it is cited again) exceeds the cost of the fine.

The most egregious systemic denials of voiding rights in Iowa slaughterhouses took place in 1999 at the 1,600-employee hog slaughter plant in Ottumwa owned by the Excel Corporation, the third-largest meat-packing firm in the

---

[12]Iowa Occupational Safety and Health Administration, In the Matter of John Morrell and Company, IOSH No. 303736573, Apr. 18, 2001.

United States, which in turn is controlled by Cargill, one of the largest food processing firms in the United States and the largest privately held corporation.[13] The degree of indignity and humiliation inflicted on workers was so intense that one worker who defecated in her pants filed an unprecedented private tort suit against the employer.

Initially, on July 21, 1999, an Iowa OSHA Compliance Safety and Health Officer (CSHO), responding to a complaint, telephoned Excel's human resources manager, Les Elders, about denial of timely access to the bathroom and faxed the OSHA sanitation standard together with Byron Orton's January 21, 1998 interpretation.[14] That complaint had alleged:

All employee[s] are having problems being allowed to go to the rest room. The shift starts at 6:00 am and employees are not allowed to go until 7:00 am, nor are employees allowed to go to the rest room for the proceeding [sic] half hour before breaks. Employees are required to have a relief person take their place on the line before leaving. Sometimes there is no relief person available for the entire shift.[15]

Elders responded on July 26:

Excel-Ottumwa operates a union facility in which all breaks and rest periods are a negotiated item. The company has no policy or practice that would restrict employees from using the restroom when requested. The company does ask the employee to notify management when it is necessary to use the restroom outside of a normal break or rest period. A relief person would be used, where possible to ensure the flow of the process continues. If no relief person is available, employees are still allowed to use the restroom.[16]

The CSHO replied the following day, recommending that the employer cover the complaint with its supervisors so that they understood the requirements. Nevertheless, two weeks later, on August 11, A,[17] a female hogpusher cooler,

---

[13]Iowa OSHA, Excel Corp., Inspection Report, Insp. No. 300375060, Nov. 22, 1999; http://www.sierraclub.org/factoryfarms/rapsheets/iowa/excel.asp.

[14]IOWA OSHA, Excel Corp., Inspection Report, "Coverage Information/Additional Comments" at 3.

[15]IOWA OSHA, Excel Corp., Inspection Report, "Coverage Information/Additional Comments" at 4.

[16]IOWA OSHA, Excel Corp., Inspection Report, "Coverage Information/Additional Comments" at 4.

[17]Throughout this account of the events at the Excel plant in Ottumwa the names of the workers denied access have been deleted and replaced with letters. The OSHA Inspection Report and inspector's notes were released to the lawyer representing the worker who

requested but never received a bathroom "let out" from a lead person because she had not gone to a supervisor, Gene Miller, and there were people ahead of her on the list because they had gone to him. She had not been trained that she was required to go to a supervisor, recalling only that the lead person gives let outs and a doctor's excuse was necessary to get a let out "whenever needed." An hour and 35 minutes later, at break time, still without having received the let out, she informed Miller that she had had an accident in her pants and had to go home to change them. Then in the middle of the kill floor "Mr. Miller said loudly to her: 'You mean you shit your pants.'"[18]

On August 19, Miller's supervisor, Jim Greinert, met with Miller, employee representative Denny Glattfelder, and Elders, concerning this incident. Elders absolved Miller of any involvement in preventing A from going to the bathroom. Elders told the group that if someone had to go to the bathroom, "they will go," and that Greinert would tell A that if she could not find a relief person, "she should go to the bathroom before having an accident." In a "personal note," the OSHA compliance officer observed that although Excel knew that there had been previous problems with toilet access, "the company chose to drop the issue," taking no further action. The OSHA officer added that Glattfelder "told me that one of the big issues in becoming the union president was to make sure that the employees at Excel-Ottumwa would be allowed to go to the bathroom when needed. This statement impressed upon me...that this must have been an ongoing problem. Mr. Glattfelder told me this was one of the major items on his platform when running for union President."[19]

---

filed a tort suit against the company; because Iowa OSHA did not redact these records, the names of the workers appear in them, but since it is not clear whether these workers agreed to have their names divulged, they have been suppressed here after telephonic consultation with Kathleen Uehling, Asst. Commissioner of Labor, Des Moines (Dec. 30, 2002).

[18]IOWA OSHA, Excel Corp., Inspection Report, "Coverage Information/Additional Comments" at 4.

[19]IOWA OSHA, Excel Corp., Inspection Report, "Coverage Information/Additional Comments" at 4-5. Glattfelder refused to be interviewed by the author on the grounds that he had in the meantime become part of management. When told that Ron Brown, the current union president, who had suggested the interview, had stated that Glattfelder's platform had included bathroom breaks, Glattfelder denied the truth of that statement and claimed that there had not been any problems with bathroom breaks. When he nevertheless insisted that the author speak to Brown, the author pointed out the self-contradiction in suggesting that he interview someone who allegedly had not told the truth; seeing this impossible situation, Glattfelder instead recommended another Excel worker and former union official. Telephone interview with Denny Glattfelder, Ottumwa, IA (Nov. 3, 2002). That person, however, also refused to be interviewed. Telephone inter-

Three weeks later, on September 8, bathroom history repeated itself at Excel as tragedy and farce. The previous day B, having hurt her hand, was assigned light duty marking export hogs with a marker, which, however, was a two-handed job; on the morning of September 8, B saw the company nurse, who gave her permission to return to her old job wiping rails in the cooler, and told her to tell Miller to call the nurse if he gave her any trouble. That morning Miller told B she would have to keep marking hogs until he could find a replacement, and, even after she had complained that the work was hurting her hands, added: "I don't care, switch hands." Later in the morning when she asked Miller for a let out, he replied: "'Well I will put you on the list.'" When she told him that "she really needed to go," he responded: "'Well in 35 minutes I will have someone here to take your place.'" When she told a co-worker she really needed to go, he told her to just go, but "Ms. Miller's [sic; must be B's] comment to this [was], 'You just don't do that with Gene.'" Her relief person did not show up for 33 minutes, but when B stood up to go, she "could no longer hold it. She defecated in her pants." The same day Miller did not allow a male worker to go to the bathroom for an hour and a half; that worker "also felt that he would be written up if he stopped the line to go to the bathroom." That same day B contacted the union, which contacted Elders, who put Miller on a fully-paid suspension until September 13, pending an investigation.[20]

On September 14, Glattfelder called in the complaint to Iowa OSHA[21] and the OSHA opening conference took place a week later. To the OSHA inspector Miller later acknowledged that he had known that B had a doctor's slip allowing her to go to the bathroom, and that since they were friends, he also knew that she was on dietary pills and "might need to go to the bathroom unexpectedly and suddenly." Miller also explained to the inspector that "through a union grievance that he had once done back when he was the union steward he knew that he had 35 minutes to get" B to the bathroom. Yet the company later told the inspector that this 35 minutes did not pertain to bathrooms, and Miller himself stated to her that he found out later from the employer that "he was to get those people who needed to use the bathroom out right away." This confusing welter of conflicting statements left the inspector "wondering what Mr. Miller thought the August 19, 1999 meeting was about" and whether he had lied to her about the 35 minutes.

---

view with Michael Larkin (Nov. 3, 2002).

[20]IOWA OSHA, Excel Corp., Inspection Report, "Coverage Information/Additional Comments" at 5-6.

[21]According to a very small article months later in the *Des Moines Register,* Glattfelder said that a female employee had lost control of her bowels after being prohibited from going to the bathroom until the next scheduled break. "Company Appeals Restroom Fine, *Des Moines Register*, Dec. 30, 1999, at 5M, col. 1.

After all, one of the workers she interviewed had reported that "Mr. Miller's 'favorite comment' was 'that will be 35 minutes.'"[22]

The OSHA inspector interviewed 12 workers, three of whom stated that Excel had not told them that they were allowed to go to the bathroom if no relief came and they could no longer hold it: "The overall tone of these interviews indicate[d] that there were definite problems with going to the bathroom. 4 of the interviewed employees indicate that they have had an accident in their pants. 2 of these employees were found just by random interviewing. The other 2 employees were part of the complaint filed by...Glattfelder. Several employees were required to wait 1.5 hours to use the toilet." During her last visit to the plant on November 12, 1999, Glattfelder told the inspector that the sanitation issue no longer existed, prompting her to issue a citation but to consider it abated. Nevertheless, the citation was to be "willful." She had contacted Helen Rogers at Federal OSHA General Industry Compliance in Washington, who told her that Federal OSHA had been citing this violation as "serious."[23] The inspector determined that the citation had been willful because Excel committed the violation intentionally and knowingly since it was aware of the law and the practice in violation thereof and did not abate the hazard; in addition: "There has been a lot of media publication on this issue concerning rights of employees to use the bathroom." The inspector's passion and compassion were poignantly on display when she insisted:

> 33 minutes...is too long. I have also documented several cases where people were required to wait 1.5 hours to use the bathroom. Having to wait this long when you need to use the bathroom must have been very painful for some of these people. Much less the loss of a primary human and animal right to use the bathroom as needed. This is harmful both physically and mentally. The loss of dignity for these people whether or not they los[t] control or had 'accidents' is very obvious to this inspector.[24]

The violations for which Iowa OSHA penalized Excel the sum of $36,000, which the company initially announced it would contest[25] but subsequently paid, included:

---

[22]IOWA OSHA, Excel Corp., Inspection Report, "Coverage Information/Additional Comments" at 5-6.

[23]IOWA OSHA, Excel Corp., Inspection Report, "Coverage Information/Additional Comments" at 6-7.

[24]IOWA OSHA, Excel Corp., Inspection Report, "Coverage Information/Additional Comments" at 7.

[25]Jeff Strait, "Excel Fined for OSHA Violation," *Ottumwa Courier*, Dec. 29, 1999, at 1, col. 1; see also "Company Appeals Restroom Fine."

Throughout the area of the Excel-Ottumwa plant where Gene Miller serves as supervisor, toilet facilities were not provided to employees.

An employee was required to wait 1 and ½ hours to use the toilet facilities. The employee asked his/her supervisor for a bathroom let out and was not allowed to use the bathroom until their scheduled break time. The employee thought they would be written up if they left without a relief person present.

An employee was required to wait 30 to 45 minutes to use the toilet facilities. The employee found their own relief person. The employee was delayed to the point that one or more of the following 3 events occurred—urinated, defecated, or a heavy menstruation in their clothing.

An employee was required to wait 1 and ½ hours to use the toilet facilities. The employee felt that they would be fired if they just left the line. The employee was delayed to the point where one or more of the following 3 events occurred—urinated, defecated, or a heavy menstruation in their clothing.

An employee was required to wait 1 hour to use the toilet facilities. The employee reports that the supervisor gave them a "hard time" about the frequency of their visits to the toilet.

An employee with a doctor's excuse slip to use the toilet facilities because of a medical problem was required to wait 33 minutes to use the toilet facilities. The employee felt if they left the line they would be written up by the supervisor. The employee was delayed to the point that one or more of the following 3 events occurred—urinated, defecated, or a heavy menstruation in their clothing.

An employee was required to wait 1 and ½ hours to use the toilet facilities. The employee did not believe they could leave the line without a relief person taking over. The employee was delayed to the point where one or more of the following 3 events occurred—urinated, defecated, or a heavy menstruation in their clothing.[26]

Iowa Labor Commissioner Byron Orton, who was personally and decisively involved in the resolution of the Excel case, attending the closing conference at the plant on November 23, explained that he had initially offered to resolve the case informally without an on-site inspection, but that the company had been very recalcitrant and refused. Although Orton characterized the supervisor who refused to let workers go to the bathroom as a "renegade" with a "Hitler complex," the commissioner insisted that management had been aware of what the supervisor had been doing. Iowa OSHA then classified Excel's violation as willful because the employer had continued to deny workers access to the toilet

---

[26]Citation and Notification of Penalty to Excel Corp., Inspection No. 300375060, (Nov. 29, 1999).

even after the agency had sent it a copy of Orton's January 21, 1998 interpretation. The company sought a settlement that would have deleted the "willful" classification because, according to Orton, like many employers, it regarded that designation as injurious to its corporate reputation. In an innovative enforcement approach, Orton agreed to reduce "willful" to "serious"—which conversion would not have helped the company financially since in both cases a repeat violation would warrant OSHA's levying the highest statutory penalty of $70,000—if Excel paid the full $36,000, and in addition made a contribution of $25,000 to the Ottumwa fire department and a seven-county hazardous materials entity.[27]

Iowa OSHA's enforcement actions in these slaughterhouse cases clearly demonstrate that this state agency has not adopted the aforementioned narrow position of Federal OSHA's director of compliance that a 30-minute wait did not rise to the level of a violation.[28]

On September 10, 2001, Linda Long, who worked at the Excel hog slaughter plant in Ottumwa, filed suit in Iowa District Court for Wapello County against Excel, Cargill Incorporated, which owns a controlling interest in and sets employee policies for Excel, Earl Gene Miller (her immediate supervisor), Jim Greinert (head of Long's department and Miller's immediate supervisor), and Les Elders (Excel's human resources manager). Long's legal claim is based on the accusation that "Miller refused to allow Plaintiff to leave her post to go to the bathroom within a reasonable period of time after Plaintiff requested to do so. As a result of Defendant Miller's conduct, Plaintiff defecated in her clothing." Nor, according to her court filing, had Long been the first victim. Despite the fact that Iowa OSHA had notified Elders about employees' toilet access complaints and advised him of state and federal regulations requiring employee access: "On multiple prior occasions, Defendant Miller and other supervisory personnel had refused to allow employees to go to the bathroom within a reasonable period of time, causing some employees to urinate or defecate in their clothing."[29] After Long had defecated in hers, Excel gave a new dimension to adding insult to injury by virtue of having "informed others, allowed word to spread, or otherwise caused knowledge of what had occurred to pass to other employees of Excel."[30]

Excel, while claiming that it did "'not deny[ ] its employees the right to go to the bathroom if they need to go to the bathroom,'" asserted in its defense that

---

[27]Telephone interview with Byron Orton, Iowa Labor Commissioner (Oct. 17, 2002).

[28]See above ch. 8.

[29]Long v. Excel Corp., Petition at Law, No. LALA 102518, ¶¶8-10, 13 (Iowa D.C. Wapello Cty, Sept. 10, 2001).

[30]Long v. Excel Corp., Petition, ¶27.

"companies like Excel have to guard against those employees who would abuse the privilege." Although the company insisted that it "does not take lightly its responsibility to allow workers to use toilet facilities," it nevertheless appeared unable to make up its corporate mind as to whether workers have a right or merely a privilege to void when they have to at work.[31]

The core of Long's claim, which is not only simple, straightforward, and commonsensical, but also innovative and even perhaps unprecedented in the annals of jurisprudence in the United States,[32] is reminiscent of the human rights approach developed by the Conseil des Prud'hommes in the Bigard case.[33] It reads:

> Human beings have a fundamental right to defecate or urinate in a reasonably private and dignified way. This right is grounded in principles of liberty and privacy.
>
> The actions of Defendants in refusing...Plaintiff reasonable access to a toilet served to deprive Plaintiff of this fundamental right.
>
> Plaintiff suffered a loss of personal dignity and experienced embarrassment and humility [sic; should be "humiliation"] as a result of Defendants' conduct.[34]

The compelling nature of Long's claim of a fundamental human right to void in a private and dignified way is powerfully confirmed by the astonishing fact that Excel admitted the truth of this allegation in its Answer.[35] The resolution of this creative litigation, which may be tried in 2003,[36] could potentially expand the legal resources—beyond passive reliance on OSHA enforcement—available to workers to defend their individual and collective autonomy in the workplace.

---

[31]Strait, "Excel Fined for OSHA Violation" (quoting Mark Klein, Excel communications manager)

[32]An appeals court did permit a General Motors worker who had defecated in his pants because his foreman had made him wait 35-50 minutes for relief to sue the latter for intentional infliction of mental stress for non-employment-related "gratuitous and intentional disclosure of plaintiff's predicament" (i.e., for having said that "plaintiff had 'crapped his pants'") to 40 co-workers after he finally let the worker go home to change his clothes. Kissinger v. Mannor, 285 N.W.2d 214, 216, 217 (Mich. App. 1979). However, a similar claim was dismissed involving a telephone operator whose AIDS medication caused diarrhea and who defecated on himself before he was permitted to leave his station and then had to sit in his soiled pants for three hours. Swatzell v. Southwestern Bell Tel. Co., 2001 U.S. Dist. Lexis 17733 (N.D. Tx. Oct. 31, 2001).

[33]See above ch. 10.

[34]Long v. Excel Corp. ¶¶20-22.

[35]Long v. Excel, Answer and Affirmative Defenses to Petition at Law and Jury Demand, ¶20 (Sept. 28, 2001).

[36]Telephone interview with Steven Lawyer, Linda Long's attorney, Des Moines (Dec. 27, 2002).

13

Bourbon and Urine Don't Mix: Jim Beam in Kentucky

"Basically, we're being asked to train our bladders and other organs to meet their needs, not ours...."[1]

"The company keeps computer spreadsheets documenting each time we go to the bathroom. It's like the potty police. It seems like they spend more time monitoring our bathroom time than we actually spend going to the bathroom. They should focus more on the business and less on the toilet"....[2]

Right now the members are just in shock that we finally won this. Most are grateful enough to curtail unnecessary breaks, but there are still that few.[3]

Trying to Break the Break Tradition

In the four and half years following the issuance of OSHA's Memorandum the only case of deprivation of voiding rights to captivate the national (and non-U.S.) news media featured the Jim Beam Brands Company bourbon plant in Clermont, Kentucky.[4] Although examination of this media attention is important because that spotlight itself helped shape the outcome of the dispute, of transcending analytic significance is the two-day OSHA appeals hearing, which offered an unprecedented public debate among labor, capital, and the state over the

[1]Ameet Sachdev, "Firm Defends Bathroom Break Rules, Jim Beam Workers Lodge Complaint," *Chicago Tribune*, Aug. 29, 2002, at 1 (Lexis) (quoting Jo Anne Kelley, president, UFCW Local 111-D).

[2]"Kentucky Bourbon King Faces Government Hearing on Denial of Bathroom Access," PR Newswire, Aug. 28, 2002 (Lexis) (UFCW press release) (quoting Anne Culver).

[3]Email from Jo Anne Kelley to Marc Linder (Oct. 13, 2002).

[4]The Canadian press, for example, the *Guelph Mercury*, carried the AP report on Aug. 28, 2002.

question of workers' right to extricate themselves from employers' control over their time and activities at the workplace and the power of governmental labor standards-setting to override capital's power to prescribe when, how often, and for how long its employees can eliminate waste from their bodies. This political-economic confrontation was all the more valuable for the supporting role played by representatives of medical (urological and urogynecological) science, whose presentations and critiques of each other underscored how malleable scientists' opinions and conclusions can be in the medium of class struggle.

Jim Beam bourbon, the world's best-selling bourbon, generated $350 million in sales in 2001 for its parent company, Fortune Brands, a firm with annual sales of $5.5 billion, which in 1994 had sold off its subsidiary American Tobacco Company. The operating company contribution of its "spirits and wine" segment as a share of net sales of alcoholic products (including whiskey, vodka, and gin) amounted to 22.3 percent in 2001 and 25.2 percent in 2000.[5]

Jo Anne Kelley, who had been president of UFCW Local 111-D—which in 1995 merged with the Distillery Workers Union, which had had a collective bargaining agreement with the company at that plant since 1947—for six years and worked at the plant for 34 years, explained that under the established practice, which went back 50 years, "everybody got a little break between breaks. It was a benefit."[6] These mini-breaks, according to the UFCW's assistant general counsel, Peter Ford, were "a longstanding past practice that the company called 'tag rotation,' under which Bottling Dept. employees essentially relieved each other while they took unscheduled breaks to smoke or hang out in the restroom, cafeteria, etc."[7]

According to data collected by Jim Beam by monitoring breaks in August 2000 and February 2001, the total average duration of these four mini-breaks taken by bottling-line workers amounted to 25.3 minutes per day.[8] Added to the two scheduled breaks of 15 minutes each, these more informal breaks—whose frequency and average duration were confirmed by Kelley[9]—gave workers about 55 minutes of rest breaks per day (beyond the 30-minute lunch period), or almost

---

[5]Fortune Brands, *Annual Report 2001*, at 23, 59, on http://www.fortunebrands.com.

[6]Telephone interview with Jo Anne Kelley, Bardstown, KY (Oct. 4, 2002).

[7]Email from Peter Ford, Asst. Gen. Counsel, UFCW, Washington, D.C., to Marc Linder (Sept. 30, 2002).

[8]Calculated as the average of 9 monitorings that Jim Beam performed on bottling lines D, G, H, and K, and that were submitted as deposition exhibits 3-12 at Dr. James Stivers' deposition on May 7, 2002, plus one of line A in Aug. 2000, a copy of which was furnished by the UFCW. The average daily breaks ranged between 17.56 and 32.4 minutes per worker.

[9]Email from Jo Anne Kelley to Marc Linder (Oct. 9, 2002).

seven minutes per hour, which is marginally more than the six-minutes per hour that the UAW, arguably the strongest mass production union in the United States, negotiated with Ford (and GM) in its 1999 collective bargaining agreement.[10] And spread out uniformly throughout the workday, such rest pauses were probably also more beneficial to the Jim Beam workers than the two lumped together in the automobile industry. In any event, Jim Beam management was determined to put an end to this considerable achievement, which it viewed as an "excessive"[11] withdrawal of labor from the task of profit-making .

That firms are driven to eliminate such nonproduction time is made plausible by customer-based market research conducted by Scott Paper Company. In touting to employers its paper towels over hot air dryers, the company stated that workers spent 12 seconds more with the latter: "Twelve seconds doesn't sound like a lot, but it can add up. The typical employee visits the bathroom 3.3 times daily while at work. If you do the math, it works out to nearly three hours per year per employee of wasted time standing in front of a hot air dryer."[12] If Scott is right in believing that eliminating such relatively small amounts of time that workers spend in the bathroom would appeal to employers, they would be considerably more interested in eliminating those bathroom visits altogether that give rise to hand-drying.

Although disputed by the workers, Beam claimed that it established a new break policy in the fall of 2001 "because some workers abused the privilege of unlimited bathroom breaks...." As Richard Griffith, the company's lawyer, described it to a television audience at the height of the conflict in August 2002, the workers had "'engaged in what is called a tag rotation. In order to fit in four unscheduled breaks a day, they would begin going out in regular patterns...so they could all get their breaks in between the start of the workday and the first scheduled break, between the first scheduled break and lunch, between lunch and the second scheduled break, between the second scheduled break and the end of the work day. [L]iterally one would come back, point to somebody and say "it's now your turn to go," and that person would go.'"[13] The use of "literally" to prompt

---

[10]In its 1993 agreement with General Motors, the UAW extracted 46 minutes of relief time per eight-hour day, or 5.75 minutes per hour. *Agreement Between UAW and the General Motors Corporation* 433 (Oct. 24, 1993). In 1999 the UAW-Ford contract provided workers with six minutes per hour. Letter from Rocky Comito, President, UAW Local 862, to Marc Linder (Oct. 17, 2002).

[11]Letter from John C. Allen, Jr. to Dr. James Stivers, Mar. 16, 2001, deposition exhibit 2 at Stivers' deposition of May 7, 2002.

[12]R. Dixon Thayer, "Everything You Ever Wanted to Know About Bathroom Habits (but were afraid to ask)" n.d. (ca. mid-1990s).

[13]"Jim Beam Bathroom Controversy Continues," Aug. 28, 2002, on http://wave3.com/Global/story.asp?=913912.

the audience to share in the employer's alleged disbelief and outrage is amusing since the term and practice of "tag relief" are widespread in industrial relations: it merely designates serial relief as opposed to "mass relief," where the production line is shut down and all workers take a break.[14]

According to Kelley, there had never been any pretense that workers had been using their mini-breaks exclusively to void; rather, they were rest breaks that also included voiding. Although Jim Beam made elaborate calculations as to how much money it was costing the company to pay "highly paid highly skilled technicians" to relieve bottling-line workers while they were in the bathroom—in fact, the technicians were paid $18.85 per hour, only marginally more than the $17.63 that the machine operators they relieved were paid, while the other bottling-line workers, the general helpers, who were paid $16.60, either did not need to be replaced or were replaced by janitors—the break system, in Kelley's estimation, did not result in loss of production and most of the time the technicians were not actively performing work anyway, since much of their workday was taken up with observing and only in the case of a machinery breakdown did they have to intervene. Kelley recounted that in recent years the company had become more and more obsessed with controlling workers and their time and capturing their movement and actions through various kinds of surveillance such as computers and cameras (but "even this employer wouldn't stoop to putting cameras in the bathrooms").[15]

Jim Beam management contended in a three-page "Summary—Restroom Break Issue" that it gave to the OSHA inspector on October 15, 2001, that it had met with the union executive board once in 2000 and numerous times in earlier years "to discuss addressing the practice of break rotation. The Union would agreed [sic] to 'try' and help reduce the obvious abuses, but clearly saw no problem with the practice. Some improvement would result after the meetings but the practice and abuses would soon return." The company also informed the inspector that during the week ending August 24, 2000 it had conducted a survey of restroom break practices in the distillery industry: "The survey confirmed that the practice of break rotation (outside scheduled breaks) was present in other industry operations to a highly varying degree. Both extremes and all points in the middle appear to exist."[16]

---

[14]Joe Ward and Bill Wolfe, "Local Ford Workers Feel Relief," (Louisville) *Courier-Journal*, Jan. 12, 2002, on http://www.courier-journal.com/business/news2002/01/12/bu011202s 136877.htm; *Agreement Between UAW and the General Motors Corporation* 433 (Oct. 24, 1993); "Auto Plants Debate How to Take Coffee Break," Tampa Tribune, Sept. 10, 1996, at 5 (Lexis);

[15]Telephone interview with Kelley.

[16]Jim Beam Brands Co., "Summary—Restroom Break Issue." This document was

Jo Anne Kelley, who noted that with one exception all distilleries in Kentucky are unionized, conjectured: " I suppose the reason the practice is so common is because in the past there were so many women who worked in the bottling facilities before all the automation. There were labor laws in the past...which were more liberal with regards to women and breaks before we were all "made equal" on the job. I suppose it had to do with...the way women were viewed as being weak and naturally needing extra breaks. ... Beam began automating and eliminating women's jobs (work traditionally performed by women) about 20 years ago.... When I went to work at Beam 34 years ago, there were 400 men and 400, women. Now there are 60 women and 185 men producing twice as much."[17]

To be sure, Jim Beam did not explain why the producer of the world's best-selling bourbon was so concerned about these breaks when they had been going on for half a century and its competitors also operated with them, but on August 28, 2000 it began "conduct[ing] data surveys to quantify the actual restroom break rotation in terms of how frequent, when, time taken, etc." Use of the expression "conducted data surveys" to cover both undertakings was misleading since what the company did in the second case was to create the data by monitoring its employees' breaks. A week after the monitoring had been completed in February 2001, senior Beam managers met "to consider proper strategy for addressing excessive restroom breaks...." Three months later senior managers met on the same matter again, this time articulating two considerations: "Try to again enlist the Union's support in 'managing out' of an abusive practice <u>or</u> approach the problem through negotiations. It was determined that we may have been asking the Union to do something it could not do. Management decided to address the problem in Contract Negotiation," which then began in early July and led to adoption of a new break arrangement.[18]

In March 2001, in support of its plans to restrict toilet access, Jack Allen, Beam's director of human resources, hired a local Louisville urologist in private practice, Dr. James Stivers, at $250 per hour,[19] to learn "what was reasonable in

---

given to OSHA at the opening conference on Oct. 18, 2001 and was attached to the OSHA Citation.

[17]Email from Jo Anne Kelley to Marc Linder (Oct.. 13, 2002). According to a Jim Beam Brands executive, tag rotation was the "vestigial" remains of the pre-automated production system when there were fewer skilled workers on the bottling line and workers going to the bathroom could be replaced by a similarly nonskilled co-worker standing next to him or her rather than by skilled workers performing other work. Telephone interview with an executive of Jim Beam Brands Co., Clermont, KY (Nov. 8, 2002).

[18]Jim Beam Brands Co., "Summary—Restroom Break Issue."

[19]Deposition for UFCW, Local 111-D of James R. Stivers, M.D., Secretary of Labor v. Jim Beam Brands Co., Occupational Safety and Health Review Commission, KOSHRC No: 3681-01, Adm. Action No: 02-KOSH-0015, at 12 (May 7, 2002).

terms of the needs to urinate in a certain time frame from a medical standpoint,"[20] and to comment on data that the company had collected on breaks and smoking in August 2000 and February 2001. Stivers, whom the company later lauded to the OSHA inspector as "a renowned urologist"[21]—an epithet that Professor Ingrid Nygaard, a urogynecologist, professor at the University of Iowa College of Medicine, and co-author of *Void Where Prohibited,* characterized as "utter nonsense"[22]—included only two publications, neither concerned with voiding, on his curriculum vitae during the more than three decades following his graduation from medical school: one was one page, written with three others in 1973, and the other two pages, written with three others in 1994.[23] Stivers came to Allen's attention as a result of the intervention of the "company doctor for Jim Beam Corporation," a "colleague" of Stivers', whom "we call a [sic] occupational physician."[24] In a letter to Stivers of March 16, 2001 on "Reasonable Restroom Breaks," Allen biased his charge to the urologist by transforming workers' right to void when they need to into a mere "privilege," which managers, bolstered by physicians, were even entitled to restrict:

As you are aware, the Jim Beam Brands Co. manufacturing operation in Clermont, Kentucky is experiencing what we believe to be an excessive pattern of employees taking restroom breaks. Management has devised a policy regarding 'break privileges' that we feel will reduce the abuse of this privilege, while assuring employees with real needs reasonable access to restroom facilities....

Prior to introducing a change of this nature, we want to be sure as possible that the policy is reasonable from all angles, especially the medical perspective. To that end we asked you to visit the plant, review the draft policy and answer the following questions:

1. Does the policy allow reasonable access to restroom facilities from a medical point of view?

2. What, if any, medical conditions should be considered for additional accommodation above those provided by the policy?

3. What, if any, temporary medical conditions should be considered for temporary accommodation above those provided by the policy?

---

[20]Deposition of Stivers at 11.

[21]"Summary—Restroom Break Issue"

[22]Email from Ingrid Nygaard to Marc Linder (Oct. 9, 2002).

[23]Curriculum vitae James R. Stivers, M.D., submitted at his deposition on May 7, 2002.

[24]Commonwealth of Kentucky Occupational Safety and Health Review Commission, Administrative Action No. 02-KOSH-0015, Secretary of Labor Commonwealth of Kentucky v. Jim Beam Brands Co., KOSHRC #3681-01, "Transcript of Administrative Hearing" at 114 (Aug. 29, 2002) (Stivers' testimony) ("Transcript of Hearing").

In addition, we provided you with data demonstrating the current patterns of restroom break activity. From your knowledge of human bodily functions, what conclusions, if any, can you draw from this data?[25]

A mere 11 days later, Stivers wrote two one-page letters to Allen. In one, Stivers seemed almost to be accusing the company of offering overly generous breaks: "It appears despite the opportunity to use the facility before and after work that there are two scheduled breaks, as well as mid-day lunch break." If these interruptions were not bad enough, Stivers observed that "almost every employee takes an additional break prior [sic] and following the morning and mid afternoon scheduled breaks. ... The average time of these breaks varies from line to line and the average duration...ranged from slightly more than five to as high as nine minutes." Stivers then asserted, without disclosing his source or that these averages included walking time to and from the toilet, that: "Normal voiding would be less than three minutes." Again, without revealing his sources, he asserted: "It is generally assumed that a normal person at a moderate temperature with average fluid intake would void between three and six times per 24 hours." Consequently: "The data does not seem to indicate a normal physiological response and it appears that the break pattern is not motivated by urination or defecation, but is most likely driven by smoking habits and other external factors." Since (the company had told him that) 60 percent of the employees smoked, Stivers suggested that a smoking cessation program might be of use to them. Although he was ignorant of OSHA's ruling that employers are obligated to make toilets available to workers so that they can use them when they need to, he offered his suspicion that "OSHA regulations are slowly marching in the direction of [sic] smoke free workplace."[26] Why the company never proposed solving the problem of smoking in the bathroom by prohibiting it,[27] Stivers failed to explain.[28] In turn, to OSHA during its inspection in October Jim Beam characterized what had merely been Stivers' speculation about what the workers were really doing in the bathroom as "his professional opinion...."[29]

In his other one-page letter of March 27, 2001, Stivers, charging $250/hour for a work product replete with incoherent syntax, misspellings, and typos, bolstered his opinion by noting that he had "reviewed OSHA regulations reference

---

[25]Letter from John C. Allen, Jr. to Dr. James Stivers, Mar. 16, 2001.

[26]Letter from James R. Stivers to Jack Allen, Jr. (Mar. 27, 2001), deposition exhibit at Stivers' deposition of May 7, 2002. In fact, Beam's monitoring included "the amount of time employees spend away from lines." Untitled data set furnished by UFCW (Mar. 18, 2002).

[27]"Transcript of Hearing" at 83 (Aug. 29, 2002) (testimony of Jo Anne Kelley).

[28]Deposition of Stivers at 49-50.

[29]Jim Beam Brand Co., "Summary—Restroom Break Issue"

[sic] agricultural workers, were [sic] a three hour availability for remote agricultural workers is the standard." Making explicit what he had implied in the other letter, Stivers asserted that the break "policy...seem [sic] to be more than reasonable and adequate." The medical conditions that "should be consider [sic] for additional accommodations on a permanent basis" included spina bifida, paraplegia, "[f]emales in the last half of a pregnancy," and "[o]lder men with prostate conditions." Other medical conditions that "can effect [sic] the need to urinate, especially more frequently," included medications, radiation treatment, and chronic infections. Finally, Stivers mentioned as temporary medical conditions that "should be considered for 'temporary accomodation [sic]" transient urinary or gastrointestinal infections, but not menstruation.[30]

Three times during Kelley's six-year presidency the company requested that she ask the workers not to use the mini-breaks for activities other than voiding; she did ask the workers to comply, and, by and large, she believed that they had complied.[31] Jim Beam proposed during contract negotiations in 2001 that the past practice of unrestricted restroom breaks on a regular basis be eliminated: "After much discussion the union agreed that we would no longer have the regular rotating cycles of breaks with the assurance that no one would be denied access to toilet facilities if needed, and we took that proposal to the membership and the membership agreed to it."[32] Kelley reported that 99 percent of the workers accepted the change because they would still be able to use the toilet.[33]

The collective bargaining agreement between UFCW Local 111-D and Jim Beam Brands Company that went into effect on August 16, 2001 provided:

> After a two (2) hour working period in the morning and in the afternoon, there shall be a fifteen (15) minute rest period. This will not affect any existing practices with regard to break periods. These periods shall not be deducted from the regular lunch or deducted from the day's wages. If overtime is for more than an hour, a fifteen minute break will be granted at the end of the hour and at two hour intervals thereafter.[34]

But the "past practice" of mini-breaks was eliminated in negotiations by a side letter"[35] of August 16, 2001, which provided: "Effective September 1, 2001, the

---

[30]Letter (# 2) from James R. Stivers to Jack Allen Jr. (Mar. 27, 2002), submitted as deposition exhibit at Stivers' deposition of May 7, 2002.

[31]Telephone interview with Kelley.

[32]"Transcript of Hearing" at 84 (Aug. 29, 2002).

[33]Telephone interview with Kelley.

[34]Agreement By and Between United Food and Commercial Workers International Union Local No. 111-D and Jim Beam Brands Co. Clermont, Kentucky, Art VIII, Sect. D (Aug. 16, 2001).

[35]Email from Ford to Linder (Sept. 30, 2002).

practice of taking unscheduled restroom breaks in a regular rotating cycle in the Clermont Bottling Department will be discontinued. Unscheduled restroom breaks will be allowed, but on an as needed basis only, with notice. Employees requiring medical consideration will be accomodated [sic]. Abuse of restroom priviledges [sic] will be managed."[36]

The union discovered what the company's real intentions were "when supervisors started explaining the new policy to employees around the 1st of September." In its introduction of the policy, according to Kelley, Jim Beam stated that the month of September would be "a breaking in period to allow you to adjust (your bladders, I guess) to the new policy."[37]

At this time Jim Beam created an internal document entitled, "Administrative Controls on Restroom Breaks," which revealed the bureaucratic solemnity that the company bestowed on its new restrictive policy, which an outsider might have been excused for imagining applied to a prison or kindergarten:

> I recommend a Supervisors [sic] daily log be maintained for a period of two months. This will assure an accurate and timely documentation of the transition from "Tag Rotation" to the negotiated "As needed" method of securing restroom relief.
>
> September will be a transition month. During the transition month, employees who take two or more unscheduled restroom breaks in a day will be counseled about the new expectations, but will not receive discipline. Starting in October, employees who have received two or more counseling's [sic] in September, will be issued a Verbal Warning on the next abuse of privilege, ... those with one counsel in September will be issued a second counsel.... And so on.
>
> At the end of each shift, each Supervisor will turn in their daily log to the Bottling Dept Secretary. Early the next workday, the Clerk will review the slips, note employees with two or more unscheduled breaks and report the information to the Human Resources Assistant....
>
> Human Resources will maintain a RR Breaks Disciplinary Spreadsheet. The HR Assistant will check the RR Breaks Disciplinary Spreadsheet to determine if counseling or discipline is required. She will fill out a counseling slip or disciplinary notice for employees as needed. RR Break notices will be walked down to the Bottling Department to be distributed and issued.
>
> The slips and notices will be issued to the supervisor assigned to supervise the employee. The Supervisor will verbally counsel the employee, sign the slip and send it to the Human Resources Department. Disciplinary Notice will be processed in the standard manner.[38]

---

[36]Side Letter of Agreement (Aug. 16, 2001) (printed on the last page of the collective bargaining agreement between Local 111-D and Jim Beam).

[37]Email from Jo Anne Kelley to Marc Linder (Oct. 5, 2002).

[38]"Administrative Controls on Restroom Breaks" is included in the OSHA Citation, where it bears the handwritten notation: "10/30/01 confirmed as policy by Jeff Conder."

The semi-public version of this policy was presented on August 27, 2001 (without disclosing the monitoring routine) as department managers' and supervisors' "bench talk" to employees who were likely to be assigned to the bottling department [39] Like other Jim Beam Brands Company pronouncements on the subject, it too transmogrified workers' voiding rights into a mere privilege revocable on management's terms. Cloaked in the neutral-sounding title, "Changes in Restroom Relief Procedures," and pretending that management too ("we all") would be subjected to the new regime, it announced a veritable revolution in the structure of time-control that the union workers had secured over a half-century:

> **Beginning the first week of September,** the practice of taking unscheduled restroom breaks in a regular rotation cycle will be discontinued.
>
> Unscheduled restroom breaks will be allowed, but on an **"as needed basis" only**, with notice. Most of the times [sic] a supervisor will be available to notify in advance and help arrange cover, if needed. If a supervisor is not available, go ahead, [sic] (get cover if needed), then notify the supervisor when you return. **You do need to notify a supervisor, one way or the other**. Failure to notify will be treated just as serious [sic] as abusing the privilege.
>
> Two or more unscheduled breaks in a day will be considered abusing the privilege. Anyone having a medical condition causing increased restroom needs may talk with Human Resources about arranging an accommodation.
>
> Through September, while we all get used to the new procedure, Supervisors will only counsel those who take two or more unscheduled breaks a day, but no discipline will be issued.
>
> Starting in October, employees who abuse the privilege will be subject to the standard discipline.
>
> **(Verbal, Written, Final Warning, Termination)**[40]

In preparation for putting the restrictive policy into effect, Jim Beam management issued to its supervisors a Bottling Department Restroom Break

---

Conder is the plant manager, who attended the OSHA closing conference on that date. According to Jo Anne Kelley, this document, which was the work Jack Allen, the human resources director, was not shared with the union until she demanded to see all documents pertaining to the bathroom break policy. Email from Jo Anne Kelley to Marc Linder (Oct. 13, 2002).

[39]Jim Beam Brands Co., "Summary—Restroom Break Issue."

[40]Jim Beam Brands Co., "Supervisor's Bench Talk: Bottling Operations: Change in Restroom Relief Procedures" (Aug. 27, 2001), submitted as documents at Stivers' deposition. Attached to the one-page document was a sheet with signature lines, which "each employee was required to sign indicating they were aware of the new policy." Email from Jo Anne Kelley to Marc Linder (Oct. 13, 2002).

Implementation Procedure, which included a "talk sheet" for first and second counselings and "standard disciplinary" forms for warnings. All of these documents, which uniformly speak of "privileges" rather than rights, poignantly convey the disciplinary system's military-type command structure, infantilization of the workers, and characterization of normal human voiding as medical abnormalities:

1st CONTACT -2 or more UNSCHEDULED BREAKS

Employee , yesterday, you were excused twice for unscheduled restroom breaks. We can and do consider one unscheduled break as reasonable, but I need to tell you that a pattern of two or more such breaks is seen as abusing the privilege. If you believe there is a medical issue that would call for an accommodation, you can talk with HR,...otherwise, you need to comply with the standard of no more than one unscheduled break in a day.

2nd CONTACT -2 or more UNSCHEDULED BREAKS

Employee , this is the second time I have had to speak with you regarding restroom privileges. My job is to clearly communicate what is expected, and yours is to comply, unless there is a medical reason preventing you from compliance. ... If not, I must warn you, continuing a pattern of two or more unscheduled breaks out side [sic] of your scheduled breaks will be cause for disciplinary action in the form of a documented Verbal Warning.[41]

The disciplinary forms seemed almost like self-caricatures modeled after organized religion's efforts at extirpating 'self-pollution':

(Name) was counseled regarding the abuse of restroom privileges on at least two occasions. On (date) he/she again abused the privilege. A recurrence of this abuse will be cause for further disciplinary action up to a Written Warning.[42]

Firing workers who made one unauthorized bathroom trip on each of six days over the course of one year was redolent of the Stalinist labor regime implemented in the Soviet Union in 1938-39, under which manual workers and white-collar employees who left 20 minutes early for, or returned 20 minutes late from, their meal break three times within a month or four times over two months were

[41]Jim Beam Brands Co., "Bottling Department Restroom Break Implementation Procedure" (undated), submitted as deposition exhibit at Stivers' deposition.

[42]Jim Beam Brands Co., "Bottling Department Restroom Break Implementation Procedure."

to be fired for truancy.[43]

On October 1, 2001, Beam put in place a new policy, which applied to about 100 bottling-line workers at the plant, which produces more than 5 million cases of bourbon annually. The new policy limited breaks of any type to scheduled halting of the line for 15 minutes at 10 a.m. and 2 p.m. and 30 minutes for lunch at noon in addition to one single unscheduled bathroom break.[44] Under the new policy, a worker was not required to ask permission to take her or his one unscheduled break, but if the supervisor was present, she or he had to notify him; if he was not there, the worker had to tell him when he returned. Failure to tell the supervisor was also subject to discipline. The same rule applied to workers with medical accommodations who were permitted to go to the bathroom as often as was necessary. At one point the union did consider having every single worker obtain a doctor's statement, but was concerned because the company had a history of not permitting employees to work if they were not 100 percent physically able to perform all available jobs, even if they would not be performing more than one job.[45] Finally, if workers had to work overtime on a bottling line and had already used up their one scheduled break, they had to work from 2 p.m. until 5 p.m. without a break.[46]

The UFCW Calls In OSHA

Kelley herself was well acquainted with the OSHA memorandum because two years earlier at Beam's distillery and warehouse in Boston, Kentucky (which

---

[43]Donald Filtzer, *Soviet Workers and Stalinist Industrialization: The Formation of Modern Soviet Production Relations, 1928-1941*, at 310-11 n. 3 (1986); Solomon Schwarz, *Labor in the Soviet Union* 102-103 (1952). From 1940 on, instead of being fired, workers were sentenced to corrective labor at their workplace for up to six months at as little as three-fourths of their pay for the crime of any unauthorized loss of work time in excess of 20 minutes, including returning from lunch break 20 minutes late even once. Schwarz, *Labor in the Soviet Union* at 106-109; John Hazard, *Law and Social Change in the U.S.S.R.* 175-76 (1953); Filtzer, *Soviet Workers and Stalinist Industrialization* at 233-53; Donald Filtzer, *Soviet Workers and Late Stalinism: Labour and the Restoration of the Stalinist System After World War II* 160-63 (2002).

[44]"Jim Beam Workers Turn Sour Over Bathroom Break Limit," The Oak Ridger Online, Aug. 28, 2002, on http://www.oakridger.com/stories/082801/bus_0828020026.html.

[45]Email from Jo Anne Kelley to Marc Linder (Oct. 6, 2002). Individual workers also testified at the hearing about their fears about their qualifications for holding or obtaining certain positions if they had sought a medical accommodation. See below.

[46]Email from Jo Anne Kelley to Marc Linder (Oct. 7, 2002).

Local 111-D has also organized) the company disciplined a worker for interrupting work to drive back from the warehouse, which lacked a toilet, to the main building to urinate. Kelley told the employer that it was violating OSHA by interfering with his right to void, and the company ultimately agreed to comply with the OSHA standard by installing a port-a-potty in the warehouse so that the worker would not have to waste so much company-paid time traveling back and forth. She was therefore well-prepared to inform Beam at Clermont in the fall of 2001 that its new policy violated OSHA's Memorandum. The company insisted that it had consulted a urologist and OSHA, an official of which, however, had merely sent Beam a copy of the Memorandum without confirming that the policy was in conformance.[47]

Jim Beam, however, had not needed that copy of the Memorandum: A company executive had received it at the time it was issued and had read and been aware of its contents. He assumed that his counterparts at other large firms had as well.[48]

Although the policy of permitting only one unscheduled toilet break per day implemented on October 1, 2001 breached the side letter agreement of August 16, 2001, which provided that "unscheduled restroom breaks will be allowed, but on an as needed basis only, with notice," the local did not grieve the breach because, as president Kelley put it: "The union chose to file a complaint with OSHA thinking that OSHA would cite the Co. and they would remove the policy. No one ever dreamed it would go this far. And frankly, I have more faith in OSHA than I do in arbitrators. Also, at the same time, I was in the process of filing two unfair labor practice charges, one was for failure to bargain in good faith as a result of the newly signed contract. This is not the only change the company implemented that was not agreed to in negotiations."[49]

As soon as the new policy went into effect on October 1, Kelley filed a complaint with the Kentucky Occupational Safety and Health Program of the Kentucky Labor Cabinet (Kentucky OSHA) on October 2, 2001. After conducting an inspection on October 15 and a closing conference with the company and union on October 30, the compliance safety and health officer (CSHO) wrote a report recommending issuance of a citation. The officer's observations attended closely to the intent of Federal OSHA's Memorandum: "The employees are limited to the number of times the restroom facilities may be used during the shift, and the number of times of voiding and/or defecation are determined by the employer, not the employee. The times chosen by the employer may also be changed arbitrarily by the employer. This does not allow for prompt access to the

---

[47]Telephone interview with Jo Anne Kelley.

[48]Telephone interview with an executive of Jim Beam Brands (Nov. 8, 2002).

[49]Email from Jo Anne Kelley to Marc Linder (Oct. 5, 2002).

restroom facilities."[50] The CSHO's internal "Instance Description" continued in a similar vein, cogently pointing out that:

> By allowing employees to access the restroom facilities only at those times scheduled by the employer for breaks and lunch, the restrooms are not available, but are indeed denied, to the employee at times other than the scheduled break rooms. It is unreasonable to expect an individual to require restroom facilities at times deemed by another party, or at times good for the production line. The length of delay in some instances could reach one hour in duration. The employee may be allowed to leave the line if necessary; however, disciplinary action may then be taken. In these instances, prompt and unrestricted access to restroom facilities is not provided.[51]

In incidents that exquisitely capture the precise way in which the employer insisted on treating the workers like children, the CSHO also related: "Employees in interviews revealed that some supervisors have, at times since the policy was enacted, refused to allow employees to use the restroom although the production line was stopped. Additional employees stated they were not allowed to use the restroom on the way to meetings, although they passed the restroom on the way to the meeting. The reason given to the employees for the denial to use the restroom was that it was not the employee's break time."[52] Although President Kelley did not hear the employees make these statement to the OSHA officer, some employees told her that "when they were moved from one line to another line they were told not to stop by the bathroom without telling the supervisor so that he could write it down. Some employees ignored this and rushed into the bathroom and voided as fast as possible and rushed back out without even washing their hands so as to avoid getting caught. This was referred to as a 'free pee.'" Kelley believed that employees in their statements to OSHA were making reference to Jim Beam's "claim that going to the restroom hurt production. Obviously, if the line is stopped anyway, production is not affected."[53]

Then on November 19, 2001, the agency issued this Citation and Notification of Penalty:

---

[50]Kentucky Labor Cabinet, Occupational Safety and Health Program, Federal Inspection # 304292311, Jim Beam Brands Co., CSHO/Optional Report # H0052/001-02, at 8 (n.d. [ca. Nov. 2002]).

[51]Kentucky Labor Cabinet, Occupational Safety and Health Program, Federal Inspection # 304292311, Jim Beam Brands Co., CSHO/Optional Report # H0052/001-02, at 11-12.

[52]Kentucky Labor Cabinet, Occupational Safety and Health Program, Federal Inspection # 304292311, Jim Beam Brands Co., CSHO/Optional Report # H0052/001-02, at 11.

[53]Email from Jo Anne Kelley to Marc Linder (Oct. 9, 2002).

29 CFR 1910.141(c)(1)(i): Toilet facilities were not provided in accordance with Table J-1 of this Section:

a) In that a newly enacted administrative policy limits the number of times an employee may use the restroom and determines the interval of time between restroom breaks at this facility. The employer may change the interval between restroom breaks arbitrarily. One unscheduled restroom break is allowed; however, any additional unscheduled restroom breaks result in disciplinary action, up to, and including, termination of employment.

b) In that some supervisors refused to allow employees to use the restroom during production shutdown because it was not the scheduled break time.[54]

Kentucky OSHA gave Beam until December 7 to abate the violation, but proposed no monetary penalty.[55] The failure to impose a penalty was based on the statutorily prescribed method of considering the gravity of the violation and then applying the penalty adjustment factors—the (small) size of the employer's business, its good faith, and its history of previous violations.[56] The most important factor, the gravity, is, in turn, determined as a combination of the severity and probability assessment.[57] The severity was deemed minimal because "the injury or illness most likely to result would probably not cause death or serious physical harm. The most serious injury or illness which is reasonably predictable as a result of an employee's exposure to this hazard would be: Increased frequency of urinary tract infections." The probability was deemed to be lesser on the grounds that the likelihood of an injury or illness was relatively low because: "An adequate number of restroom facilities are available, and in proximity to the working area; allowances are made for those with disclosed medical conditions requiring frequent use of restroom facilities."[58] To be sure, by this logic, as the

---

[54]Kentucky Labor Cabinet, Occupational Safety and Health Program, Citation and Notification of Penalty, Jim Beam Brands Co., Clermont, KY, Inspection No. 304292311, 11/19/2001.

[55]Kentucky Labor Cabinet, Occupational Safety and Health Program, Citation and Notification of Penalty, Jim Beam Brands Co., Clermont, KY, Inspection No. 304292311, 11/19/2001.

[56]29 USC sect. 666(j). The term "penalty adjustment factors," application of which may reduce the "gravity-based penalty" by as much as 95 percent, is explained in OSHA, *OSHA Field Inspection Reference Manual* CPL 2.103, sect. 8, ch. IV, C.2.i. (1994), on http://www.osha.gov/Firm_osha_data/100008.html.

[57]OSHA, *OSHA Field Inspection Reference Manual* CPL 2.103, sect. 8, ch. IV, C.2. Even an other than serious violation can be assessed a $1,000 to $7,000 fine if it is "determined to be appropriate to achieve the necessary deterrent effect." *Id*. at C. 2. g. (5).

[58]Kentucky Labor Cabinet, Occupational Safety and Health Program, Federal Inspec-

inspector herself conceded, it is difficult to imagine how a monetary penalty would ever be imposed in a citation for denial of access[59] or how this result would be compatible with the monetary penalties assessed in 12 other cases, including one in Kentucky.[60] The failure to fine the company is especially remarkable in light of the fact that of all the employers cited for restricting access, Jim Beam had implemented the most highly formalized policy, which was written, general, and issued by the top of the corporate hierarchy—expressly mentioned by the citation policy in the OSHA Memorandum as an exacerbating factor.[61]

While creating no financial deterrence, the citation was significant for consciously extending the reach of the law beyond the grotesque but obvious situations, in which an employer's denial of access causes a worker to void on herself, to the empirically much more prevalent situations in which workers are 'merely' disciplined for going to the bathroom or deterred from going when they

---

tion # 304292311, Jim Beam Brands Co., CSHO/Optional Report # H0052/001-02, at 14. The employer, which was not eligible for any size adjustment because it had 1000 employees nationwide while the regulatory ceiling is 250, was given the maximum 25-percent penalty reduction for its good faith in having all required programs, monitoring, and a safety committee, and its history of previous violations. *Id.* at 6. The *OSHA Field Inspection Reference Manual* ch. 8 sect. IVC.2.i.(5)(c) mandates a reduction of 10 percent if an employer has not been cited for any serious, willful, or repeated violation in the last three years. However, Beam had been assessed $1,300 for a serious violation at Clermont on Mar. 9, 2001. Insp. No. 303752596. Questioned about this inconsistency, the inspector stated that at the time she was writing her report for the citation she had been unaware of this earlier citation because it had not yet been posted on the OSHA website. However, she also insisted that the omission would not have mattered since no monetary penalty would have been imposed anyway for the reasons explained above. Telephone interview with Diane Marraccini, Frankfort, KY (Oct. 14, 2002). According to Jo Anne Kelley, Beam had been cited for violating 830 KAR 2:015 section 3(1): "Employees working on unguarded working surfaces which are elevated 10 feet or more above a lower level were not secured by safety belts and lanyards, lifelines where necessary, or shall be protected by safety nets. This one is quite near and dear to me considering it involved a job which I performed for 17 years and complained the whole time to no avail. The loading and unloading of tanker trucks without protection. The worst was the export tankers where the ladder was approximately 10" wide and straight up and down the back of the tanker. Tankers are loaded and unloaded in rain, sleet, snow, whatever. The Co. has since spent a great deal of money building loading and unloading stations. I am just thankful that I never fell." Email from Jo Anne Kelley to Marc Linder (Oct. 13, 2002).

[59]Telephone interview with Diane Marraccini, Frankfort, KY (Oct. 14, 2002). For an overview of the penalty calculation process, see *Occupational Safety and Health Law* 223-45 (2d ed. Randy Rabinowitz ed. 2002).

[60]See above ch. 11.

[61]See below Appendix II.

need to go. OSHA's declaring unlawful this particular form of autocratic managerial control over workers' bodies and time was doubtless considerably more worrisome to some employers than the prospect of any trivial monetary penalty.

However, why, in spite of having—given its preferred management style—good reason for regarding this kind of government intervention as anathema, Jim Beam decided to contest this non-monetary citation[62] on December 10 and gave the UFCW to understand that it was determined to litigate the interpretation of the OSHA toilet standard to the highest court, remains unclear.[63] In retrospect, the company conceded that it had been "asking for trouble" in contesting a zero-dollar citation.[64] Even though this dispute was to be conducted within the Kentucky OSHA system and any administrative or judicial decision would therefore be of no direct precedential value for any other state-plan program or Federal OSHA, the UFCW was nevertheless correct in asserting: "If Jim Beam wins the right to restrict workers' access to the restroom, the OSHA regulations protecting that right will be challenged. Millions of workers across the country could be affected."[65]

The texture of labor-management relations at the Jim Beam plant at this juncture was captured by Kelley's later testimony that she had posted on the union bulletin board a copy of the citation, of Beam's appeal, and of a letter stating that the union had requested third-party status to represent workers, but that Jack Allen, the human resources director, had unlocked it and removed the copies, telling her that it had no place on the union bulletin board and that from then on she was not to post anything without his permission.[66]

Dueling Urinary Scientists

As the appeal got underway, Dr. Stivers, at his deposition, taken by the UFCW on May 7, 2002, went so far as to call Beam's new, much more restrictive toilet access policy "more than physiologically adequate to accommodate a nor-

---

[62]An abbreviated version of the inspection report is available on the federal OSHA website: http://155.103.6.10/cgi-bin/est1xp?i=304292311 as well as in the Lexis-Nexis OSHAIR file.

[63]Telephone interviews with Ana Avendano and Peter Ford of the UFCW general counsel's office and Robyn Robbins, assistant director, UFCW Occupational Safety and Health Office during the spring and summer of 2002.

[64]Telephone interview with a Jim Beam executive (Nov. 8, 2002).

[65]"If You Gotta Go, You Gotta Go at Jim Beam" (Aug. 2002), on http://www.ufcw.com/worker/internal.cfm?subsection_id=189&internal_id=799.

[66]"Transcript of Hearing" at 88-90 (Aug. 29).

mal person"—indeed, even "physiologically and medically excessive."[67] Stivers then asserted, based on no research or review of the medical literature, that since "we know that a normal person most likely does not have to get up to go to the bathroom while they're sleeping unless they've ingested excessive amounts of fluid with or without alcohol...as the benchmark you can say that six to eight hours a person can go without having to urinate."[68] (In fact, an important study of 300 asymptomatic women—i.e., those without urinary tract symptoms—in the process of being published as he was being deposed revealed that 44 percent voided during the night.)[69]

Stivers testified that on average people produce one quart of urine daily (within a range of three-fourths to one and one-third quarts); with an average bladder capacity of 250 to 500 cubic centimeters, he reckoned that a normal person "should urinate between three and six times in a 24-hour period...."[70] Not only did Stivers not take menstruation into account, but he asserted that it was "[n]ot likely" that a menstruating woman would use the bathroom for a longer time than a non-menstruating woman because younger women tend to use tampons, while older women tend to use sanitary napkins and when they urinate, they would probably change their napkin but not a tampon.[71] Stivers also admitted that his claim that normal voiding took less than three minutes did not take into account walking to or from the bathroom, waiting in line, undressing or dressing, menstrual needs, bowel movements, or hand washing.[72]

Stivers' total inadequacy as an expert on whom Beam could have reasonably relied to help formulate an appropriate workplace voiding policy was underscored by his admission that he had reviewed OSHA's field sanitation standard, despite the fact that it was not relevant, because "there weren't any OSHA regulations that applied to people on assembly lines in terms of voiding breaks."[73] Moreover, Stivers' inability to understand the crux of the problem generated by the employer's unilaterally fixed voiding breaks was painfully on display in his admission that he did not "think that an individual is assigned a time to go to the bathroom. It's not like every day at 10:15 you take a break and every day at 2:15 you take a break...." Nor had Beam ever told him whether the company or the worker

---

[67]Deposition of Stivers at 13.

[68]Deposition of Stivers at 15.

[69]M. FitzGerald, U. Stablein, and L. Brubaker, "Urinary Habits Among Asymptomatic Women," *American Journal of Obstetrics and Gynecology* 187(5):1384-88 at 1386 (Nov. 2002).

[70]Deposition of Stivers at 15.

[71]Deposition of Stivers at 40-41.

[72]Deposition of Stivers at 44-45.

[73]Deposition of Stivers at 54.

chose the time.[74]

On June 19, Jim Beam's lawyer telephonically deposed Dr. Ingrid Nygaard, who set out six separate grounds showing that the company had failed to permit reasonable access:

> First, many healthy women who have no bladder complaints need to void more frequently than every two hours during the daytime. Second, women who have bladder complaints comprise a sizable minority of the population of women, and those women have to void more frequently even than asymptomatic women who don't have bladder complaints. Third, delaying the urge to urinate predisposes women to urinary tract infections, painful urgency and urinary incontinence. Fourth, voiding infrequency predisposes women to recurrent urinary tract infections. Fifth, menstruating women require restroom breaks to change menstrual hygiene products that may not be at every two-hour cutoff, and finally women with urinary incontinence who require the use of protective pads also require restroom breaks that may be outside of the every two-hour interval to change those pads to avoid problems from the pads.[75]

Nygaard also rebutted Stivers' testimony by citing the results of the theretofore most reliable study[76] revealing that the median number of voids during waking hours among 300 asymptomatic women was eight with a range of four to 18 and 95 percent of the women voiding fewer than 13 times; thus more than half of the women voided more than eight times.[77] To present a more concrete picture, Nygaard also observed that studies have shown that 20 to 50 percent of women fall into her second category of those with bladder complaints, which include urinary incontinence, urinary frequency, and overactive bladder.[78] In addition, menstruating women in her fifth category may have to change their hygiene products as frequently as every 30 minutes on the first day of their period and up to every

---

[74]Deposition of Stivers at 58.

[75]Commonwealth of Kentucky, OSH Review Comm., Secretary of Labor, Commonwealth of Kentucky vs. Jim Beam Brands Co., Deposition of Dr. Ingrid Nygaard 34-35 (June 19, 2002). See also letter from Dr. Ingrid Nygaard to Ana Avendano Denier, assistant general counsel, UFCW (June 21, 2002).

[76]Mary FitzGerald, Ursula Stablein, and Linda Brubaker, "Voiding Patterns Among Asymptomatic Women," *Obstetrics and Gynecology* 99(4)(Supp):111S (Apr. 2002) (abstract).

[77]Deposition of Nygaard at 37-38.

[78]Deposition of Nygaard at 49-50. One study of women between the ages of 16 and 64 revealed that the incidence of urinary incontinence ranged between 10 and 30 percent; another study of women 18 to 44 years of age found a prevalence of 24.5 percent, with 6.6 percent of respondents reporting incontinence daily. Sheila Fitzgerald, Mary Palmer, Susan Berry, and Kristin Hart, "Urinary Incontinence: Impact on Working Women," *AAOHN Journal* 48(3):112-18 at 113.

eight or 12 hours on later days; the most typical range is every one to three hours.[79] Nygaard testified that she had not seen any study showing an average of as low as three to six voids per day, as Stivers had asserted; and in contrast to his claim that the average voiding volume amounted to three-fourths to one and one-third quarts per day, Nygaard reported that studies in fact showed the much higher range of 1,332 to 1,759 cubic centimeters (whereby a quart equals about 946 cubic centimeters).[80]

In a follow-up letter two days later to the UFCW's assistant general counsel (a copy of which was also given to Jim Beam's counsel), Nygaard significantly expanded her analysis, explaining why recent cutting-edge urological research bolstered scientific support for an at-will workplace voiding regime:

> Until recently, the symptom of urinary frequency has been defined as voiding eight or more times in a 24 hour period. This cut-off is based on one study, done in Scandinavia in 1988 in which 151 white women with no bladder complaints completed a bladder diary (a prospective record of all voids) for 48 hours.[81] The number of voids per 24 hours ranged from 3-11 and 92% of women voided fewer than eight times daily. Of note, in the 65 women who reported one working and one free day, voiding frequency was higher during work days (6.2 vs 5.8).
>
> Since that study, there have been fewer than a dozen studies identified in the English language medical literature that address the question of how often women without symptoms of bladder dysfunction void. Interpretation of these studies is problematic for one of several reasons: the number of women queried was small, the population was not typical of the population in general (for example, some studies queried hospital employees only), the age range of the participants was limited to mainly one age bracket, or the participants were not residing in North America, and thus not representative of women here.
>
> Because of the limitations in the medical literature, FitzGerald, Stablein, and Brubaker recently conducted a study to assess voiding habits in asymptomatic women residing in the U.S. This study was presented at the 2002 American College of Obstetrics and Gynecology meeting.[82] The researchers recruited 300 ambulatory non-pregnant women with no bladder complaints. Women were excluded from participating if they had any symptoms of urinary incontinence (urinary leakage) or voiding difficulty, pelvic organ prolapse or pelvic pain, current or past use of medications with anticholinergic effects (which can affect the bladder), neurological disorders that could affect the bladder, prior

[79]Deposition of Nygaard at 58-59.

[80]Deposition of Nygaard at 72-76.

[81]G. Larsson and A. Victor, "Micturition Patterns in a Healthy Female Population: Studied with a Frequency/Volume Chart," *Scan. J. Urol. Nephrol* 1988; supp. 114:53-57.

[82]M. FitzGerald, U. Stablein, and L. Brubaker, "Voiding Patterns Among Asymptomatic Women," *Obstetrics and Gynecology* 99:111 S (2002). This publication is an abstract of M FitzGerald, U Stablein, and L Brubaker, "Urinary Habits Among Asymptomatic Women," *Am. J. Obstetrics and Gynecology* 187:1384-88 (Nov. 2002).

surgery for incontinence or prolapse, or worked primarily at night. Thus the results of this study can be generalized to a healthy population of women of various ages and ethnic backgrounds. (Women with urinary incontinence and pelvic organ prolapse are likely to void more often than women without these conditions.) In this study the median age was 40 years, with a range of 18-91. Self-assigned race was 39% African-American, 39% Caucasian, 12% Hispanic and 9% Asian. The median number of voids during waking hours was 8 (range 4-18). 95% of the women voided fewer than 13 times. The number of voids increased with age and fluid intake and did not differ according to race, weight, or having borne children or having had a hysterectomy. The volume that women voided in a 24 hour period averaged 1759 cc, which is not considered abnormal according to the International Continence Society, which defines polyuria (excessive urine output) as > 2.8 liters of urine per 24 hours in adults.[83]

Based on these results, the use of fewer than 8 voids to indicate a normal voiding pattern is inappropriate.[84]

In clinical practice, a common method used to describe a reference value (also known as "normal" values) for a test or characteristic is to use the interval that embraces 95% of the observations. A minimum of 300 subjects is recommended to describe this interval.[85]

As noted, the FitzGerald study included asymptomatic women. However, a large number of women in the U.S. general population suffer from either urinary incontinence or overactive bladder syndrome, conditions that generally increase the need to void. Studies from many different populations of healthy community-dwelling women found prevalence rates of urinary incontinence ranging from 20-50%.[86] Wide ranges are found

---

[83]P. Abrams, L. Cardozo, M. Fall et al., "The Standardization of Terminology of Lower Urinary Tract Function: Report from the Standardization Sub-Committee of the International Continence Society," *Neurourology and Urodynamics* 21:167-78 (2002).

[84]See M. FitzGerald, N. Butler, S. Shott, and L. Brubaker, "Bother Arising from Urinary Frequency in Women," *Neurourology and Urodynamics* 21(1): 36-40 (2002), discussion at 41: "We reviewed records of 200 women who had completed 24-hour frequency-volume charts, and had indicated their degree of bother with urinary frequency.... The degree of bother was correlated with daytime and nighttime voiding frequency, maximum functional capacity, mean voided volume, and demographic variables. Among 200 women, 180 (90%) indicated at least a minor degree of bother with urinary frequency. A voiding frequency of eight or more times in 24 hours was reported by 166 (83%) of women. Among the 34 women voiding fewer than eight times/24 hours, 26 (76%) reported bother with urinary frequency. There was large variation in the degree of bother reported at a given voiding frequency. Our study suggests that the currently utilized cutoff value of eight daily voids to define urinary frequency may not be helpful in the management of women in this country. A racially diverse study of the voiding habits of asymptomatic North American women is mandated."

[85]J. Henry, Todd-Sanford-Davidsohn, *Clinical Diagnosis and Management by Laboratory Methods* 53 (17th ed. 1984).

[86]C. Payne, "Epidemiology, Pathophysiology, and Evaluation of Urinary Incontinence and Overactive Bladder," *Urology* 51(2A Supp.):3-10 (1998).

because of diverse definitions of incontinence, differing research designs and population samples and a variety of often unvalidated measurement instruments. In women with urinary urge incontinence, two studies reported an average number of voids per 24 hours of 9.5 (the 95% interval ranging from 5.5 to 14.0 voids per 24 hours)[87] to 10.6 (daytime voids averaging 8.8).[88] In treatment studies of individuals with overactive bladder syndrome, the mean number of voids in the 1529 participants was roughly 11 (10.9-11.3 in each of three treatment groups at baseline), with a range of 2.0-51.3 voids per day.[89] Thus, if about a third of women have a bladder condition and half of them have overactive bladders, then about one in six are likely to void more frequently. In a large population-based study of people in six European countries, 16.6% of 16,776 individuals ages 40 and over had overactive bladder symptoms (frequency, urgency or urge incontinence).[90]

Thus, the new company policy is detrimental to the physiologic need to void in both healthy women and a large number of women with bladder complaints.

Further, while exceptions to Jim Beam's policy may be requested for pregnancy, women in the early stages of pregnancy may not be aware that they are pregnant or may want to keep this information confidential. However, 40% of young women (median age 23 years) in early pregnancy (median 11 weeks) needed to void more than 7 times during the daytime in one study.[91] Since urinary frequency increases before women are visibly pregnant, should they be forced to choose between losing their job over frequent bathroom breaks and telling the doctor and thus the boss that they are pregnant in order to receive an accommodation?

Voiding infrequently or restricting voiding when voiding is desired increases the risk of urinary tract infections. Urinary tract infections (bladder or kidney infections) are

---

[87] G. Larsson, P. Abrams, and A. Victor, "The Frequency/Volume Chart in Detrusor Instability," *Neurourology and Urodynamics* 10:533-43 (1991).

[88] H. Van Melick, K. Gisolf, M Eckhardt, "One 24-Hour Frequency-Volume Chart in a Woman with Objective Urinary Motor Urge Incontinence Is Sufficient," *Urology* 58:188-92 (2001). By showing that one voiding diary for one day was reproducible, the study also undercuts possible criticism of the FitzGerald et al. study on the grounds that measuring urine only for one day is insufficient.

[89] P. Van Kerrebroeck, K. Kreder, U. Jonas, et al., "Tolterodine Once-Daily: Superior Efficacy and Tolerability in the Treatment of the Overactive Bladder," *Urology* 57:414-21 (2001).

[90] I. Milsom, P. Abrams, L. Cardozo, et al., "How Widespread Are the Symptoms of an Overactive Bladder and How Are They Managed? A Population-Based Prevalence Study," *BJU Int* 87:760-66 (2001). This study, which was funded by a drug company, used as the definition of frequency >8, which Nygaard believes encompasses some women who are normal. On the other hand, only five percent of the respondents had only frequency, with no other symptom (thus, about 12% had frequency plus urgency and/or incontinence). Therefore the proportion with bladder disorder is still high. The authors note that the definition of >8 voids was chosen arbitrarily, based on the Larsson study.

[91] A. Cutner, L. Cardozo, C. Benness, "Assessment of Urinary Symptoms in Early Pregnancy," *British J. Obstetrics and Gynecology* 98:1283-86 (1991)

common in women and affect 20%[92] to 50% of women.[93] Women who void infrequently have a higher risk of having recurrent urinary tract infections. In a study which queried asymptomatic women (as opposed to those seeking care for infections), women who voided infrequently were more likely to have three or more urinary tract infections per year than women who voided more frequently (6.2% vs 1.6%).[94]

Delaying voiding or drinking less to avoid having to void also increases the number of women who suffer from urinary tract infection. A study of public school teachers found that 21.2% of women who reported making a conscious effort at work to drink less had a urinary tract infection in the preceding year, compared to 9.9% of women who drank as much as they wanted to.[95] In one study, 61% of young women with recurrent urinary tract infections gave a history of deferring voiding for periods of one hour or longer (i.e., not urinating when they felt the need) compared with 11% of women without recurrent infections.[96]

The average volume that asymptomatic women urinate over 24 hours ranges from

---

[92]L. Nicolle and A. Ronald, "Recurrent Urinary Tract Infection in Adult Women: Diagnosis and Treatment," *Infect Dis Clin North Am* 1:793-806 (1987) (at least 20% have an urinary tract infection and 3% have recurrent urinary tract infections—the population that restricting bathroom access really hurts).

[93]T. Hooton, D. Scholes, J. Hughes, "A Prospective Study of Risk Factors for Symptomatic Urinary Tract Infection in Young Women," *New England J. Med.* 335:468-74 (1996). The study found that among 348 healthy university women (mean age 23) the incidence of urinary tract infection (culture confirmed) per person-year was 0.7 and in a healthy HMO cohort (age 29) 0.5. At baseline, 28% and 58% respectively had at least one urinary tract infection and 14% and 38% had two or more. In the six months of the study, 98 urinary tract infections occurred in the 348 university women and 82 in the HMO women. Thus in a group of 100 working women, in the course of a year the group will have had 50-70 urinary tract infections (the lower number being more probable among Jim Beam workers); thus the odds are high that at any given time, someone has urinary tract infection symptoms since they usually last three to seven days.

[94]A. Nielsen and S. Walter, "Epidemiology of Infrequent Voiding and Associated Symptoms," *Scan. J. Urol Nephrol* 157:49-53 (suppl.) (1994). The cut-off point of three voids or fewer per day was chosen arbitrarily and it is impossible to tell from the data whether there are differences in women who voided more than three times but less often than some other arbitrary number. The population was large (1048) but limited to Danes under age 46; the data were collected by questionnaire and not by prospective recording. Of interest, the number of urinary tract infections in the last year (21% among women voiding three or fewer times and 10.4% among those voiding four or more times) is very similar to the finding in I. Nygaard and M. Linder, "Thirst at Work—An Occupational Hazard?" *Int Urogynecol* 8:340-43 (1997).

[95] Nygaard and Linder, "Thirst at Work—An Occupational Hazard?"

[96]K. Adotto, K. Doebele, L. Galland, et al., "Behavioral Factors and Urinary Tract Infection," *JAMA* 241:2625-26 (1979).

1332cc (+/-59 cc)[97] to 1759 cc (+/- 797 cc) in the large study of American women by FitzGerald et al. Thus, normal women void more than " a quart of urine" per day, as stated by Dr. Stivers (one quart being equivalent to 946 cc).[98]

As a urogynecologist, Nygaard was not prepared to deliver a professional opinion on men's urinary problems; despite being a urologist, Stivers failed to deal with them. In order to present a more comprehensive view of the voiding needs of a labor force consisting of males and females, it is crucial to point out that many, especially older, men also have special urination problems that may require frequent bathroom visits during the workday. According to one British study, more than one-half of men over the age of 50 have enlarged prostates (benign prostatic hyperplasia or BPH), as a result of which a significant proportion of male workers must also urinate more often than do men without prostate problems:

[A]s the prostate grows and BPH begins to restrict urine flow, men may experience the following symptoms: weak flow; a need to push or strain to start urine flow; intermittent urine stream (starts and stops several times); difficulty stopping urination; "dribbling" or leakage after urination; a feeling of being unable to empty the bladder completely.

As the bladder muscle works to push urine through a narrowed urethra, the bladder wall thickens and is less able to stretch. As a result, the bladder can't hold as much, causing a need to urinate more often. These symptoms can include:

frequent urination, especially at night, disrupting sleep; an urgent need to urinate that can't be postponed; urge incontinence: an inability to get to a bathroom in time when the urge to urinate occurs.[99]

According to a large study of more than 1,000 men at autopsy, the prevalence of BPH rises from five percent among men under 30 years of age, to 23 percent between the ages of 41 and 50, 42 percent between 51 and 60, and 71 percent

---

[97]Y. Homma, O. Yamaguchi, S. Kageyama, S. Nishizawa, M. Yoshia, and K. Kawabe, "Nocturia in the Adult: Classification on the Basis of Largest Voided Volume and Nocturnal Urine Production," *J. Urology* 163:777-79 (2000). Although the mean number of voids (8) was similar to that found in FitzGerald et al., the small numbers (32 women) studied and the non-generalizable population (Japanese) did not make it sufficiently scientifically valid to support the FitzGerald study. It is nevertheless included here as one of the few to report the 24-hour voided volume. Despite being the lowest reported voided volume, it was still higher than Dr. Stivers' estimate.

[98]Letter from Dr. Ingrid Nygaard to Ann Avendano Denier (June 21, 2002).

[99]http://www.prostatecanceruk.org/bhp.html

among those 61 to 70 years old.[100] Another study of men age 50 to 78 found that men with moderate and severe symptoms of BPH voided 2.0 and 2.5 times more per day, respectively, than those without symptoms.[101]

The Hearing Begins

Shortly before the Kentucky Occupational Safety and Health Review Commission administrative hearing, at which Jim Beam contested issuance of the citation,[102] the Kentucky Labor Cabinet filed a motion with the hearing officer to exclude Beam's medical experts' testimony on the grounds that the employer was trying to "relitigate medical issues that have been fully researched and addressed in the promulgation of standard 29 CFR 1910.141(c)(1)(i)." The Labor Cabinet therefore requested that the hearing officer "take judicial notice...that it is generally known that no one can ascertain with complete accuracy when each and every individual will have the sense of urgency to void. Therefore there is no factual issue to be addressed. Medical testimony is not needed to appreciate that 'when you gotta go, you gotta go!'"[103] Although it is difficult to imagine how this fundamental fact of exclusive sensory self-perception could possibly be refuted, the hearing officer summarily rejected the request.[104]

On August 28, as the two-day hearings got underway, the *Wall Street Journal*'s closely watched weekly "Work Week" column unleashed a flood of media attention by devoting most of the piece to the Beam dispute. The reporter, Carlos Tejada, quoting Jack Allen, Beam's director of human resources as bemoaning that "'[t]here's not a lot of guidance to go by with regards to what's reasonable and what's not,'" portrayed the hearing as possibly giving "both workers and companies a better sense as to how much employers can control bathroom access." Despite this lack of guidance and specificity, Allen was certain that Beam's policy "fits the bill." Tejada then quoted the author in refutation: "'Peo-

---

[100]Stephen Berry et al., "The Development of Human Benign Prostatic Hyperplasia with Age," *Journal of Urology* 132:474-79, tab. 2 at 475 (1984) (percentages calculated based on absolute numbers).

[101]Marco Blanker et al., "Normal Voiding Patterns and Determinants of Increased Diurnal and Nocturnal Voiding Frequency in Elderly Men," *Journal of Urology* 164:1201-1205, at 1203 (Oct. 2000).

[102]803 Ky Adm. Regs. ch. 50 (2002).

[103]Commonwealth of Kentucky, Occupational Safety and Health Review Commission, Adm. Action No. 02-KOSH-0015, Motion in Limine and to Take Judicial Notice on Behalf of the Secretary of Labor (Aug. 15. 2002).

[104]Email from Peter Ford to Marc Linder (Aug. 23, 2002).

ple's bladders don't operate on clock time.'"[105]

In the same vein, later that day the Associated Press quoted President Kelley capturing the essence of Beam's policy: "'They're saying you go to the bathroom when we say it's time to go to the bathroom, not when your body says it's time to go to the bathroom. It's like training kindergarten kids to get onto a schedule." Kelley also explained how workers tried to cope with the six-step termination process: while some workers "urinated on themselves because they were afraid to leave the line" and others began wearing protective undergarments, still others "feigned illnesses to go home and avoid getting violations." By the time of the hearing, 40 workers had been counseled, five warned, and one was one violation short of being fired. That employee, Krystal Ditto, a 36-year-old single mother of two children who was healthy and therefore had not sought a medical waiver, confirmed the kindergarten analogy when she told the AP that a Beam official had "suggested she practice going to the bathroom every two hours at home to get into the habit."[106]

Since no other employer in the United States had proceeded so far in the appeal of a citation for having denied workers permission to stop work to go to the bathroom, the provocative questions posed by lawyers for OSHA, Jim Beam, and the UFCW[107] and the hearing testimony they elicited represent by far the most extensive public discussion of OSHA's intervention. The broad array of witnesses included the OSHA inspector, Jim Beam's plant manager and human resources director, the president of the union, several workers who had been adversely affected by the policy, two local urologists hired by Jim Beam, and Dr. Ingrid Nygaard, junior author of *Void Where Prohibited*, who traveled to Kentucky to testify on behalf of the union and workers.

The Kentucky Occupational Safety and Health Review Commission hearing took place in the Bullitt High School Auditorium in Shepherdsville, Kentucky, a town of 6,000, 30 minutes south of Louisville and just a few miles from Fort Knox and the Jim Beam plant in Clermont. The brief but pithy opening statement of Karen Triplett, counsel for the Kentucky Labor Cabinet, aptly expressed OSHA's aggressive biology-based enforcement position: "Utilizing a toilet is not a privilege: it is a physical necessity."[108]

---

[105]Carlos Tejada, "Work Week," *Wall Street Journal*, Aug. 28, 2002, at B3, col. 1.

[106]"Jim Beam Workers Turn Sour Over Bathroom Break Limit."

[107]Peter Ford, assistant general counsel of the UFCW, entered the litigation late, having replaced his colleague, Ana Avendano.

[108]Commonwealth of Kentucky Occupational Safety and Health Review Commission, Administrative Action No. 02-KOSH-0015, Secretary of Labor Commonwealth of Kentucky v. Jim Beam Brands Co., KOSHRC #3681-01, "Transcript of Administrative Hearing" at 12 (Aug. 28, 2002) ("Transcript of Hearing").

Jim Beam's attorney, Richard Griffith, sought to radicalize the proceedings with hyperbole calculated to convince the world that OSHA had rendered employers powerless to deal with workers who fraudulently availed themselves of the Memorandum's mandate to taunt management by spending more time in the bathroom than working: "The Secretary of Labor has taken the very extreme position that employees have the absolute statutory right to use the restroom any time they want and that any restriction on that is a violation of OSHA standard [sic]. That position has been taken despite a very clear history of abuse of restroom breaks at the bottling line.... The position has been taken despite a complete absence of published regulatory or medical support for it. The position has been taken though neither the Secretary of Labor nor the expert witness that the union has hired to support it's [sic] position in this case, has offered any suggestion of any reasonable alternatives that might be available to restrict access to restrooms. ... The regulation does not speak to what we will do if the employee simply says they need to go every five...minutes. I need to go a hundred times a day, what would you do?"[109] Griffith went on to claim that the union had "conceded in negotiations" that in the days of tag rotation: "people went sometimes and didn't need to go. ... Nine...times in an eight and half hour...period they were having access to restroom facilities. [B]ecause it had been a longstanding practice or for whatever other reasons, it [the company] didn't change the practice...."[110]

The first witness was Dr. Nygaard, who declared unequivocally that "an at-will policy is the most medically tenable policy for voiding."[111] Under questioning by Peter Ford, counsel for the complainant representing UFCW Local 111-D, she laid out a statistical overview of the distribution of conditions affecting frequency of urination among women:

> If you started with that hypothetical group of a hundred women, or just a hundred employees, then based on over all [sic] weights of conditions, we would assume that about...20% of them or...20 women have incontinence or an over active [sic] bladder, abnormal bladder. That about...3% of them would have a recurrence of bladder infections. That about one of them would be pregnant. If we assume that half of those hundred are before menopause, then we assume that while normal menses is about five days, that the heavier flow might be just for the first two days. And on any given day about...3 women are menstruating. Then if we assume that about...5% of them have medical problems that cause needing more frequent bathroom access like diabetes or high blood pressure, they are on medications for those conditions. We add all those together, we come up with...about...64 people that are left healthy, no bladder problems, no medical com-

[109]"Transcript of Hearing" at 12-14 (Aug. 28).

[110]"Transcript of Hearing" at 17-18 (Aug. 28).

[111]"Transcript of Hearing" at 86 (Aug. 28).

plaints. And then we know that half of those people need to urinate more often than an average of every two hours. So when we put all that together, it's only about a third of [women] that would not be affected on a daily basis by this bathroom policy.[112]

Nygaard went on to discuss the results of the aforementioned study by FitzGerald et al.—in her opinion the heretofore most reliable investigation of the frequency of female urination—which found that half of women without medical problems had to urinate more than eight times during waking hours: "And that finding didn't surprise me because it agrees with my clinical practice. When I ask people how often they urinate...it's generally somewhere between every one hour to every three hours and these are healthy women without problems."[113] Finally, Nygaard stressed that urinary retention leads to distension of the bladder wall, while blood vessels become so thinned out they no longer supply blood to bladder, resulting in infections. If the bladder does not empty completely, bacteria can grow leading to infection.[114]

In cross-examining Nygaard, Jim Beam's lawyer sought to buttress its claim that workers do not need to urinate more often than every three hours by reference to OSHA's field sanitation standard for farmworkers, which does not require employers to provide toilets to workers who work fewer than three hours.[115] However, Griffith undermined his own position by telling Nygaard, literally in the same breath, that OSHA also excluded from its protection workers on farms with fewer than 11 employees.[116] Yet Congress and OSHA chose to exclude such workers not because they are somehow biologically immune to medical problems related to retention of urine, but, as already explained,[117] in order to benefit certain farm employers at the expense of their workers. And regardless of the political background of OSHA's pretextual claim in the 1980s that there was no need to impose financial burdens on agricultural employers for the sake of farm workers' health, because fixed-site industrial employers do not provide short-term toilets, no analogous financial burden could apply to them to justify depriving workers of voiding rights for periods of less than three hours.

Jim Beam's lawyer then attempted to manipulate Nygaard into acquiescing in his campaign to characterize *Void Where Prohibited* as authority for Jim

---

[112]"Transcript of Hearing" at 38-39 (Aug. 28).

[113]"Transcript of Hearing" at 41 (Aug. 28).

[114]"Transcript of Hearing" at 43-44 (Aug. 28).

[115]29 CFR sect. 1928.110(c)(2)(v).

[116]"Transcript of Hearing" at 65 (Aug. 28); 29 CFR sect. 1928.110(a) ("This section shall apply to any agricultural establishment where eleven...or more employees are engaged on any given day in hand-labor operations in the field").

[117]See above ch. 5.

Beam's policy of toilet breaks every two hours:

Q. The book that you...co-authored...with Professor Lender [sic].... Would it be fair to say that he has been very well published and outspoken about his strong conviction that workers' need to have strong rights to get to go to the restroom whenever they want to?
A. I think that's clear from his writing in the book. You have to ask him otherwise.
Q. And you wrote some parts of this book?
A. I wrote the chapters that specifically dealt with the medical aspect of waiting. It's a small part of the book.
Q. Your book is called "Void Where Prohibited. Rest Breaks and the right [sic] to Urinate on Company Time: by Mark [sic] Linder and Ingrid Nygaard. Before the book was published and was put out with your name on it, did you have the opportunity to read the book and assess whether there was anything you thought medically was not valid?
A. I did not read the whole book. I read the parts that were related to medicine and skimmed other parts but I certainly didn't check the whole book.
Q. I'm not trying to trap you in any way but I just don't understand what your participation was. You had the opportunity to read all of this, and...if there was something that is medically incorrect, you would have had the opportunity to correct it?
A. I would hope that I had that.
Q. There's a statement in the book where the authors, and perhaps this was Professor Linder who was writing the bulk of this, is arguing for regular restroom breaks, mandated statutory restroom breaks. And in it, he proposes that there be allowed a...12 minutes [sic] break every two hours so you can go to the restroom. Do you recall that argument by him that that would be good social legislation?
A. As opposed to having no breaks, yes.
Q. What he said is quote "...48 minutes of rest per...8 hour work day could for example be taken as four 12-minute blocks every two hours." Do you recall that statement in the book?
A. I recall that statement in the book.
Q. And you didn't object that such a policy adopted by legislation would allow twelve-minute restroom breaks every two hours?
A. Yeah, if you read the book he made it clear that at-will voiding is the policy of choice.
...
Q. But you don't dispute the book which advocates strongly the workers' right to urinate proposed in this and its possible legislative remedy,...12 minute breaks every two hours. You don't dispute that that's what it says.
A. I don't dispute that that's what it says.[118]

In fact, whether Griffith, carried away by an advocate's zeal, negligently misread the passage or intentionally misled and deceived the witness and the hearing officer, his claim that the book "proposes that there be allowed a...12 minutes [sic] break every two hours so you can go to the restroom" is a complete fabri-

---

[118]"Transcript of Hearing" at 71-74 (Aug. 28).

cation. The proposal has absolutely nothing to do with "the restroom" and everything to do with rest: this part of the tripartite proposal (the introduction of meal breaks and rest breaks under the Fair Labor Standards Act and at-will urination under OSHA) deals with rest breaks, not toilet breaks:

> Workers' needs for rest, nutrition, excretory autonomy and well-being, and free time for self-development and community activities should be addressed by a comprehensive re-orientation of the national wage and hour statute. [T]he FLSA should be amended to create a framework for setting minimum standards for the duration and scheduling of meal and rest periods and voiding breaks. ...
>
> Grounded on the fundamental nonwaivable and nonpurchasable right to restore their physical and mental powers through rest, the statute should at a minimum entitle workers to a forty-five-minute meal period and the equivalent of six minutes of freedom from work per hour. ... The forty-eight minutes of rest per eight-hour workday could, for example, be taken as four twelve-minute blocks every two hours. Employers who failed to provide hourly rest breaks would be prohibited from disciplining worker who had to use the toilet more frequently than during the scheduled breaks. ...
>
> As coordinated measures, the OSHA general sanitation regulations should also be amended to make express the employer's duty to permit workers to urinate as often as necessary, while the ADA regulations should incorporate a threshold for urinary disability in terms of voiding frequency.[119]

Having scored his fraudulent debating point, Griffith then shifted his focus, suggesting that OSHA's directive rendered employers helpless to detect fraudulent urinators:

> Q. In accessing [sic] that concept of at-will voiding, have you taken into consideration at all, what could be done? In your consultation with Professor Linder, what could be done if, in fact, an employee wished to claim that he or she needed to go when, in fact, he or she really didn't need to go? How do you deal with that as an employer? ... Let's say hypothetically...an employee wishes not to work any particular day so every...20 minutes they say they need to go to the bathroom. Would you agree with me that because you're not going to go into the stall with the person, there's really no way to regulate or deal with that risk...."[120]

After failing to prompt Nygaard to concede that the book had failed to "factor into your position...the risk of any sort of abuse of restroom privileges if you have an at-will rule,"[121] Griffith unsuccessfully sought to induce Nygaard to abandon her criticism of the inadequacy of Jim Beam's medical accommodation:

---

[119]Linder and Nygaard, *Void Where Prohibited* at 164-65.

[120]"Transcript of Hearing" at 74 (Aug. 28).

[121]"Transcript of Hearing" at 75 (Aug. 28).

A. [Y]ou told me that if somebody with...leakage of urine...takes it upon herself to go to a physician and get a letter, and then share that letter with her supervisors, in essence opening herself up to embarrassment and ridicule, that person can be accommodated. Now, that same person who has urinary leakage or painful urgency, under a system at which she can void at will, can self accommodate....
Q. And what you're arguing is the reason why a void-at-will is better [sic] way to go than some other alternative?
A. It is a better way to go because the line between healthy and illness is very, very...blurred. And for urinary incontinence, part of what keeps people from being healthy...is specifically that they can't get to the bathroom. So, simply by providing the bathroom, they can remain healthy.[122]

Nygaard thus objected to Beam's medical accommodation policy because it required workers to come forward and tell "[a]bout embarrassing medical problems and...it becomes a medical problem only when they are denied bathroom access. Otherwise it may not be a medical problem."[123]

If the weakness of Jim Beam's case was reflected in the lack of expertise of the expert medical witnesses it hired, its predicament may have been dictated by the possibly limited universe of qualified physicians willing to testify on behalf of an employer's profit-driven preference for a more restrictive urination policy than that mandated by the national occupational safety and health agency. Its first witness, Dr. Tamara James, who declared three times that Beam's policy of letting workers void every two to three hours was "reasonable" and "adequate,"[124] admitted under cross-examination that prior to reviewing the literature that Nygaard had cited in her deposition she had never read it because her work was "purely clinical"—she did not keep up with research in the areas of workplace urination or urinary incontinence: "That is not my forte."[125] Her dogmatism and closed-mindedness was vividly on display in her statement that, despite Nygaard's deposition and hearing testimony that some women need to change their feminine hygiene products more often than every two hours, reasonable minds could not differ on the issue that if "someone needs to change their tampon every hour to hour and a half they need to be evaluated" and could therefore get a medical accommodation to use the bathroom more often.[126]

Kentucky OSHA's only witness was the officer who had conducted the inspection at Jim Beam, Diane Marraccini, an industrial hygienist in the agency's compliance division, with almost ten years of experience there.[127] She testified that

[122]"Transcript of Hearing" at 76-77 (Aug. 28).

[123]"Transcript of Hearing" at 78 (Aug. 28).

[124]"Transcript of Hearing" at 106-108 (Aug. 28).

[125]"Transcript of Hearing" at 112-13 (Aug. 28).

[126]"Transcript of Hearing" at 105-108, 110 (quote) (Aug. 28).

[127]"Transcript of Hearing" at 126-27 (Aug. 28).

she had determined that Beam had violated the OSHA standard because, "by allowing the employees to frequent the restroom only during scheduled breaks, that were determined by the employer and not the employee...at the times other than those scheduled breaks, the employer was, in fact, denying access to the restroom...." In other words, the policy limited employees not only in the number of times they could frequent the bathroom, "but at pre-determined intervals that were determined by the employer and not the employee."[128]

Marraccini explained that she had classified the violation as "other than serious" "[b]ecause the...most likely resulting injury or illness would not result in permanent disability, death or serious physical harm to the employee." The severity assessment was "minimal," while the probability assessment was that the most likely injury or illness was a urinary tract infection. The assessment of no monetary penalty was based on the table in OSHA's field operations manual with predetermined base-penalty amounts arrived at through a combination of severity/probability; the probability was "lesser" because "they did have an adequate number of water closets available, and they did allow the exemption for those with medical conditions."[129]

Just as Griffith and his client remained impervious to the fact that workplace voiding was a right and not merely a privilege, it was not clear that he understood that Marraccini was correcting his misunderstanding of the wording and meaning of the regulation itself, which had served as the crucial basis for Federal OSHA's expansive Memorandum. When he asked Marraccini: "The sanitation standard explicitly says that the employer has to *have* an adequate number of toilet facilities based upon the number of employees, correct?" she replied: "It indicates that they need to *provide* employees with those facilities."[130] On the other hand, Griffith was able to make one point—that the only medical article cited by the OSHA Memorandum as support for the argument that involuntary urine retention may lead to urinary tract infection defined infrequent voiding as three or fewer voids per day.[131] The claim, as Nygaard later remarked, "is true, and...is a problem for the scientific basis of the opinion."[132] What this gap also reveals is that a study of urinary tract infections using a higher threshold of voids is an important occupational health research desideratum.

A particularly valuable colloquy unfolded between Jim Beam's lawyer and Marraccini concerning the logic of the at-will voiding regime created by the Memorandum and employers' power lawfully to police the truthfulness of workers'

---

[128]"Transcript of Hearing" at 144-46 (Aug. 28).

[129]"Transcript of Hearing" at 146-48 (Aug. 28).

[130]"Transcript of Hearing" at 157 (Aug. 28) (italics added).

[131]"Transcript of Hearing" at 162-63 (Aug. 28). The article is Nielsen and Walter, "The Epidemiology of Infrequent Voiding and Other Symptoms."

[132]Email from Ingrid Nygaard to Marc Linder (Oct. 6, 2002).

claims that they need to void:

Q. What I'm understanding your testimony to be is that any restriction by an employer where an employer says there is a specific part of the day that you can't go...that would be a violation of the OSHA standard, correct?
A. What I am saying is that if the employer denies them going to the restroom when they have a physical need to go to the restroom, yes, that is a violation of the standard.
Q. Okay, then what if they say they've got a physical need. Is it your interpretation of the standard that a simple statement, I have the urge to go gives them the regulatory, statutory right to go to the restroom...?
A. I know of no other way to determine if that's a true statement or not. So, yes, I would have [to] say they would have to accept it at face value.
Q. ...So, if a person were to say every half hour they need to go to the restroom...—and they tell their employer that and they really mean it—I really need to go. I don't have a medical condition, I just need to go every half hour, I really need to go all day. There is no way an employer could address that use, correct? ... Your position would be that someone...saying they needed to go every half hour, whether or not they needed it, that that would be a protected activity under OSHA...?
A. If they have a need to use the restroom and indicate that to the employer then yes, they must be allowed access to the restroom.[133]

To be sure, in her final response, Marraccini sidestepped the crucial issue by assuming what presumably was the principal question—namely, whether workers "need" to void or are merely saying that they do.[134]

In responding to a hypothetical question posed by Griffith, Marraccini stated that if an employer gave an employee access to the bathroom 20 times a day but dictated each of those times, it would be necessary to look at the question on a case by case basis because it is possible that those 20 breaks might include times when the employee might actually need to use bathroom. Griffith then asked: "So, is there a number where it becomes reasonable for an employer to set a number and say you can only go this many times?" To which Marraccini replied: "Twenty...may not be okay in some instances. I don't think that question is answerable."[135] Marraccini also undermined Jim Beam's effort to inflate the number of toilet breaks it provided by observing that, since the employer has no control over employees' going to bathroom immediately before and after shift and therefore is "at that point providing them no access," it was not relevant to the reasonableness of the policy from OSHA's perspective that, including those two times, workers could go to the bathroom six times in 8.5 hours.[136]

---

[133]"Transcript of Hearing" at 166-67 (Aug. 28).

[134]For further discussion of the issue of abuse, see below ch. 17.

[135]"Transcript of Hearing" at 169 (Aug. 28).

[136]"Transcript of Hearing" at 180-81 (Aug. 28).

The remainder of the first day of the hearing was devoted to testimony by two female bottling-line employees. The first, Krystal Ditto, 36 years old, who had worked at Jim Beam for seven years, testified that she considered herself "normal" and urinated 10-15 times during waking hours. Because she had just undergone a full physical examination and been found normal and healthy, she did not want to apply for a medical accommodation; she was also reluctant to state to management that she was not 100 percent able to work because the company "can send you home until you are...."[137] The only employee to have received a final warning and thus to have been facing termination the next time she exceeded her one daily discretionary bathroom break, she wrote on that warning that she disagreed with it because she had been "using the restroom to relieve myself when needed and nothing more. Why is it not stated on this warning that I'm using the restroom instead of excessive personal needs time."[138] When she received her final warning she met with Jack Allen, the human resources director, and the bottling manager, who "asked me if when I went home at night, and on the weekends did I try using the restroom every two hours." She said no and asked him whether he did and he said no.[139] Later, when one day she had already used up her one unscheduled break and still had to go to bathroom, she had to ask to be permitted to go home early to use the bathroom, and she lost pay for the time. She stressed that it was humiliating to have to "justify the need that you had to use the restroom" to a supervisor; or to sit here at hearing and "discuss...your period...."[140]

Ditto's experience in leaving early for the day to avoid termination was not unprecedented. Thirty years earlier, just as OSHA was being established, a meat-processing firm in Wisconsin entered into an agreement with the Amalgamated Meat Cutters and Butcher Workmen (which later became part of the UFCW) under which unscheduled bathroom relief breaks were granted on an "emergency" basis only. At the arbitration of its grievance over the employer's overly strict interpretation of this standard, the union argued that the company was actually causing more absenteeism: some employees "requested the rest of the day off and went home on the premise of being ill rather than taking the few minutes to engage in an emergency relief period with the result of being written up...."[141]

Margaret Boone, a Class B technician who had worked at the plant for 34 years, testified that she sometimes had a "urination bladder problem" and seepage, and urinated 10-12 times during waking hours: "I had some seepage problems before the policy and I still have them. I don't think I'm ill. I have been to the doctor, and I think I'm just a normal fifty-four year old woman. My bladder is

---

[137]"Transcript of Hearing" at 186-87 (quote), 198 (quote) (Aug. 28).

[138]"Transcript of Hearing" at 191, 196 (quote) (Aug. 28).

[139]"Transcript of Hearing" at 199 (Aug. 28).

[140]"Transcript of Hearing" at 200-201, 203 (quote) (Aug. 28).

[141]Jones Dairy Farm, 72-2 Labor Arbitration Awards (CCH) ¶8639 at 5256 (1973).

fifty-four years old just like the rest of my body. ... It's somewhat humiliating to have to ask a young male supervisor if I can go to the bathroom. ... It makes me feel like I'm back in grade school."[142] The first day's hearing ended with Boone's testimony that on days when it was "just sprung on" workers at 3:00 or 3:30 in the afternoon that they were being "drafted" for three hours of overtime, they would have no scheduled break between 2:00 and 5:00; thus, if they had already used their one discretionary bathroom break that day, they would have to wait three hours to void.[143]

Half-Time Media Events

Midway through the hearings, Richard Griffith, Beam's lawyer for the OSHA appeal, told a Louisville television station that "'[a] reasonable restriction on the use of a restroom is perfectly appropriate. And what you cannot have is a restriction that leads to extended delays.'"[144] He was not even completely correctly interpreting the Memorandum in the abstract,[145] but he totally failed to explain how Beam could possibly have satisfied that reasonableness criterion by permitting workers to void when they did not have to and prohibiting them from (or sanctioning them for) voiding when they did have to.[146]

As the tide of print, radio, and television attention—excluding the national newspaper of record, whose labor reporter had deemed it newsworthy to mention that foremen in clothing sweatshops in the Northern Mariana Islands "often limit the number of bathroom breaks"[147] or that immigrant workers speak up "about not being allowed to take bathroom breaks" at an AFL-CIO forum,[148] but not to report

---

[142]"Transcript of Hearing" at 205, 207 (quote), 212 (quote), 216 (quote) (Aug. 28).

[143]"Transcript of Hearing" at 221-23 (quote at 222) (Aug. 28).

[144]"Jim Beam Bottlers Want More Bathroom Breaks," Aug. 28, 2002, on http://www.thelouisvillechannel.com/lou/news/stories/news-163862420020828-180831.html.

[145]See above ch. 6.

[146]Beam's impermissible fixed-break system is scarcely unique; indeed, it is not even the worst of its kind. For example, according to an informant, at the unionized Austrian-owned MIBA bearings plant in McConnelsville, Ohio, the one unscheduled break was merely at the supervisor's discretion. Moreover, the fixed-breaks were shorter—only 20 minutes for lunch and 10 minutes for the two rest breaks; since it took four minutes to go back and forth to the bathroom, the 100 men on each shift could not all use it during the six minutes remaining. Email from Colette Carr, Columbus, OH to UFCW (Aug. 29, 2002).

[147]Steven Greenhouse, "Suit Says 18 Companies Conspired to Violate Sweatshop Workers' Civil Rights," *N.Y. Times*, Jan. 14, 1999, at A9, col. 2 (Lexis).

[148]Steven Greenhouse, "Immigrants Flock to Union Banner at a Forum," *N.Y. Times*,

on a dispute as mundane as the one at Beam[149]—rose on August 28,[150] Beam's public relations department was forced to issue a "Media Statement" that afternoon:

> Early in 2001, and in prior years, Jim Beam and the Union recognized there was a problem with too many people taking too many unscheduled breaks. Their discussions did not correct the problem. Jim Beam was charged with finding a solution to the problem.
>
> The new break policy is fair, reasonable and respectful to the 150 employees who work on the bottling line...; we do care about their safety and comfort in the workplace.
>
> The break policy allows for four breaks each day, or about one every two hours: in the morning, during the lunch hour and after lunch. One other break may be taken anytime during the day. Employees with medical conditions are immediately granted special accommodations and are allowed to take as many breaks as they need. Everyone seeking accommodation has been granted it.
>
> It is important to understand that a bottling line is a unique work setting that requires a more formal, organized procedure for taking breaks than most other work settings. Upon leaving the line to take a break, the employee's position on the line must be replaced in order to allow everyone to effectively perform his or her job.
>
> Before implementing the policy, we consulted with doctors and sought guidance from OSHA and other experts. We believe the policy balances the needs of our employees with the need to get our work done properly.[151]

Beam's use of the passive construction—"was charged with finding a solution"—left readers with the impression that the union had agreed to give the employer unilateral control of formulating a voiding policy. Beam also misled the public by calling its bottling line "a unique work setting," whereas in fact thousands of assembly-line factories do or must implement relief systems, as even OSHA's Memorandum observed.[152] Since Beam already had a relief system in place, when it asserted that its restrictive "policy balances the needs of our employees with the need to get our work done properly," by "properly" it meant not

---

Apr. 2, 2000, sect. 1, at 29, col. 1 (Lexis).

[149]When the newspaper did mention the humiliation felt by workers in a foundry "forced to urinate in their pants," it linked the oppressive policy to a whole series of such practices at "one of the most dangerous employers in America...." David Barstow and Lowell Bergman, "At a Texas Foundry, An Indifference to Life," *N.Y. Times*, Jan. 8, 2003, at A1, col. 1-2, at A14, col. 2 (nat. ed.).

[150]After being quoted in the *Wall Street Journal*, the author himself was called that day by several newspaper reporters, was interviewed live on a New York City radio program (which in keeping with the tone set by discussions in the media, played a flushing toilet in the background), and received an inquiry from CNN.

[151]"Jim Beam Issues Media Statement Regarding Distillery Break Policy," PR Newswire, Aug. 28, 2002 (Lexis).

[152]OSHA, Interpretation of 29 CFR 1910.141(c)(1)(i): Toilet Facilities.

that unscheduled voiding caused bottling-line stoppages, but that profits were reduced by the need to pay the workers giving relief.[153] As the plant manager, Jeff Conger put it: "'Obviously, breaks cost money.'"[154] The costs of not taking breaks, being borne by the workers and thus 'obviously' not appearing on Beam's income statement, needed no discussion.

On the second day of the hearing, August 29, Harry Groth, vice president of Jim Beam Brands Company, issued a memorandum to all employees at Clermont and Boston, which was a variation of the "Media Statement" and the core of which read:

I know that many of you feel this is a sensitive issue and are not interested in talking publicly about it. I certainly want to respect that. However, since there has been so much media attention and numerous factual inaccuracies, I want to be sure you understand our perspective.

Early in 2001, and in prior years, Jim Beam recognized there was a problem with people abusing the established break practice. Jim Beam completed over a year of research in developing what we believe is an appropriate policy, which was implemented in October 2001.

We believe this new break policy is fair, reasonable and respectful to all of you who work in the bottling department. As you know, a bottling line requires formal training and an organized procedure for taking breaks. Our policy allows for four breaks each day, or about one every two hours: in the morning, during the lunch hour and after lunch. One other break may be taken anytime during the day. Employees with medical conditions are immediately granted special accommodations and are allowed as many breaks as they need. One-hundred [sic] percent of those seeking accommodations have received it.[155]

The obfuscatory effect that such corporate public relations can exert was dismayingly on display in the following editorial from the Iowa State University student newspaper, which contained almost as many factual and logical errors—virtually all of them identified elsewhere in this book—as it had assertions and in its densely uncritical acceptance of a powerful entity's self-interested press release bodes ill for the powers of discernment of the next generation of journalists and editors:

---

[153]Beam calculated the yearly "Cost of Hours Off Line" based on the amount of time that "both the mechanic and sanitation person spent relieving for restroom breaks," the "estimated hours spent away from the line," and the employees' hourly pay rate plus benefits. Jim Beam data, faxed by UFCW to author (Mar. 18, 2002).

[154]"Jim Beam Workers Turn Sour Over Bathroom Break Limit."

[155]Harry Groth, Memorandum on Break Policy (Aug. 29, 2002) (copy faxed to author by Local 111-D).

A Jim Beam bourbon distillery plant in Clermont, Ky., is under fire because of a policy that limits the number of trips an employee can make to the restroom.

The policy helps maximize quality control and productivity while allowing employees a reasonable amount of restroom breaks.

Workers on the bottling line are restricted to four trips per 8 and a half hour work shift. Three of the breaks are scheduled and the fourth can be taken at any time. Additional trips may lead to reprimands and after six violations employees may be fired.

Workers are allowed to use the restroom about every two hours and never have to wait more than three hours.

The local union, United Food and Commercial Workers, is complaining on behalf of about 100 employees who find the restrictions to be outrageous.

The union claims some of the employees have been incontinent in fear of being punished for leaving the line, whereas others wear protective undergarments, similar to diapers, to avoid leaving the line at all.

Union representatives are calling it a "shame" to make people choose between soiling themselves or securing their job. But the union has failed to mention what might cause a company to implement such a policy -- workers abusing unlimited bathroom breaks.

The policy was put in place last October as a solution. It is the only policy of its kind in the company, so it is fair to say the workers brought this upon themselves. Those who did not abuse the original system should be upset with those who did, not management.

The policy sounds as if the company is causing harm to people's bodies as an effort to increase already outrageous profits. There was a problem. In the business world when there is a problem, it has to be fixed.

And four trips to the restroom is plenty. The company consulted a urologist before putting the policy into effect. If an employee has a medical condition they can exempt themselves with a note from a physician.

There is not a good reason people should be urinating themselves or wearing diapers. If it is that bad, they can get out of it.

The workers and the union are simply complaining over something they need not complain about. The company has a reasonable policy and tolerates a number of violations.

Grade school teachers get even less of an opportunity to use the restroom at work, yet teachers wetting themselves is not a common occurrence, at least not to the point where it requires a national news story.

Perhaps name recognition is playing a bigger role in this complaint then anyone cares to mention. If Jim Beam were not Jim Beam, there would be no issue at all.[156]

The Hearing Ends

The second day of hearings on August 29 saw testimony from four additional employees (including the union president), the employer's paid urologist, Dr. Stivers, and its plant manager and human resources director. Patricia Culver, who

---

[156]"Jim Beam Has That Effect on Everyone," *Iowa State Daily*, Sept. 4, 2002 (Lexis).

had worked at the plant for 23 years, was a Technician B. She considered herself "[p]erfectly" "normal" and urinated 8-10 times during waking hours. After her first counseling she reduced her intake of water and called in sick three times while having her period, losing wages. The policy had been "humiliating.... I feel like I've been stripped of privacy." She did not apply for a medical accommodation because she feared she would be disqualified from her job, which is "the highest paying ladies job at Jim Beam. I was afraid that they say okay you need these special accommodations so we're going to put you over here...where it doesn't take someone to replace you to use the facilities." But she did get a medical note (despite being healthy and normal) because "us females have other needs, special needs."[157]

Bobby Hill, a 44 year-old general helper who had worked at the plant for 10 years and considered himself "normal," urinated 10-12 times during waking hours. He had received two counselings and an oral and written warning, but did not believe he needed to have the medical accommodation form filled out "because I've been using the restroom ever since I was born, and nobody's told me when I needed to go. I know when I need to go to the restroom...."[158]

Angela Mattingly, a 25 year old who urinated 8-10 times during waking hours, had received two counselings and an oral and written warning. She wrote on her warning that she had been told to get a doctor's note because of using the bathroom too much: "I didn't know that mother nature was a medical problem. ... It just irritates me to have to get a paper to use the rest room and it bothers me to be told to go to Human Resources when I'm on my period." Mattingly went to a nurse practitioner, who did a urinalysis, which was clear. She tried to explain to the nurse that "sometimes when I go it just don't land on break time. ... I was trying to explain to her that I just can't time myself to wait until 9:30." The nurse filled out her accommodation as "overactive bladder."[159]

Jo Anne Kelley had worked at Beam for 34 years, the last six of which as president of UFCW Local 111-D, which represented 248 workers at the Jim Beam plants in Clermont and Boston—200 of those employees working in Clermont, including 100 on the bottling line.[160] Kelley testified that the past practice of taking unscheduled restroom breaks on a regular basis had been in effect at least as long as she had been working there. During collective bargaining negotiations in 2001 Jim Beam had proposed that it be eliminated: "After much discussion the union agreed that we would no longer have the regular rotating cycle of breaks with the assurance that no one would be denied access to toilet facilities if needed, and we took that proposal to the membership and the membership agreed to it, that regular

[157]"Transcript of Hearing" at 8 (quote), 14-15, 17, 18 (Aug. 29).

[158]"Transcript of Hearing" at 29, 32, 37 (quote) (Aug. 29).

[159]"Transcript of Hearing" at 42, 44, 48, 49, 51, 52 (Aug. 29).

[160]"Transcript of Hearing" at 59, 61 (Aug. 29).

cycle of rotating unscheduled breaks is longer in place."[161]

The union, however, did not agree to the new bathroom restriction policy. Kelley testified that if Jim Beam had wanted to discipline workers for "taking unauthorized rest breaks when they did not actually need to use the restroom,"[162] it could have had recourse to three rules set out in the collective bargaining agreement, which had probably been in effect for 50 years Two warranted discharge without notice: "Absence from duty without notice to and permission from Supervisor or Manager" and "Changing working places without orders or prowling around the works away from assigned places." Warranting discharge with notice was a violation of the rule that: "All employees shall, during working hours, devote themselves diligently to their assigned duty."[163]

Returning to his favored theme of OSHA's deprivation of employers of their power to discipline slackers, Griffith engaged Kelley in the following dialog on cross-examination, which Kelley adroitly and playfully turned against her inquisitor:

Q. [Y]ou testified there was a long-standing practice...concerning taking unscheduled breaks. [T]he way it worked, was that employees took typically four unscheduled breaks a day, isn't that true?
A. That probably the way it worked the majority of the time.
Q. [I]f...an employer were to have a void-at-will policy, you heard that term, did you not when you heard Dr. Nygaard testify?
A. Yes
Q. She said she thought a void-at-will policy is the right policy. Secretary of Labor has taken the same position. If an employer has a void-at-will policy, wouldn't you agree that there is no way to monitor whether any given employee is taking unnecessary trips to the restroom?
A. I would say that I have no knowledge of any way to monitor.
Q. So, if in fact someone were doing that, there is no way for Jim Beam to discipline any such employee in the contract...isn't that true?
A. No, that's not true.
Q. Well, they have to know that the person had actually gone unnecessarily in order to be able to do it, isn't that true?
A. Well----.
Q. Isn't that true?
A. Not necessarily.
Q. They could discipline them, even though they did not know whether the person needed to go...is that your testimony?

---

[161]"Transcript of Hearing" at 83, 84 (quote) (Aug. 29).

[162]"Transcript of Hearing" at 85 (Aug. 29) (question by Peter Ford).

[163]"Transcript of Hearing" at 86 (Aug. 29) (Kelley citing Agreement, Article XXX Supplement 'B" Working Rules H, J, and 1).

A. Not necessarily.
Q. That's a very unresponsive answer in my judgment, Ms. Kelley.
A. I could elaborate if you wish.
Q. You go ahead.
Q. Alright. Under the collective bargaining agreement, the employer can and does, too frequently in my opinion, discipline individuals for things they may not necessarily have done, and at that point it's the union's place to grieve, and try to prove that they did not do that.
Q. So, your position would be that the employer would have the ability to speculate that the person, when they said they were going to the restroom really wasn't going...?
A. They do have that, and they do do that."[164]

Picking up Griffith's theme and seeking to undermine the company's claim that the OSHA Memorandum deprived the employer of its power to discipline—which a Beam executive stated was not the purpose or intent but might be the outcome of the OSHA Memorandum[165]—the union's lawyer elicited from Kelley on re-examination that nothing in the collective bargaining agreement would prevent the employer from requiring an employee to undergo medical evaluation if the employer contended that he was using bathroom too often or from announcing its intent to ban smoking in the restroom.[166]

Q. If the company wanted to monitor just what was going on inside the restroom, is there anything in the contract...that would prohibit them from at least attempting to monitor what goes on inside the restroom, try for example, having an attendant in the restrooms?
A. Contractually, I see no reason why that they couldn't have a monitor. I see other legal ramifications for having surveillance cameras or something of that nature, but I think they probably could, and have had monitors in the restroom.[167]

Nor, Kelley testified, would the contract prohibit the company from requiring workers to notify their supervisor when they go on and return from unscheduled bathroom breaks.[168]

At this stage in the proceedings, before calling its star urological witness, Jim Beam moved for summary judgment on the grounds that "the Secretary of Labor failed to prove the existence of [sic] unsafe condition that would give this court jurisdiction to hear this case under the OSHA statute. And also it's failed to prove violation of any OSHA Standard." However the hearing officer found that there were sufficient factual disputes between the parties to preclude granting such a

---

[164]"Transcript of Hearing" at 102-105 (Aug. 29).
[165]Telephone interview with a Jim Beam Brands executive (Nov. 8, 2002).
[166]"Transcript of Hearing" at 106-107 (Aug. 29).
[167]"Transcript of Hearing" at 106-107 (Aug. 29).
[168]"Transcript of Hearing" at 107-108 (Aug. 29).

motion: "I think I have enough testimony that if I accept as true would be sufficient to support Labor's case: therefore I'm going to overrule the motion for summary judgment."[169]

Despite the debacle of his deposition testimony, which had revealed Dr. Stivers to be both dogmatic and ignorant of fundamental aspects of urination, Jim Beam chose to give him an opportunity to rehabilitate himself and to take the risk of exposing him to withering cross-examination. Stivers, who had retired from active practice earlier that year,[170] told Griffith that when Jim Beam had initially contacted him, the company doctor had asked him whether he would talk to management and how much he would charge. A Jim Beam official then came to Stivers' office, talked to him about "what they perceived to be a problem situation," and asked him to come to the plant and "try to educate them as to what was reasonable, what was medically necessary, what was physiologically normal in terms of urination."[171] Stivers then testified that the first paragraph of his (aforementioned) letter to Jim Beam "indicated that we did review what existing OSHA regulations were present, and they were for agricultural workers, and it recommended...3 hours as a time that people should not exceed in terms of having access to a bathroom."[172] To be sure, as already noted,[173] that policy decision was rooted not in medicine, but in farmers' lobbying power, but Stivers nevertheless tried using it to lend credence to his "conclusion that...3 to...6 voidings is what we would call normal in a...24 hour period. [T]hat conclusion is derived from the fact that almost every study of medication that's used to treat conditions that are associated with frequent urination uses the bottom level of acceptance into a study of a medication to alter urination [a]s...8 voids per...24 hours." While one or two would be "very rare," 3-6 was "in the range for [sic] normal person.... And therefore it seemed that the...4 scheduled breaks would be more than adequate to address ninety to ninety-five percent of employees."[174]

Commenting later in the year on this testimony, Dr. Nygaard stated that Stivers' claim about medication studies "is true: the recent very large drug company studies done to evaluate and promote drugs for 'overactive bladder' tend to use >8 voids as an acceptance criterion into the study. This was chosen based on the Larsson study and because it makes way more people look like they have overactive bladder which drives the stock prices up of the new drugs. This in no way relates to the next sentence which, obviously from my depo[sition], I disagree

---

[169]"Transcript of Hearing" at 109-10 (Aug. 29)

[170]"Transcript of Hearing" at 112 (Aug. 29).

[171]"Transcript of Hearing" at 114-15 (Aug. 29).

[172]"Transcript of Hearing" at 118 (Aug. 29).

[173]See above ch. 5.

[174]"Transcript of Hearing" at 118-19 (Aug. 29).

with."[175]

Asked for his professional opinion on Nygaard's testimony (based on Fitz-Gerald at al.'s study) that in her judgment a normal healthy woman could be expected to void as many as 12 times a day, Stivers, without explaining the basis of his answer, replied: "I would suspect that anybody voiding...12 times a day would be making an appointment to see their doctor or see a urologist. That's excessive. That usually interferes with most peoples [sic] ability to carry out their normal activity."[176] In later commentary, Nygaard observed:

> In FitzGerald's study, there were women who considered themselves normal who voided 12 times a day. The problem is that doctors only see the women that present for care, not all the ones out there who may have the same symptom as the women that present for care but who are not bothered by it. For something subjective like urinary frequency, it becomes a medical diagnosis rather than a normal fact of life when the person is bothered and wants treatment. Restrictive work rules can take an asymptomatic person and turn them into someone who is seeking medical treatment—on the one hand, you could say that this makes it a medical diagnosis/condition for that person, but on the other, why should they have to pay $80/month and sustain side effects for a drug that helps them pee less at work?[177]

Stivers testified that since his deposition he had read the abstract of Fitz-Gerald's paper and, without revealing any criticism, opined that "the paper is subject to criticism. The author criticizes the paper, the data herself."[178] (To be sure, on cross-examination, he admitted that since he had not read the full paper, he was not in a position to comment on it.)[179] Again, Nygaard later dismissed Stivers' assertion as unfounded: "Any good researcher tries to explain the data. In no way did FitzGerald criticize it. I think he is talking about the last sentence of the abstract: 'The differences (ie the fact that the frequency was more than [in] Larsson's study) may be due to the differing racial composition of the study sample or they may simply be due to higher fluid intake.' Abstracts have a small word limit, so one can't expand much. Clearly the other thing that may explain the difference is that this study with a larger population and less homogeneous population may be closer to the 'truth.'"[180]

Stivers argued that at an average of 8 ounces per void, 12 voids would amount to 96 ounces or twice as much as a normal person puts out (1250-1500 cc's). Since there is, according to him, no reason to believe that more than a third of that

---

[175]Email from Ingrid Nygaard to Marc Linder (Oct. 9, 2002).

[176]"Transcript of Hearing" at 121 (Aug. 29).

[177]Email from Nygaard to Linder (Oct. 9, 2002).

[178]"Transcript of Hearing" at 122 (Aug. 29).

[179]"Transcript of Hearing" at 140 (quote) (Aug. 29).

[180]Email from Nygaard to Linder (Oct. 9, 2002).

amount (500 cc's) should be occurring during the time of this work shift, and the kidneys make 60 cc's an hour, a normal bladder should be able to store and expel it in an average of two-hour intervals.[181] In contrast, Nygaard stated: "Many people void more often in the morning than later in the day. Younger people don't make as much urine at night as in the daytime, so dividing a 24 hour total up into thirds doesn't work."[182]

Asked directly by Griffith, Stivers allowed as how even three-hour intervals would still be a reasonable time between breaks.[183] And prompted by the company's lawyer, he sought support once again in OSHA's toilet standard for farmworkers, about the contents of which he was manifestly ignorant and confused: "I believe there was a ruling by OSHA. I reviewed this ruling, that agriculture workers did not have to be provided with a bathroom. They just needed to be provided access. So, if they're in a field, and there was no facility available, that...3 hours was the maximum time that they should be without." To Griffith's question as to whether a three-hour period would be "consistent with your knowledge about human physiology," Stivers replied: "It should be adequate, especially considering the fact that most agriculture workers are under a hot sun. They're going to lose more of their fluid by sweating than they are by urine production."[184] Nygaard, once again, later corrected this claim by pointing out that farmworkers are "supposed to be drinking way more to make up for the sweat plus some."[185]

As at his deposition, Stivers seemed almost to fault Jim Beam for being overly lenient: its policy of letting workers go every two hours "is more than generous, more than physiologically adequate, and the accommodation is anybody that needs an accommodation for some reason. It seems to cover all bases. It should be a policy. It should be something that would accommodate ninety-five, ninety-eight percent of the employees."[186] Nevertheless, on cross-examination he testified that in spite of being paid $250 per hour, he had not gone to Beam "with a preconceived notion."[187]

Instead of rehabilitating his expert witness, Griffith committed the tactical error of enabling the union to delegitimate Stivers as a urologist who advocates the view that holding it builds strong bladders and character:

Q. Would you agree that in general...employees should urinate as frequently as necessary?

---

[181]"Transcript of Hearing" at 122, 125 (Aug. 29).

[182]Email from Nygaard to Linder (Oct. 9, 2002).

[183]"Transcript of Hearing" at 125 (Aug. 29).

[184]"Transcript of Hearing" at 126 (Aug. 29). On the field sanitation standard, see above ch. 5; for the text of the field sanitation standard, see below App. III.

[185]Email from Nygaard to Linder (Oct. 9, 2002).

[186]"Transcript of Hearing" at 129 (Aug. 29).

[187]"Transcript of Hearing" at 130, 132 (quote) (Aug. 29).

A. Not as stated that way, no.
Q. What do you disagree with, about that statement?
A. Well, I think you're trying to put some [sic] into my mouth. And if you want me to agree that someone should go to the bathroom when they have the urge to go to the bathroom, yes, under most circumstances we recommend that.
Q. [Under] what circumstances would you not agree with that?
A. Well, there are a lot of situations where it's not possible for someone to go to the bathroom if they have the urge to. Having done lots of long surgeries, I personally have had to suppress the desire to go to the bathroom. And, I didn't go to the bathroom every time that I wanted to. But, I made sure that the surgical field was stable, and then I excused myself. ... You don't just drop everything and run out to go to the bathroom because you feel like you need to. Truck drivers do the same thing. They're coming down the highway and get the urge to urinate, they suppress it for a few minutes, and then it goes away. So, they don't stop at the next place.[188]

To confront this allegation with the real world of work, Nygaard observed: "My patients who are truck drivers (and I've had at least 10) say that they pee as soon as they get the urge—in the truck. There are gizmos sold to drivers for this purpose."[189] Moreover, as a bus drivers union has pointed out, "involuntary urine retention also represents a safety hazard to the riding and driving public since an operator who is forced to 'hold it' is distracted and more likely to make driving mistakes."[190]

Conceding that there may not be anyone to replace a surgeon, the UFCW's lawyer then elicited from Stivers an admission that he in effect had little if any sympathy for the entire purpose of the OSHA Memorandum:

Q. How about in a case of someone working on a production line, where there are other employees who could relieve,...would you agree that if there was someone available to relieve them that it would be advisable for the employee who needed to go [to] the restroom to do so?
A. I don't have an opinion about that. ... My personal opinion is, that is not reasonable for everyone who has a transient urge to go to the bathroom, to immediately go to the bathroom. And that's not the way the world turns around. ... I don't think it's a medical

[188]"Transcript of Hearing" at 136-38 (quote) (Aug. 29).

[189]Email from Nygaard to Linder (Oct. 9, 2002).

[190]Amalgamated Transit Union Division 757, Portland, OR, "Medical Implications of Urine Retention" (document filed in In the Matter of the Informal Conference on Citations Issued to C-Tran, Employer, of Vancouver, Washington, by the Washington State Department of Labor, Health and Safety Division (Jan. 31, 1995). See also In the Matter of the Informal Conference on Citations Issued to C-Tran, Employer, of Vancouver, Washington, by the Washington State Department of Labor, Health and Safety Division, Transcript of Proceedings 33-34 (Jan. 31, 1995) (statement of Susan Stoner, General Counsel, ATU Div. 757).

problem. I don't think it's medically appropriate. It may be socially appropriate, but to say medically appropriate, no. I don't think it falls into the realm of medically appropriate....[191]

The most that Stivers could say in favor of OSHA's rule that workers need prompt access to toilets was: "I think it would be nice." And although he felt constrained to concede that he did not disagree with Triplett's statement that "it's the healthiest environment for a worker to have prompt access to toilet facilities,"[192] he quickly reverted to form, apparently leaving even the UFCW's lawyer Peter Ford incredulous (and Jim Beam management doubtless ruing its decision to afford the "renowned urologist" an encore):

Q. Dr. Stivers, someone who has a gastro-intestinal problem, would you agree that they should, at least ideally, they should have prompt access to the restroom?
...
A. I would hope that anybody that has a GI problem would have access to the problem [sic].
Q. Would you agree that if they didn't have prompt access, someone in diarrhea could very well soil themselves if they had to wait any appreciable length of time?
A. It's a social problem, not a medical problem.
Q. Pardon?
A. That's a social problem, not a medical problem.
Q. Your [sic] not answering my question, by saying that.
A. I'm not answering your question the way you want me to answer the question. The answer to the question is, of course someone with a GI condition should under every hopeful circumstance have access to a bathroom when they need to go there. But if they have an accident, it's a social problem, not a medical problem.[193]

Even if an administrative tribunal or court were ever to accept Stivers' startling opinion that a worker's urinating or defecating in her pants is a social rather than a medical problem, since Congress enacted OSHA "to assure so far as possible every working man and woman in the Nation safe and healthful working conditions,"[194] OSHA's administrators would still be operating within their powers in prohibiting employers from, and fining them for, causing employees to urinate, menstruate, or defecate in their pants on the grounds that it creates health and safety problems (such as chafing, rashes, contact of fecal matter with an open sore, and greater possibility of accidents and injuries attendant on workers' being inattentive

---

[191]"Transcript of Hearing" at 138-39 (Aug. 29).
[192]"Transcript of Hearing" at 155 (Aug. 29).
[193]"Transcript of Hearing" at 159-60 (Aug. 29).
[194]29 USC sect. 651(b).

and preoccupied with their wet, soggy, and smelly pants).[195] One state-plan program has already gone beyond the physical-medical to the social-psychological: in 1998 OSHA in Washington State cited an employer for having failed to provide an employee with access to toilet facilities by virtue of having failed to inform his employees that they were authorized to use the toilet in the owner's home next door to the place of employment, a repair shop, which lacked a toilet because construction had been held up by a lack of a permit. The hazard to which the employer had exposed the worker, who was expected to work eight hours a day "without being informed as to the availability of toilet facilities" for two weeks, was "physical harm or acute embarrassment."[196]

Jeff Conder, the Jim Beam plant manager, testified that from 1996 on tag rotation "was taking so much time of the sanitation person, and that Tech A, the person we were investing all that training in...would relieve the Tech B and the sanitation person would relieve almost everyone else, and these...2 folks were spending...2 to...2 ½ hours a day to relieve."[197] Tag rotation had been in place at least during the 18 years he had worked there, but "we" did not "really start taking notice of it" until the mid-90s "when the frustration levels really started getting high on the lines for the supervisors because of the number of times that employees were away from their lines." Jim Beam, according to Conder, had "hoped and expected" that workers were using tag rotation only to use the restroom, but "that wasn't the case."[198] To be sure, Jo Anne Kelley later commented that this claim "is definitely not true. The co[mpany] started expecting that when he became plant manager and did not like what he saw."[199] Moreover, Kelley stated unequivocally that workers and the union had never claimed that the mini-breaks were used only for urinating: they were a rest-break benefit that the workers had used for whatever they wanted to do.[200]

As Conder related the history of the company's opposition to mini-breaks, management in the mid-90s began working with stewards on the floor: "since we had the [union] executive board's agreement that they'd help us...they'd keep things in check"; "I have to admit we did it minimally"; "we asked them to...make sure that if people are taking a break, that they really need to be taking a break, etc. That really back-fired on us. I think we really created a lot of hard feelings out of

---

[195]As an enforcement matter this position was confirmed in a telephone interview with Elizabeth Slatten, OSHA Austin Area Office Assistant Area Director (Nov. 8, 2002).

[196]Milton Summers dba Summers Garden Center, Insp. No. 115323784 (July 29, 1998). WISHA faxed a copy of the WISHA Violation Report.

[197]"Transcript of Hearing" at 166-67 (quote) (Aug. 29).

[198]"Transcript of Hearing" at 170, 172 (quote) (Aug. 29).

[199]Email from Jo Anne Kelley to Marc Linder (Oct. 9, 2002).

[200]Email from Marc Linder to Jo Anne Kelley (Oct. 28, 2002) and from Jo Anne Kelley to Marc Linder (Oct. 29, 2002).

that. We set the supervisors up by failure because we let the seven...supervisors go out, and with their o[w]n discretion try to determine whether somebody was going to the bathroom or they weren't, or they were smoking. So, it was incredibly difficult for a supervisor to do. And it created hard feelings among the employees because some supervisors were strict: some were not so strict."[201]

Kelley later subjected Conder's version to strong criticism:

> It made for good testimony, but was basically pure bull. Mr. Conder began noticing it in the mid 1990's because that is when he became plant manager. 1992 I believe. I am sure the frustration level became high with the supervisors when...Conder...put more and more demands on them for production and quality. It was an easy out to blame the union employees for taking too many breaks because this is exactly what...Conder...wanted to hear. There were supervisors who complained to Tech A's for production being down or the bottling line not running good. The Tech A's were not going to take the blame for it, so they used the excuse that they were spending all their time relieving. It was too easy for the supervisor to run to Mr. Conder and tell him that production was down because the Tech A's could not do their job because they were relieving Tech B's. It was a matter of everyone trying to cover their own ass and the Tech B's were at the bottom of the heap and got the blame for taking breaks. Nothing was ever said about the Tech A's['] breaks or the fact that they stood around in the shop smoking or hitting on women when they weren't relieving. Maybe Mr Conder truly believes it happened the way he tells it, but it just shows how out of touch he is with reality.[202]

In talks with the union executive board in 2000, it became clear to Conder that "rotation was not the issue with the union." The problem was twofold: how to make efficiencies and manage the business and how "to figure out what needs the employees actually had to go to the restroom. And we certainly weren't the people to do that." Beam made a survey of its industry and other industries and found "an incredibly broad range of break practices, from very liberal to very strict. ... And we were really trying to get some kind of strategy at this point...but again we knew very little about the medical side of it. ... So, that's [when] we decided to talk to Dr. Stivers. But, we didn't have all the experience that was need[ed] to make a determination on what was really abusive or not."[203]

As far as the actual operation of the policy was concerned, Conder contended that "if someone claims to have" a temporary condition like diarrhea, "what happens is they identify to their supervisor that they're sick that day. And they continue to go to restroom as needed. That documentation floats up to Human Resources, and has that documentation that those employees...had diarrhea or whatever. So that...there's just no progression on discipline on that. It's just so noted

---

[201]"Transcript of Hearing" at 174 (quote) (Aug. 29).

[202]Email from Jo Anne Kelley to Marc Linder (Oct. 28, 2002).

[203]"Transcript of Hearing" at 177-78, 185 (Aug. 29).

and excused for that particular day. So, there's no progression if someone says 'I've got diarrhea today,' and so I go to the bathroom five...times."[204] Conder also praised the employees on whom Beam had conferred medical accommodations: "I think we owe that group a big compliment" because 75 percent of them were taking just 0 or 1 extra unscheduled break."[205]

Inadvertently shedding light on the real-world impediments to the OSHA Memorandum's ability to stop employers from interfering with employees' need to void, Conder tried to dispel the notion presented by the union's lawyer that the August 2001 collective bargaining agreement had put an end to the mini-breaks that workers had been taking for half a century

Q. [T]he tag rotation system...was in effect for a long time...until company and union agreed to end it last August....
A. That practice...did not come to an end just because we said that it would. I mean, you said it ended in August, well it ended when we implemented the policy and controlled it.
Q. Well, in fact, it ended when the union and company entered into a letter of agreement dated August 16th, 2001...did it not.
A. How does a practice end with a signed piece of paper? No it didn't: it did not end at that time.[206]

The plant manager thus admitted that without the imposition of coercive disciplinary measures—which were in no way mentioned or permitted by the agreement—the company did not believe that workers could be weaned of the mini-breaks they had been taking for decades. As applied to the OSHA Memorandum, the analogous lesson would presumably be: OSHA should not have assumed that, without the accompanying imposition of coercive financial penalties, a piece of paper would stop employers from continuing their long-term policy of prohibiting workers from going to the bathroom outside of scheduled breaks

In her cross-examination of the plant manager, Kentucky OSHA's attorney was able, by focusing on the employer's diarrhea doctrine, to uncover the deep self-contradiction inherent in Beam's bathroom policy. To begin with, Triplett prompted Conder to state that five or six employees had fallen under that doctrine, by which "[t]hey get to counseling and they get to go, but it doesn't count as any progression from that point." In response to Triplett's question as to whether it "[w]ould...surprise you to know that a lot of the employees weren't even aware of that policy?" Conder replied: "You know what,...that's not good."[207] She then raised with him the crucial issue underlying the entire structure and purpose of

[204]"Transcript of Hearing" at 191-92 (Aug. 29).

[205]"Transcript of Hearing" at 195-96 (Aug. 29).

[206]"Transcript of Hearing" at 201 (Aug. 29)

[207]"Transcript of Hearing" at 211 (Aug. 29)

management restriction of worker autonomy at the workplace: "[I]f a person says they have diarrhea, how do you make sure that they're actually using the facilities for diarrhea?" to which the plant manager surprisingly replied. "Trust them."[208] With this notion of trust as a baseline operating principle, the OSHA lawyer then asked Conder the logically compelling follow-up question: "Well, wouldn't it be the same if someone said they had to go to the toilet [to urinate]?" This time, however, instead of giving a straightforward and pithy answer, Conder descended into evasiveness: "Again, we consulted, we've consulted to tell us what is reasonable. That's not my opinion, but what others have said. And, if they're going more than that, then we question...."[209] Triplett then made one last effort to achieve clarity on the matter:

Q. So...what you're saying is, now...is that when someone says they really have to use the toilet, as long as it's in the B-word, they can use it?
A. They [sic] think you are stretching that there. We said if someone had a condition that they're identifying that they're sick, and they're not just lying to go because they've got [sic] go for whatever reasons on their mind, if they're identifying to us that they're sick, then we're going to take their word for it, we're going to document it, and we're going to let them go, and it's not going to count against them.
Q. But how do you know they're really sick? Isn't that the same as when you testified earlier, you couldn't tell when they [sic] really using the toilet to void?
A. I guess I don't.[210]

With Jim Beam's final witness, the Clermont (and Boston) plant human resources director, John (Jack) Allen, Jr., Triplett returned to the issue of diarrhea, this time branching out in a different direction and ultimately provoking him into uttering the most poignant non sequitur of the entire proceedings. After Allen had stated that the company had accommodated all cases of diarrhea "[o]r any temporary illness that appears to be throwing the individual out of sync," she asked:

Q. So does a female that has an onset of menses, has heavy flow, are they accommodated?
A: [W]e have not yet had such a case arise where that has been identified as the reason....
Q. Would that be accommodated?
A. If they were to identify that that was the particular problem, we would certainly consider it.
Q. So when a woman starts her period, she has to go to personnel to get approval?
A. No. ...
Q. So, what are you saying if a woman has a—
A. I'm saying that if an individual were to have a particularly difficult time where they

---

[208]"Transcript of Hearing" at 212-13 (Aug. 29)
[209]"Transcript of Hearing" at 213 (Aug. 29).
[210]"Transcript of Hearing" at 215-16 (Aug. 29).

believe that they have additional needs, and they are temporary, that's something that we would entertain. But...it isn't an automatic situation, it would be a case by case basis. ...
Q. If a woman has a menstrual accident because they didn't realize they had started, but they knew they had to go to the restroom fast. Now, I understand as a male, you don't understand just what that feels like. But you have to make that very rapid trip so as to prevent embarrassment, and so you get your counseling and you're saying if you explain to the gentleman in personnel...during the counseling, they'll be accommodated and that won't lead to the six...in a series of a corrective action. [S]he has to go to her male supervisor, it's a bit personal, don't you think?
A. Yes, it is. There's no question about it, but if I were at the last point in my disciplinary action...I'm sure I can communicate that in a way that even a man would understand without having to get terribly embarrassed. These are adults that are working here.[211]

After having imposed a regime that all its victims regarded as infantilizing, its author found it convenient to remind his charges that they were, after all, adults. Recalling her reaction to this testimony as she sat at the hearing, Kelley commented: "I thought I would throw up with disgust at this man. ... This man actually thought that diarrhea would qualify but a woman would have to have an accident to qualify."[212]

In fact, Allen's assurances to the contrary notwithstanding, this particular cause of workplace embarrassment is well known in industrial relations. One arbitrator, dealing with a employer's rule that required (mostly female) employees to register with their foremen when leaving for and returning from relief so that management could "exercise proper control of personal time" to avoid the "disastrous" consequences of employees' "taking too much time off...particularly in view of the wage increases granted during the recent contract negotiations,"[213] spoke of

the virtually inevitable affront to the sensitivities of these workers which administration of the rule entails. Certainly it cannot be presumed that in seeking employment in the manufacture of gaskets and similar products a woman worker has submitted herself to rules which may require disclosure to her foreman of information normally reserved for discussion with her physician. If she refuses the detailed "explanation" demanded of her, does she then become subject to discipline for insubordination?[214]

---

[211]"Transcript of Hearing" at 242, 245-46 (Aug. 29).

[212]Email from Kelley to Linder (Oct. 8, 2002).

[213]Detroit Gasket & Mfg. Co., 27 Labor Arbitration Reports (BNA) 717, 718, 719 (1956).

[214]Detroit Gasket & Mfg. Co., 27 Labor Arbitration Reports (BNA) at 721.

Aftermath and Lessons

A week after the hearings, as a result of the unprecedented adverse national and international multi-media publicity—including and especially a joke about the similarity between the color of Jim Beam bourbon and urine on the Saturday night Jay Leno television program[215]—and "[a]fter an outcry from employees and labor officials,"[216] expertly promoted by the UFCW, whose website urged people to email Jim Beam, the company, deciding not to carry on this debate in the media, abandoned its appeal despite its alleged belief that the hearing officer would have decided in its favor.[217] The company announced that:

> effective immediately, it has voluntarily discontinued a former policy that limits breaks for bottling-line workers at its Clermont, Ky., distillery.
>
> "Our former policy was intended only to manage excessive breaks, and we believe it provided appropriate flexibility," said Rich Reese, CEO. "However, we've listened to the concerns of our employees and have changed our policy. We will work with the local Union to find a mutually acceptable solution for managing breaks on the bottling-line."[218]

In a postmortem lamentation about the achievement of "unlimited bathroom breaks," not based on any knowledge of the case, an employer-side labor law firm empathized with Jim Beam: "Ever see a co-worker take a newspaper into the stall next to you, only to emerge an hour later having read it cover to cover? ... Imagine how much money this costs an employer. Perhaps this type of behavior is what Jim Beam Brands Company hoped to avoid by establishing a 'bathroom break' policy."[219]

The president of Local 111-D, surmising that "'reason has prevailed and they have come to their senses,'" expressed the hope that "'now we can get back to the business of making...bourbon.'"[220] As of early October, the company had still

---

[215]"Jay Leno Joke, Along with Public Outcry, Helps Bring Bathroom Pass to Jim Beam Workers (9/02)," on http://www.ufcw.org/worker/internal.cfm?subsection_id=189&internal_id=799; telephone interview with Jackie Nowell, director of occupational safety and health, UFCW, Washington, D.C. (Oct. 17, 2002). A Jim Beam executive confirmed the role of the Leno joke in the decision to abandon the appeal. Telephone interview (Nov. 8, 2002).

[216]"Jim Beam Distillery Ends Restroom Limiting Policy," *Chattanooga Times*, Sept. 7, 2001, at C5 (Lexis).

[217]Telephone interview with a Jim Beam Brands executive (Nov. 8, 2002).

[218]"Jim Beam Issues Statement Regarding Reversal of Break Policy for Bottling-Line Workers," PR Newswire, Sept. 6, 2002 (Lexis).

[219]"Unlimited Bathroom Breaks Touted As Victory" (Sept. 10, 2002), on http://www.kollmanlaw.com/quickseptember2002.shtml.

[220]Goetz David, "Jim Beam Ends Limit on Bathroom Breaks," (Louisville) *Courier-*

refused to meet with the union. Indeed, despite its press release, it had also still not announced to the workers that it had rescinded its year-old policy, although it was apparently no longer enforcing it and the union instructed the workers to ignore it and to use the bathroom when necessary, but to inform the supervisor before they went. Reaching a new agreement was delayed over the issue of Beam's removing from its files the disciplinary actions meted out for having violated the policy that it publicly rescinded.[221]

Although the denials of toilet access at the Excel Corporation hog slaughter plant in Ottumwa, Iowa were far more injurious to workers' dignity than those inflicted by Jim Beam (as suggested by OSHA's imposition of a $36,000 penalty on Excel), the whole world's heart (or bladder) seemed to go out to the bourbon workers, whereas few outside of Ottumwa ever learned of the events there. Such are the vagaries of the media and how they shape popular consciousness.

By this point it should have become obvious that both more and less was at stake than voiding rights at Jim Beam and that the media's involvement was neither fortuitous nor without consequences. Since there was never any pretense by workers or the union that employees were using tag rotation solely to void, the question arises as to why Jim Beam went to the effort and expense of monitoring bathroom breaks, generating spreadsheets, and hiring a urologist to try to prove that human beings simply do not urinate as much as the logs indicated. Formulated differently: Why was this dispute over how much time would be spent working/not working on the bottling line framed in terms of voiding when both sides knew that voiding took up only a part of the disputed time? Why did Jim Beam management not simply demand that the workers work more and spend less time off the bottling line not working? This question also occurred to Karen Triplett, the Kentucky Labor Cabinet counsel, but she, too, was unable to gain any insight into the employer's motivation for implementing a strategy that attacked only part of the problem it perceived, but, perversely, in such a way that the company inadvertently touched off a national and international groundswell of public support for workers who had been deprived of their right to urinate—support that with certainty would not have been forthcoming if Jim Beam had, instead, focused on the 6-7 minute hourly rest breaks.[222]

Ironically, Jim Beam's management and attorney could have benefited from pondering one of the fates of *Void Where Prohibited*—the disproportionate attention that has been paid to its treatment of toilet breaks, which takes up only two of its nine chapters. Although toilet breaks are merely a subset of rest breaks,

---

Journal, Sept. 7, 2002, at 1B (Lexis).

[221]Telephone interview with Jo Anne Kelley, Bardstown, KY (Oct. 4, 2002); Email from Peter Ford to Marc Linder (Sept. 30, 2002).

[222]Telephone interview with Karen Triplett, General Counsel's Office, Kentucky Labor Cabinet, Frankfort (Oct. 11, 2002).

few reviewers focused on this more general category, and its discussion of rest breaks has had incomparably less impact on public policy and opinion, which can be outraged by stories of adults forced to defecate or urinate in their pants, but would have remained indifferent to employers' oppressive practices depriving workers of respites during the workday.[223]

In the absence of privileged access to Jim Beam's management, no empirically verifiable answer is available. Nevertheless, when asked, one company executive sought to undermine the legitimacy of the question by challenging the claim that the union had never made a pretense that the mini-breaks were used solely for voiding. He believed that he had refuted it by mentioning that Kelley had once told him that if the company eliminated the mini-breaks, employees would not be able to go to the bathroom when they need to. The author pointed out that these two positions were not irreconcilable since some workers had in fact used some proportion of their tag-rotation breaks to void, but the official insisted that the union had used (or, in a word suggested to him by the author, "hijacked") the OSHA Memorandum for political purposes to preserve their non-bathroom breaks, although he hastened to exculpate OSHA from any complicity.[224]

The UFCW's attorney was himself puzzled and could only "speculate":

> Apparently Beam management had already decided to implement the policy before the parties entered into the side letter, so despite the fact that the side letter ended the tag rotation practice, Beam went full speed ahead with its policy. ... It really doesn't make sense that Beam adopted its policy rather than take other measures to ensure that workers weren't wasting time on unscheduled restroom breaks - e.g. they could have had restroom attendants, and they could have enforced more aggressively their longstanding work rules that prohibit loitering, etc..... I think this is a case of...hardheaded management with too much power...refusing to back down once challenged, until they were embarrassed into doing so by lots of media scrutiny.[225]

The secretary-treasurer of UFCW Local 227, which has 20,000 members in Kentucky and provided strong support to Local 111-D during the dispute, offered

---

[223]The chapter on regulation of rest breaks outside of the United States was, however, reprinted in a scholarly journal: Marc Linder and Ingrid Nygaard, "Rest in the Rest of the World," *Comparative Labor Law and Policy Journal* 20:105-24 (1998) (reprint of ch. 8). As a result of the editor's negligence, readers were not informed that what formally appeared to be an article without any context was in reality a chapter from a published book. Letter from Matthew Finkin to Frances Benson and Marc Linder (Oct. 15, 1998) (promising to reprint the chapter "with a suitable headnote"); email from Matthew Finkin to Amy Winger, permissions manager, Cornell U. Press (Oct. 28, 1998) ("We'll prepare a headnote giving attribution").

[224]Telephone interview with a Jim Beam Brands executive (Nov. 8, 2002).

[225]Email from Peter Ford to Marc Linder (Oct. 1, 2002).

a complementary explanation. Gary Best conjectured that some consultant had just sold Jim Beam management on the policy of a blanket ban on all unscheduled bathroom breaks as a solution to the problem of worker abuse of tag rotation instead of disciplining the violators. Since management had been used to dealing with the weak distillery workers union that had been in place for decades before the merger with the UFCW a few years earlier, the company, he speculated, thought it could get away with it. But, fortunately for the workers, the UFCW lucked into national media attention, which is always reinforced by the public's strong interest in any events that combine food or drink production with voiding or toilets.[226]

Without knowing anything about the situation other than what a reporter told him and what he had read in news accounts, a consultant with a doctorate in psychology at a "absence management" consulting firm was easily able to intuit that: "'The bathroom break rule is an example of what we see too often—where managers attempt to control all aspects of employee behavior. ... What went wrong was trying to solve a problem by attempting to have control of employees' bladders....'" But when the consultant sought to explain what an employer in Beam's situation should have done, he instinctively recommended an approach that failed to engage the workers' purposes—"'finding out why employees weren't more invested in getting the product out the door nor more committed to getting the job done.'"[227] Because the whole point is to "'engage...not punish your employees,'" Beam could not understand that its employees "feel disconnected. It's a control battle between the workers and the management instead of the focus on customers.'"[228] What the absence manager failed to understand was that the workers at Jim Beam apparently wanted less, not more, connection with customers, management, and production work. Instead, they wanted to regain their decades-long vested right to six- or seven-minute mini-rest periods every other hour. When supplied with these further details about Beam and especially about its position that the OSHA Memorandum deprived it of its managerial powers to discipline its employees, the consultant, who after 17 years of observing management was no longer surprised by its irrational actions, acknowledged that dealing with a practice of 50 years' standing was not easy, but nevertheless insisted that companies have to deal with their human capital as they do with their other kinds of capital.[229]

---

[226]Telephone interview with Gary Best, Louisville, KY (Oct. 2, 2002).

[227]Carol Kleiman, "Restroom Off Limits? Workers Merit a Break," *Chicago Tribune*, Nov. 10, 2002, Business sect. at 5 (Lexis) (quoting Mike Scofield of Nucleus Solutions).

[228]MaryBeth Matzek, "Employees Say Those Are the Breaks," (Appleton, WI) *Post-Crescent*, Sept. 17, 2002, at 8D (Lexis) (quoting Scofield).

[229]Telephone interview with Michael Scofield, Nucleus Solutions (Nov. 19, 2002).

In an undated letter sent in October 2002, Jim Beam finally informed the UFCW that it had "rescinded the Bottling Department Restroom Break Policy and withdr[a]w[n] its Notice of Contest to the OSHA citation. This has paved the way to achieving a settlement of this matter. In addition to withdrawing the break policy, be advised that we will be expunging from the employees' personnel files the disciplinary actions and related counseling which arose from the enforcement of this policy."[230] Nevertheless, the Jim Beam story still lacks a happy ending. As of the autumn of 2002, Jo Anne Kelley reported, "the company still keeps a spreadsheet on employees who take bathroom breaks. The supervisor keeps track on a daily basis and turns a report in to H[uman]R[esources] each day." Consequently, it remained her "belief that the co[mpany] is still trying to build a case to defend their position and try again at some point."[231] In November it confirmed that it was monitoring workers' bathroom use and that workers were aware of that activity. An official stated that it "will manage excessive bathroom use on a case by case basis," although no fixed number of visits or interval between visits counted as "excessive": the numbers are vague and neither workers nor management knows what they are.[232]

In the pre-Memorandum years some labor arbitrators forthrightly condemned precisely this kind of "onerous and humiliating timekeeping procedure," refusing to approve it, even where the employer alleged increased production, "at the cost to employees in loss of dignity and embarrassment."[233]

Despite the international media-mediated uproar over Jim Beam's policy, other companies continue to impose similar restrictions on their employees. For example, according to the UFCW, a specialty meats food processor in Buffalo issued a break policy on August 14, 2002, which, in addition to a 15-minute coffee break at 9:30 a.m. and a 30-minute lunch break at noon, limited workers to one 5-minute break between 7:00 a.m. and 9:30 a.m., a second between 9:45 a.m. and noon, and a last one between 12:30 p.m. and 3:30 p.m. With the notice itself declared to be a "verbal warning to all employees!," the employer announced: "If you have left your work station without permission, the following actions will be taken by management: 1st Offense - Written warning, 2nd Offense

---

[230]Letter from J. C. Allen, Jr., Director, Human Resources, Jim Beam Brands Co., to George Orlando, Int'l Vice President, UFCW (undated [before October 18, 2002]).

[231]Email from Jo Anne Kelley to Marc Linder (Oct. 5, 2002).

[232]Telephone interview with Jim Beam Brands executive (Nov. 8, 2002)..

[233]Schmidt Cabinet Co., 75 Labor Arbitration Reports (BNA) 397, 400 (1980). Cf. Elgin Instrument Co., 37 Labor Arbitration Reports (BNA) 1064, 1068 (not terming it unreasonable, but nevertheless doubting its soundness in retaining employee morale); Gremar Mfg. Co., 46 Labor Arbitration Reports (BNA) 215, 220 (1965) (ordering employer to discontinue the practice, but permitting its reinstitution with safeguards).

- 3 Day Suspension without pay, 3rd Offense - Termination of Employment."[234]

Unions and workers have at times engaged in creative guerrilla warfare against such managerial incursions into their ability to respond to calls of nature. In an innovative application of a work-to-rule tactic, UNITE reacted to a similar rule prohibiting bathroom visits outside of scheduled breaks at a textile plant in North Carolina by arranging for all the workers to hold their urine until break-time; because the long line far exceeded the capacity of the bathroom, by the time the 15-minute break had ended, many workers were still waiting in line and continued to stand there until their turn came. Management quickly rescinded the rule.[235]

One ironic product of the publicity surrounding the Jim Beam case is the potential for the proliferation of opportunistic employer responses. For example, the Manager of the Employment Standards program of the Washington State Department of Labor and Industries reported that at the time of pervasive media interest in Jim Beam he happened to be meeting with employers—many of whom give their workers only scheduled breaks—who asked him whether, if they had to start letting employees go to the bathroom, they could deduct that time from the 10-minute rest break that Washington State law requires employers to give employees every four hours.[236] The agency had still not yet issued an opinion letter on the question by the beginning of 2003, but it was, despite predictable employer opposition, preparing a ruling that the practice was unlawful.[237]

Coda: How the OSHA Memorandum Has Constituted Progress Compared with Arbitration and Litigation

The restrictive toilet-access policy imposed by Jim Beam was by no means unprecedented. Other unionized employers, also alleging that "'too many employees [were] absent from their work positions too long and too often for personal reasons, toilet, smoking, gossiping, just plain resting, etc.,'" had, years

---

[234]Battistoni [Bison Products Co.], Break Policy (Aug. 14, 2002) (copy faxed by UFCW).

[235]Telephone interview with Willie Jones, State Director for UNITE, Montgomery, AL (Nov. 25 and Dec. 16, 2002), according to whom the events took place at Cone Mills in North Carolina around 1984-85.

[236]Wash. Adm. Code sect 296-126-092(4) (2002).

[237]Telephone interview with Richard Ervin, Employment Standards Manager, Washington Dept. of Labor and Industry, Olympia, WA (Oct. 21 and Nov. 1, 2002, and Jan. 9, 2003).

before the Memorandum was issued, implemented rules prohibiting workers from leaving their work stations, except during scheduled breaks, without supervisors' permission. In 1967, Jones Dairy Farm, a sausage, bacon, and ham processing firm in Wisconsin, after years of trying to prevent workers from "remaining unnecessarily long from their work positions for personal reasons,"[238] and one year after Local P-1236 of the Amalgamated Meat Cutters and Butcher Workmen of North America (a predecessor of the UFCW) had begun representing the employees there,[239] informed them that "[s]upervisors were to grant such permission only for emergencies" and that "requirements for bodily relief on a regularly scheduled basis would not be considered an emergency. The requirement for permission was strictly enforced." After this policy and an "honor system" vesting discretion in the workers had both failed to satisfy the employer, the parties agreed on a new policy during the 1970 bargaining negotiations, stating that the then current emergency relief policy was unreasonable and that emergency relief periods had been "abused." Under the new rule, beyond a 15-minute rest period during the first three hours of work, "[a]dditional relief periods will only be granted on an emergency basis. The past practice of continual rotation of additional relief periods will be discontinued. Absences away from an employee's work station on a regular basis and/or pattern is [sic] not an emergency. ... In all cases, an employee must receive authorization before leaving his work station...." The only specificity that the parties injected into the policy was the guideline that the emergency relief periods "shall not be used for the purpose of loitering in the rest rooms" or "granted for the purpose of smoking." Without further definition, employees with "special problems" were to be given "consideration."[240]

Despite having agreed to the policy, by 1972 the union realized that it disagreed with what it regarded as the company's "much too restrictive and...unreasonable" "interpretation of an emergency as being a rare and non-recurring incident...." There do not appear to have been mass disciplinary actions, but that year Jones Dairy Farm sent one employee a reprimand letter "because of your excessive amount of 'emergencys [sic]' regarding going to the rest room during working hours." Informing the worker that if there was a "temporary physical reason" for his rest room visits during unscheduled breaks, he should get a doctor's statement verifying his condition, the employer warned him that further violation would subject him to suspension. Because he "again went to the rest room during working hours" three and a half months later, the company suspended him

---

[238]Jones Dairy Farm, 72-2 Labor Arbitration Awards (CCH) ¶8639 at 5254 (1973).

[239]Jones Dairy Farm, 245 NLRB 1109, 1113 (1979). UFCW Local 538 currently represents the workers at Jones.

[240]Jones Dairy Farm, 72-2 Labor Arbitration Awards at 5254-55.

for three days, provoking a grievance.[241]

During the 27 months since the policy had gone into effect, the grievant had gone to the bathroom on emergency relief nine times, each time having requested and been given permission by his foreman. The employer took the position that the fact that each of the employee's emergencies had taken place within an hour and a half of starting work, in combination with "the odor of liquor on the grievant's breath indicated...a pattern that was caused by the grievants [sic] conduct prior to reporting for work. ... Specifically, we are referring politely to [D's] liquid intake." The company argued that "[s]elf-imposed problems due to an unwillingness to conduct one' self [sic] during off-work hours in such a way as to prepare for work in accordance with policies, work rules, and schedule, does not constitute an emergency or special consideration within the intent of the agreement." In sharp contrast, the union insisted that "what an employee does away from work is 'none of the Company's business.'" Moreover, it argued that the company's interpretation was "unworkable" and "could result in employees being disciplined and discharged for something that is beyond the reasonable control of the employee" (though it unclear whether the union alleged that circumstance in this case).[242]

The arbitrator did not grapple with any of the principled issues raised by these severe restrictions imposed on workers' freedom to void. Declaring himself without authority to engage in any further definition of emergency relief periods, he confined himself to applying the policy to the facts of the case. Without explanation, but presuming that the employer had not even tried to determine whether any of the worker's bathroom visits were attributable to legitimate emergencies, he merely found that nine relief periods over 27 months failed to form a sufficient basis for concluding that "a pattern exists independent of and beyond the control of the employee."[243]

Without any doubt, this policy, which in its vagueness could be and was interpreted by the employer far more strictly than the Jim Beam one-a-day unscheduled void regime, would, on its face, have been a violation of OSHA's toilet standard as subject to the 1998 Memorandum. This narrowly legalistic grievance-arbitration procedure on behalf of a single worker—whose case presumably offered neither typical nor especially compelling facts—is much inferior as a means of combating employer overreaching, which the union was in part responsible for facilitating by having agreed to the "emergency" system in the first place, to an OSHA complaint.

This claim will now be tested, against the background of Jo Anne Kelley's

---

[241]Jones Dairy Farm, 72-2 Labor Arbitration Awards at 5255.

[242]Jones Dairy Farm, 72-2 Labor Arbitration Awards at 5256-57.

[243]Jones Dairy Farm, 72-2 Labor Arbitration Awards at 5257.

statement that she had more faith in OSHA than in arbitrators, with reference to a more recent arbitration of a dispute—also involving a union that became part of the UFCW—which bears an uncanny resemblance to the Jim Beam case.

Cagle's Poultry and Egg Company, which is headquartered in Atlanta, owned five chicken processing plants in the late 1970s, one of which was located in Macon, Georgia, where Local 315 of the Retail, Wholesale and Department Store Union (RWDSU) represented about 200 workers pursuant to a collective bargaining agreement, which ran from October 1977 to October 1980.[244] The RWDSU also represented workers at two of the other plants in Georgia, at one of which the National Labor Relations Board found that Cagle's in the latter part of the 1970s had committed numerous unfair labor practices after the workers there had selected the union as their exclusive bargaining representative.[245] Almost all the 150 production-line workers at the Macon plant were black women.[246] Although one of the Cagle's lawyers later called them a particularly "militant" group of workers, applying the same epithet to the civil rights movement of the period in Macon,[247] the workers' own lawyer dismissed those characterizations, but did stress that the black women were very poor and had been deeply dissatisfied with the discriminatory treatment to which they were subjected at the plant.[248]

Some sense of the rancor that must have prompted and, in turn, been intensified by the dispute was conveyed by the remarks of Cagle's chief attorney almost a quarter-century later. A retired founding and name partner in a prominent Atlanta law firm and former visiting professor at Emory Law School, he retained a vivid memory of the case, referring to some of the workers as "troublemakers" who had claimed that the company had not furnished them relief and had been watching them in the bathroom. He also allowed as how a couple of the women "were really bitches," and felt compelled to mention the "bad attitude with the brothers," who believed that "everything the white man did was to put us down." Finally, he observed that the union, being headquartered in New York City, was "communist," but, at the same time, "weak."[249]

The toilet access dispute was purportedly triggered by the company's dis-

---

[244]Cagle's Poultry and Egg Co., 73 Labor Arbitration Reports (BNA) 34, 35 (1979).

[245]Cagle's Inc. and Southeast Council, Retail, Wholesale & Department Store Union, 234 NLRB 1148 (1978).

[246]Cagle's Poultry and Egg Co., 73 Labor Arbitration Reports at 37, 38.

[247]Telephone interview with Victor Cavanaugh, attorney, Elarbee, Thompson, Sapp & Wilson, Atlanta (Jan. 23, 2003).

[248]Telephone interview with Linda Mabry, Atlanta (Jan. 28, 2003). Reverend Henry Ficklin, who had been the head of the Southern Christian Leadership Conference in 1979, also denied that the civil rights movement in Macon had been militant. Telephone interview with Ficklin, Macon (Feb. 4, 2003).

[249]Telephone interview with Warner Currie, Atlanta (Jan. 28, 2003).

covery in 1978 that the Macon plant was operating less efficiently than some of the other Cagle's plants. The company alleged that one reason for this inefficiency was that "frequently, the production line was slowed down and an excessive number of relief employees had to be provided because employees were visiting the rest room too frequently and staying too long. On occasions thirteen...or fifteen...employees were found in the rest room at the same time while the line was running. There were twelve...relief employees manning the line. The Company estimated that it should not be necessary to provide more than four...or possibly even two...relief workers."[250] A quarter-century later, Cagle's CEO, who had no difficulty recalling the dispute, characterized the union back then as "reasonably strong," observing that the workers had just walked off the line "at-will" to go to the toilet. Moreover, he asserted that they had gone to the bathroom not to void, but to smoke, which was permitted at the time in the restrooms.[251] The lead plaintiff in the lawsuit that arose out of the dispute, Margie Bagley, and one of the other named plaintiffs, Alma Oliver, both stated in refutation that many if not most of the women, including themselves, did not even smoke.[252]

In order to determine whether Plant Rule No. 39, "Abuse of emergency rest room privilege," was being violated, Cagle's decided to do a three-week time study, beginning on January 8, 1979. Remarkably, the company assigned the monitoring of the bathroom near the production line to the chief union steward and assistant steward, whose job it was to record the time each employee entered and left. (They did not want to perform this task, but "were not allowed to refuse"; nevertheless, they also did not file a grievance.) The results, according to the company, revealed that in the aggregate workers spent 188.38 hours per week in the bathroom on unscheduled breaks (including 2.25 minutes for the trip to and from the bathroom), which, at an average labor cost of $3.78 per hour, amounted to a "cost" of $712.08 per week; adding on the weekly wages ($1,209.60) of eight excess relief workers, the company calculated "the total cost for the unscheduled rest room use" at $1,921.68 per week[253] or $99,927.36 annually. Cagle's also reported that practically all 148 of the employees in seven departments made emergency bathroom visits, averaging more than two per day, each one lasting on

[250]Cagle's Poultry and Egg Co., 73 Labor Arbitration Reports at 39.

[251]Telephone interview with Doug Cagle, Atlanta (Jan. 28, 2003). When told of this explanation, the company's attorney at the time of the dispute responded that it made sense. Telephone interview with Victor Cavanaugh, Atlanta (Jan. 30, 2003).

[252]Telephone interviews with Margie Bagley and Alma Oliver, Macon (Jan. 29, 2003). Reverend Henry Ficklin, a civil rights activist who aided the workers in 1979 confirmed that most of the women were churchgoers and did not smoke. Telephone interview with Ficklin.

[253]Cagle's Poultry and Egg Co., 73 Labor Arbitration Reports at 39.

average 7.34 minutes including travel time.[254] The company and union met in February to discuss the issue, but management did not respond to a request for the names of the workers "who might be abusing the privilege" so that the union could "'try to correct the problem,'" and on March 5 the company issued a memorandum laying out the guidelines that supervisors would be following in order to comply with Rule No. 39.[255] The spirit underlying the new regime was manifestly driven by the world view that "the price of tyranny is eternal vigilance."[256]

> Any employee will be allowed to visit the bathroom eight...times in any four...week period with a five...trip limit in any two...week period of the four...weeks. Maximum time from job is five...minutes.
>
> 1st offense will be a verbal warning.
> 2nd offense will be a written warning.
> 3rd offense will be three...days off without pay.
> 4th offense will be discharge.
>
> Anyone taking medication which requires [the] employee to visit the bathroom more than the above stated limits will be required to furnish the Company with a doctor's statement of said medication. The Company will then give the employee three...weeks of excessive bathroom privileges in order to correct their problem. If after three...weeks the problem is not corrected, the said employee will be given a Leave of Absence to get the problem corrected.
>
> Any employee who has been on continuous medication prior to 1-1-79 which requires additional bathroom privileges will be handled on an individual basis by their supervisor.
>
> ...
>
> Each employee will be issued a Bathroom or "B" Card to be placed in the racks next to the Time Clock outside of the Bathrooms. It will be the responsibility of each employee to punch their card before they enter the Bathroom and again as they exit. Any employee failing to do so will be severely reprimanded. First offenders will be given a disciplinary layoff. Second offenders will be terminated from their employment with Cagle's. Any deviation from this will be done so in good faith by the employee's immediate supervisor.[257]

The Cagle's regime was far more restrictive than Jim Beam's, limiting unscheduled breaks to an average of two per week, expressly capping the length of the visit, requiring only four steps to discharge, putting the burden on workers to

---

[254]Cagle's Poultry and Egg Co., 73 Labor Arbitration Reports at 42, 43.

[255]Cagle's Poultry and Egg Co., 73 Labor Arbitration Reports at 39.

[256]Richard Dunn, *Sugar and Slaves: The Rise of the Planter Class in the English West Indies, 1624-1713*, at 246 (1973 [1972]).

[257]Cagle's Poultry and Egg Co., 73 Labor Arbitration Reports at 36.

overcome their medical problems, and adding a bathroom time clock as a means intimidation. But it was similar to the rule in effect at that time at the Sanderson Farms poultry plant in Laurel, Mississippi, which granted workers only three weekly toilet breaks.[258] The huge change brought about by the new system was highlighted by the fact that, according to Annie Kitchens, the shop steward who had worked at the Macon plant since the day it had opened 21 years earlier, until 1979 no one had ever been "discharged for abusing the bathroom privilege."[259]

However, in order to appreciate fully the rule's oppressiveness, it is necessary to consider that the company had been systematically depriving the workers of one of their two contractually guaranteed breaks. In addition to an unpaid one-hour lunch period, according to Article X, Paragraph 31: "One paid ten (10) minute rest period shall be provided for all employees during the first four...hours of normal shift work and one paid ten...minute rest period during the second four...hours of normal shift work and a fifteen...minute paid rest period after ten...hours total of normal shift work."[260] Nevertheless, as an arbitrator found later in 1979: "The Company has interpreted Article X, Paragraph 31 to mean that the second rest period would only be provided if and when employees worked more than four...hours in the afternoon. Thus, no rest period was provided if employees worked four...hours in the afternoon. Thus, no rest period was provided if employees worked four... hours or less in the afternoon. The Union has never objected to this practice."[261] Consequently, on days when a shift lasted no more than eight hours, workers had only the one scheduled rest break and lunch during which they could void.

---

[258]David Moberg, "Puttin' Down Ol' Massa: Laurel, Mississippi, 1979," in *Working Lives: The Southern Exposure History of Labor in the South* 291-301 at 292 (Marc Miller ed. 1980 [1979]).

[259]Brief on Behalf of the Union at 3, 4, In the Matter of Arbitration Re: Cagle's Poultry and Egg Company, Inc., and Retail, Wholesale and Department Store Union, Local 315 (FMCS No. 79K-12882). Isach James, the then Atlanta-based representative (and later president) of Local 315, who confirmed Kitchens' statement, added that the shop stewards had told him at the time that some employees had "abused" breaks; the example he gave, however, was of women who wanted to leave the line to void soon after returning from lunch. Telephone interview with Isach James, Atlanta (Feb. 5, 2003).

[260]Cagle's Poultry and Egg Co., 73 Labor Arbitration Reports at 35-36.

[261]Cagle's Poultry and Egg Co., 73 Labor Arbitration Reports at 41. According to Cagle's then attorney, the reason for this practice was that the length of the workday was not known in advance—the shift ended when the last chicken had been processed. Thus not only did the company not know whether the second half of the shift would last four hours, it would have been unusual for it to have lasted exactly four hours rather than less time (in which case no rest period was granted) or more time (in which case instead the overtime rest period kicked in). Telephone interview with Cavanaugh.

Unsurprisingly, the same day that Cagle's issued the memorandum the union filed a grievance requesting discontinuance of the bathroom clock and of Rule No. 39 itself. Efforts to resolve the grievance during the next few days were unsuccessful and the guidelines were put into effect on Monday March 12.[262] Three days later the banner headline on the front page of *The Macon Telegraph* read: "Cagle's Employees Fired Over Restroom Rules." On March 14, some of the workers, who considered the new rule "'harsh' and even 'ridiculous,'" had asked to speak to company officials about the right to use the bathroom and the three-day suspension of Margie Bagley without pay (for failing to use the bathroom clock), "and ended up being fired." Already during the first three days under the rule six workers had received warnings and four had been fired and rehired because they had refused to clock in when going to the bathroom. Defiantly, one of the workers declared that the new rules "'can't control your body.'" The newspaper reported a "mass-firing" of 175 workers, and although it could not confirm that production had been interrupted, reportedly all the chickens at the plant had been moved to a different slaughter plant.[263]

As an arbitrator later depicted the week's events: "[T]he production line was disrupted several times as the employees displayed resentment to the guidelines and the time clock which had been installed at the restroom."[264] In the midst of these wildcat strikes "in protest" against what *U.S. News & World Report* called the company's "edict," Cagle's sought to settle the grievance by agreeing to "liberalize its policy."[265] On Thursday March 15, the approximately 175 workers standing outside the plant's locked gates "still considered themselves fired for not wanting to abide by the company's new rule...." In the course of the day, Isach James, Local 315's representative from Atlanta, met with the company and then told the workers outside that Cagle's officials had increased the number of permissible bathroom visits to 16, but that otherwise they were not "going to change," and that the workers should "'go back to work Monday and let the union handle it.'" The message was met with "outspoken disapproval" by most of the workers, who complained of race discrimination because, unlike the virtually all-black and -female line workers, the all-white office workers were not subject to the rule.[266] The same day, Cagle's posted a revised memorandum, incorporating

[262]Cagle's Poultry and Egg Co., 73 Labor Arbitration Reports at 39.

[263]Yvonne Shinhoster, "Cagle's Employees Fired over Restroom Rules," *Macon Telegraph*, Mar. 15, 1979, at 1A, col. 1-4 (Home ed.). Alma Oliver stated 24 years later that the workers themselves had not been fired. Telephone interview with Oliver.

[264]Cagle's Poultry and Egg Co., 73 Labor Arbitration Reports at 39.

[265]"Henhouse Spat," *U.S. News & World Report*, Apr. 2, 1979, at 65.

[266]Yvonne Shinhoster, "Restroom Rule Eased at Cagle's," *Macon Telegraph*, Mar. 16, 1979, at 1A, col. 5, at 10A, col. 1-4 (Home ed.).

the 16-visit maximum, raising the two-week limit to 10, and introducing two- and three-trip maximums for two- and three-day weeks. In addition, it omitted the last paragraph of the original memorandum altogether. If the offer was unacceptable to the union, the company requested immediate arbitration.[267]

The dispute sharpened on the weekend when the shop steward, Annie Kitchens, and the two assistant stewards, Dorothy Stanford and Alma Oliver, received letters informing them that they had been "discharged immediately" for violating the contract provision prohibiting employees from taking part in a walkout or slowdown as well as for having failed to carry out their duties as shop stewards. That same day about 30 workers met with the local chapter of the Southern Christian Leadership Conference (SCLC), the black civil rights organization, whose president, Reverend Henry Ficklin, declared that it supported the workers "100 percent." Ficklin also had the presence of mind to announce that SCLC would contact the U.S. Department of Labor and OSHA to determine whether Cagle's had violated any federal regulations.[268] (Ficklin's recalled that he did contact OSHA, but that its response was "nebulous" and that the agency never did anything.)[269] Ficklin stated that the organization's primary concern was "'the complete disregard for human rights. The worst thing that comes to mind...is after you have used all of your allotted times for restroom visits, what do you do—lose your job, or wet your clothes?'"[270] By this point reporting was intensive, with television cameras present, and public sentiment heavily in favor of the women, whose dignity was seen as under attack by an absurd policy.[271] Press coverage had also become national with the *Washington Post* and *Chicago Tribune* both printing wire service reports.[272]

At a meeting on Sunday night at Ficklin's Mt. Vernon Baptist church, Robert Steele, an American Civil Liberties Union lawyer who had filed a discrimination

---

[267]Cagle's Poultry and Egg Co., 73 Labor Arbitration Reports at 39-40.

[268]Sidney Hill, "Cagle Fires 3 Union Stewards," *Macon Telegraph*, Mar. 18, 1979, at 1A, col. 3-5, at 6A, col. 3-6 (Home ed.). See also Yvonne Shinhoster, "Arbitrator May Handle Cagle Case," *Macon Telegraph*, Mar. 26, 1979, at 1B, col. 4. A quarter-century later, Alma Oliver noted that she, Kitchens, and Stanford had in fact told the workers that it would be best to let the union try to work things out; they were reinstated with back pay after six months. Telephone interview with Oliver.

[269]Telephone interview with Ficklin. Ficklin is still the head of the local SCLC and has been a member of the Macon city council ever since 1979.

[270]Hill, "Cagle Fires 3 Union Stewards" at 6A, col. 3-4.

[271]Telephone interview with Yvonne Shinhoster Lamb, Washington, D.C., Feb. 6, 2003). Shinhoster was the *Macon Telegraph* reporter who covered the dispute.

[272]"Firm Settles Dispute Over Restroom Visits," *Washington Post*, Mar. 17, 1979, at A5, col. 3 (AP report); "ACLU Hits Company's Employe [sic] Restroom Limit," *Chicago Tribune*, Mar. 19, 1979, at sect. 1, at 6, col. 5-6 (UPI report).

complaint on behalf of the workers with the Equal Employment Opportunity Commission, advised the 130 assembled workers to return to work so that he and the union could continue the fight against the bathroom restrictions. Insisting that even the revised rule was "'inconceivable'" and that they "'would be opposed to any restrictions,'" Steele assured them that three labor attorneys had told him that Cagle's "could not impose [a] rule regulating restroom use." However, both he and Kitchens argued that the no-strike clause in the contract meant that a successful outcome required them to resume work.[273] Though they were "unhappy" about the firing of the three shop stewards, on whose behalf charges of unfair firing had been filed with the National Labor Relations Board, most of the workers did return to work on Monday. As before, Cagle's refused to comment on the events.[274] That day's editorial page cartoon in *The Macon Telegraph* suggesting that the company had laid an egg may explain Cagle's studied silence.[275] That the company nevertheless intended to continue wielding a big stick was confirmed on March 21 when it fired Bagley for having forgotten again to clock in and out of the bathroom.[276]

Although the union had asked for arbitration on March 16,[277] it told workers to return to work on Monday, March 19,[278] but the company's offer did not satisfy them. As early as March 19, the *Chicago Tribune* reported the ACLU lawyer Steele as saying that the "'whole thing is degrading' and held racial connotations because most of the affected employees are black." He added that "'[t]hese women have no skills and...really need the job.'"[279]

Then on March 28 three women workers (with Bagley as lead plaintiff)[280]

---

[273]Yvonne Shinhoster, "Workers to Return to Jobs at Cagle's Today," *Macon Telegraph*, Mar. 19, 1979, at 1A, col. 1-4 (Home ed.).

[274]Yvonne Shinhoster, "Cagle's Workers Return," *Macon Telegraph*, Mar. 20, 1979, at 1B, col. 6.

[275]*Macon Telegraph*, Mar. 19, 1979, at 4D, col. 3-5.

[276]Yvonne Shinhoster, "Woman Fired for Violating Toilet Rule," *Macon Telegraph*, Mar. 22, 1979, at 1B, col. 5.

[277]Brief on Behalf of the Union at 2, In the Matter of Arbitration Re: Cagle's Poultry and Egg Company, Inc.

[278]"ACLU Hits Company's Employe [sic] Restroom Limit." According to Isach James, the representative of Local 315, the new terms had been reached through negotiations on March 15. "Firm Settles Dispute Over Restroom Visits."

[279]"ACLU Hits Company's Employe [sic] Restroom Limit." The article erroneously stated that the lawsuit had already been filed.

[280] Margie Bagley, Willie Mae Pearson, and Elmyra Lester filed the suit, and on April 17, Annie Mae Kitchens, Dorothy Stanford, and Alma Oliver were added as plaintiffs. Kitchens' name was misspelled McKitchens in Bagley v. Cagle's Poultry and Egg Company, Inc., and Retail Wholesale and Department Store Union, Local No. 315, AFL-

CAGLE'S POULTRY
BATHROOM USE RESTRICTION
BARTLETT 1979
THE MACON TELEGRAPH AND NEWS

filed suit under section 301 of the National Labor Relations Act in U.S. District Court for the Middle District of Georgia, Macon Division, against Cagle's for breach of the collective bargaining agreement and against their own union for breach of its duty of fair representation.[281] The workers alleged that the guidelines, which were "directed only toward plaintiffs and other production and maintenance employees, who are predominantly Black and females," were "racially discriminatory and obviously unreasonable," and constituted an attempt to "limit a natural biological function.... If said natural biological function is not deferred within the limits of plant rule #40 [sic; should be 39] guidelines, plaintiffs are left in the unenviable position of suffering a loss of pay and/or employment with defendant company."[282] Alleging that the modification of the guidelines to which the union had agreed was "still discriminatory and unreasonable,"[283] the plaintiffs stated that the union had "conspired" with Cagle's "to allow the guidelines" to "remain intact." Moreover, the grievance negotiations "were spurious, and deliberately designed to give the false impression that a sincere effort was being made by defendant union to resolve the grievance."[284] In conclusion, the plaintiffs, who, as their attorney explained to *The Macon Telegraph*, "'wanted the rule abolished, not amended,'"[285] requested that the court enjoin Cagle's from "adhering to the guidelines...."[286] Cagle's senior attorney in effect lent credence to the plaintiffs' account of the union's role by noting years later both that the nub of the complaint had been that the company either dominated or was in cahoots with the union and that the union representative was someone "you could run over...you could get a better deal from him than you should have."[287]

CIO, Amendment to Complaint, ¶4, C.A. No. 79-71-MAC (M.D. Ga., Apr. 13, 1979), and on the docket sheet, which the Deputy Clerk of the District Court made available.

[281]Bagley v. Cagle's Poultry and Egg Co., Complaint, ¶1, C.A. No. 79-71-MAC (M.D. Ga., Mar. 28, 1979).

[282]Bagley v. Cagle's Poultry and Egg Co., Complaint, ¶7.

[283]Bagley v. Cagle's Poultry and Egg Co., Complaint, ¶9.

[284]Bagley v. Cagle's Poultry and Egg Co., Complaint, ¶10. Local 315's representative from Atlanta, Isach James, who was later its president for many years, when asked why the workers had sued the union, asserted that the ACLU lawyer had done it because he wanted money, intimating that the workers had not even realized that they had also sued their own union. In a similar vein, James also related that at the April 19 hearing before Judge Owens, Steele and the union's lawyer, Morgan Stanford, got into a shoving match in the courthouse. Telephone interview with James. In turn, Ficklin of the SCLC regarded the union as akin to a company union. Telephone interview with Ficklin.

[285] Alan Sverdlik, "Cagle, Union Conspired over Rule, Suit Charges," *Macon Telegraph,* Mar. 29, 1979, at 1A, col. 5-6, at 2A, col. 2 (Home ed.) (quoting Jerry Boykin).

[286]Bagley v. Cagle's Poultry and Egg Co., Complaint, at [5] (Prayer).

[287]Telephone interview with Currie. On November 21, 1979, the union was dismissed

Two weeks later the plaintiffs filed an amended complaint alleging that the bathroom-access rule, which violated their "right to privacy," was racially and sexually discriminatory because it applied to the production-line workers, 97 percent of whom were black and female, but not to office, supervisory, or maintenance personnel, all of whom were white.[288] The amended complaint also charged that Cagle's was engaged in a pattern and practice of race and sex discrimination in employment and promotions by denying the black women on the production line "the opportunity to advance into the office, maintenance or supervisory positions."[289] The point of the amendments, according to one of the workers' lawyers, was to attack a whole series of sex-discriminatory practices at the plant—which the union had not effectively opposed—that prevented women from securing the same production jobs that black men held and office jobs that white women held.[290]

On April 19, the court held a preliminary hearing, well attended by the press,[291] on the issues raised by the alleged breach of the collective bargaining agreement, at which Judge Wilbur D. Owens, Jr. (a Nixon appointee) suspended Plant Rule No. 39 until the arbitration proceeding could be completed, but in the interim "required that every employee clock in and out of the bathroom."[292] According to the account the next day in *The Macon Telegraph*:

> Speaking to the Cagle employees sitting in the courtroom, Owens said that in the event any employee fails to clock in and out, the company still has the authority to deal with them. The judge added Cagle management also has all "the authority it has always had to deal with what it perceives to be an abuse of the bathroom privilege."[293]

It was perhaps this kind of tough talk that prompted one of the Cagle's attorneys to recall the judicial proceeding as having been favorable to the company. In

---

from the action with prejudice. Bagley v. Cagle's Docket Sheet. The union continued to represent the workers at the Macon plant until it was shut down and sold in 2001. Telephone interview with Doug Cagle. In retrospect Alma Oliver was of the view that the union did try to help: it was a good, but not a great union, simply overpowered by a much stronger company. Telephone interview with Oliver.

[288]Bagley v. Cagle's, Amendment to Complaint, ¶¶14-18.

[289]Bagley v. Cagle's, Amendment to Complaint, ¶¶21.

[290]Telephone interview with Mabry.

[291]Isach James related that a group of "obnoxious" reporters at the courthouse had followed him into the bathroom and tried to interview him, holding microphones up to him as he stood at a urinal. Telephone interview with James.

[292]Bagley v. Cagle's, Conset Order and Decree at 1-2.

[293]Yvonne Shinhoster, "Cagle Toilet Rule Suspended," *Macon Telegraph*, Apr. 20, 1979, at 1B, col. 4.

particular he remembered that Owens had declared at the hearing that he was taking judicial notice of the fact that he drove some six hours to South Florida without stopping to use the bathroom and that he therefore had no sympathy for workers who wanted to go every hour or so.[294] (When asked about his comment, Judge Owens, on senior status, admitted that he was plain-spoken, but had no recollection of the case, and could only express nostalgia for the days when he could hold it that long. Since he remembered nothing about the dispute, it was, unfortunately, not possible to discover whether, with his now shortened urinary time-horizon, he would have viewed the workers' complaint differently.)[295]

However, the judge's sympathies did not rest solely with the company, as witnessed by this question that he posed to the plant's general manager: "[W]hy don't you deal with the individual case here instead of just passing some blanket rule that's inflexible?"[296] The plaintiffs' attorney also had reason for viewing the judge's order as a "partial victory," since Owens also ruled in open court that

> employees who have a "genuine need" to go to the restroom at times other than their break or lunchtime should feel free to do so, if there is a relief person available to take their place on the production line.
>
> "In the court's best judgement [sic] this is a fair arrangement pending arbitration. That is, the company is in position to run the plant, and you all are in position to work and to go to the bathroom when you really need to do so."[297]

Despite his acceptance of Cagle's unilaterally and self-regardingly structured relief system, Judge Owens appeared to harbor some skepticism about the basis of the company's determination of "the number of times a worker should be allowed to go to the restroom outside of company-paid breaks [sic] and lunchtime." When the plant general manager replied that he had both conducted a time and motion study of the bathroom trips and asked other poultry plants how they maintained efficiency, Owens was clearly dissatisfied: "'What is your scientific or medical basis for saying you can expect a human being, a normal human being, to refrain from going to the bathroom two hours after lunch? What knowledge

---

[294]Telephone interview with Cavanaugh. Mabry, one of the plaintiffs' lawyers, confirmed the substance of the judge's remarks, which she and her clients had regarded as odd. Telephone interview with Mabry.

[295]Telephone interview with Judge Owens, Macon (Jan. 26, 2003).

[296]Brief on Behalf of the Union at 5, In the Matter of Arbitration Re: Cagle's Poultry and Egg Company, Inc. (quoting court transcript at 39-40). Unfortunately no such transcript was found in the case file housed at the federal archive in Atlanta. Telephone interview with Denise Partee, Deputy Clerk, U.S. District Court for the Middle District of Georgia, Macon Division (Feb. 6, 2003).

[297]Shinhoster, "Cagle Toilet Rule Suspended."

do you have of human habits?'"[298]

Cagle's, whose Answer to the plaintiffs' Complaint denied virtually all the allegations and offered as one of its defenses that it had promulgated the bathroom rule "out of 'business necessity' to insure an adequate number of employees on the production line for safe and efficient operation of the plant,"[299] had made the most of the brief period of 24 working days before the court suspension took effect to enforce the rule, disciplining 22 employees one or more times, including six who were suspended for three violations (none of which involved exceeding the five-minute visit rule).[300] The spirit in which the company conducted this campaign was captured by the lecture that the plant manager gave a worker after she had received a verbal warning for exceeding the permissible number of trips. After informing her that "she had to train herself to go to the bathroom before she came to work in the morning and during the ten-minute break," he added that he knew that she "could train herself, because he had a dog who stayed in his house seven or eight hours a day, and when he got home, he would let the dog out of the house to relieve himself."[301] In order to assist the women with their urinary obedience training, Cagle's radically reduced the number of relief workers from twelve to four, thus increasing substantially the amount of time that a worker who signaled the need for relief had to wait before being allowed to leave the line.[302]

In the Answer that Local 315 filed with the court, the union denied the plaintiffs' allegations concerning its role, but rehearsed the same facts about the bathroom rule in its Cross-Claim against Cagle's seeking "a permanent inunction restraining Cagle's from discriminatorily maintaining and enforcing rules which restrict employees' use of bathroom facilities."[303]

At the arbitration, the RWDSU raised a number of objections, in opposition to the company's new rule, alleging contract violations. First, it was unreasonable for the company to "change its longstanding bathroom rule which required that the individual employee who abused the rule should be dealt with." The new rule was at odds with "our democratic, industrial society" in which "guilt should

---

[298]Shinhoster, "Cagle Toilet Rule Suspended." Unfortunately the article ended at that point without reporting the plant manager's answer.

[299]Bagley v. Cagle's Poultry and Egg Co., Answer at 3, C.A. No. 79-71-MAC (M.D. Ga., Apr. 24, 1979).

[300]Cagle's Poultry and Egg Co., 73 Labor Arbitration Reports at 40.

[301]Brief on Behalf of the Union at 14-15, In the Matter of Arbitration Re: Cagle's Poultry and Egg Company, Inc.

[302]Brief on Behalf of the Union at 8-9, In the Matter of Arbitration Re: Cagle's Poultry and Egg Company, Inc.

[303]Bagley v. Cagle's Poultry and Egg Co., Answer and Cross-Claim of Defendant Retail, Wholesale, and Department Store Union, Local No. 315, AFL-CIO, at 6, C.A. No. 79-71-MAC (M.D. Ga., Apr. 24, 1979).

be individual and not by association, so that the individual employee who abuses the legitimate rights of an employer should be penalized instead of penalizing the whole group of employees." Next, Local 315 accused the company of having "seized on the pretext of abuse of the bathroom rule in order to get more production at the Macon plant." The union adroitly noted that it was "manifestly unfair" for Cagle's to limit bathroom breaks—which was "undignified" in its own right and "not in keeping with...biological variances"—while it was in breach of the contract by denying workers their afternoon rest break. Finally, the RWDSU raised the specter of racism by pointing out that "[w]ith the office staff being 100% white and subject to the old bathroom rule of individual abuse, and with the production employees being over 90% black, it could easily be argued that the change in the bathroom rule was racially motivated."[304]

For its part, Cagle's insisted at arbitration that uniform application of the rule to everyone in the bargaining unit was more just and fair-minded than case-by-case procedures. Otherwise, it focused on the need to "curtail the abuse of the bathroom privileges afforded by the Company" in light of the drop in production and the lack of any definition of abuse in the old rule. Ratcheting up the rhetoric, Cagle's insisted that the "lack of production in the Macon facility" had made it necessary to "establish[ ] order" there. Finally the company declared that "[i]n addition to providing an adequate number of times an employee can take emergency leave from the production line, there are five other times during the working day [during] which employees may visit the rest room."[305] This total derived from including not only, as Jim Beam would do more than two decades later, use of the toilet before and after work, but also the afternoon rest break that Cagle's denied the workers.[306]

These six opportunities to void were within the range specified by Dr. Theopholius Worrell, a general practitioner—several of whose patients worked at the plant[307]—who had offered expert testimony that "[a] normal person uses the bathroom for elimination purposes four...to six...times a day, primarily in daylight hours" and that "[u]nder normal circumstances, it would be reasonable for a person to be allowed to go to the restroom five...times between 6:30 a.m. and 4:00 p.m." The doctor focused on a number of factors that would heighten the need for women and blacks, who formed the vast majority of the line workers, to urinate. The medication prescribed for hypertension, high blood pressure, and diabetes, which are "frequently found in blacks, may cause a need to use the bathroom more frequently." In addition to pointing out that women tend to go to the

---

[304]Cagle's Poultry and Egg Co., 73 Labor Arbitration Reports at 37.

[305]Cagle's Poultry and Egg Co., 73 Labor Arbitration Reports at 37-38.

[306]Cagle's Poultry and Egg Co., 73 Labor Arbitration Reports at 46.

[307]Telephone interview with Theopholius Worrell, Macon (Jan. 30, 2003).

bathroom more often than men during or after their menstrual cycle, Worrell testified that "[r]estrictions on the time used for urination by females...may require them to hasten the process which could cause urinary tract infection." With special reference to conditions prevailing at the poultry plant, the expert also observed that "[w]orking around water may have [the] psychological effect of requiring use of the bathroom more frequently" and that "[s]tress or fear of punishment or loss of job could psychologically cause biological variances which would require changes in bathroom habits."[308]

The normal range of daily voids suggested by Worrell was almost identical to that presented by Stivers and much lower than that indicated by Nygaard 23 years later. Ironically, although Worrell testified on the workers' behalf, his testimony served to support the arbitrator's decision to uphold the company's final offer of one unscheduled void per day, which, together with the morning rest break, lunch break, voids on the workers' own time before and after work, and the restoration of the mid-afternoon rest break, brought the total to the top of the "normal" range of six daily voids.

In his decision, the arbitrator, George Roberts, who had spent 35 years as a personnel director and in industrial relations with the U.S. Postal Service, Willys Motors, and Babcock and Wilcox,[309] first turned to the issue of the company's failure to provide the contractually guaranteed afternoon rest period, to which practice the union had "never objected." After taking notice that work on the continuous production line was "confining and monotonous" and that "[t]he protective clothing and working conditions are not conducive to long periods of peak personal efficiency," Roberts concluded that since it was "obvious" that the contractual rest periods were "for the purpose of reducing fatigue and satisfying the personal needs of the employees," Cagle's failure to provide the second one "defeats the fundamental purpose of the rest periods." Excoriating the company's interpretation for "completely disregard[ing] the human needs of the employees," the arbitrator declared that it was "ridiculous to think that the negotiators of this Agreement intended to provide a rest period only if employees worked more than four...continuous hours...." In addition to ordering Cagle's to cease violating the provision, Roberts rebuked the union for its complacency about the company's contract breach until the issue of bathroom breaks had arisen.[310]

The arbitrator's discussion of toilet access was circumscribed by his acceptance at face value of the language of Rule No. 39—"Abuse of emergency rest room privilege." Since an "emergency" is an unforeseen combination of circum-

---

[308]Cagle's Poultry and Egg Co., 73 Labor Arbitration Reports at 40.

[309]*Labor Arbitration: Cumulative Digest and Index Covering Volumes 71-80,* at 998 (1984).

[310]Cagle's Poultry and Egg Co., 73 Labor Arbitration Reports at 41.

stances calling for immediate action, the term itself suggests that, as the antithesis of a normal situation, it was designed to encompass, for example, sudden and unexpected diarrhea, but certainly not the worker's normal quota of workday urinary voids, which apparently in the company's and his view could be satisfactorily taken care of during the mid-morning rest break and lunch. Given this understanding, Roberts upheld both Cagle's right to perform an analysis of "emergency use" of the bathroom, regardless of how well or ill founded the causal connection was to perceived inefficiency at the plant, and its decision, since "the abuse was...out of hand," to "set uniform standards for all employees before taking individual disciplinary action." Indeed, the arbitrator was so deferential to managerial prerogatives that, despite the absence of any evidence that the bathroom use study was credible, he declared that the company's "conclusion that the emergency use of the restroom was being abused was a judgement [sic] decision of the Company and must be accepted as factual...." Apart from the definitional implication that emergencies do not occur (twice) daily, neither Cagle's nor the arbitrator offered any evidence at all that there was anything 'abusive' about voiding twice outside of two scheduled breaks. Ironically, Roberts lamented that the employer had not made greater use of the results of the study to explain the limitations in the guidelines to the employees. He faulted Cagle's for articulating only one conclusion from the study—"that the cost of the emergency use of the restroom was excessive"—and then drawing the ultimate conclusion of abuse from it. Roberts raised the possibility that if the company had informed the workers that virtually all the line workers made two emergency trips a day, each averaging 7.34 minutes (including travel time), "it would have possibly dispelled the impression that the decision to issue the guidelines was made arbitrarily and capriciously and that it was 'pretext—in order to get more production.'"[311]

It is extraordinarily revealing of the arbitrator's own bias that he could even have imagined that workers whose employer was disciplining them for voiding more than twice a day would somehow undergo a conversion and share Cagle's outrage over their "abusive" efforts to void four times a day if only they knew how often and for how long they were going to the bathroom.

Roberts then dealt with the union's two charges of unreasonableness directed at the guidelines to Rule No. 39 concerning the time clock and the limit on the number of bathroom visits. He summarily upheld the use of the clock on the grounds that the permissible purpose of the guidelines was to prevent abuse and that there was "no better way to determine who, when, and how long an employee uses the restroom."[312] He was equally dismissive of the union's claim that the five-minute limit on bathroom visits was also unreasonable, especially since the

---

[311]Cagle's Poultry and Egg Co., 73 Labor Arbitration Reports at 42-43.

[312]Cagle's Poultry and Egg Co., 73 Labor Arbitration Reports at 44.

company had failed to consult with medical authorities and testimony had indicated that "'biological variances of individuals' made it "virtually impossible to set a maximum time limit...." Roberts' sole basis for rejecting this claim hinged on his literalist acceptance of the company's designation of the breaks as "emergency restroom privileges": he asserted that if 10 minutes was reasonable for a scheduled rest break, which could also be used for going to the bathroom, a lesser amount of time was "[c]ertainly" reasonable for emergency use. Moreover, even conceding that there were circumstances—for which he did not even purport to have evidence to call "rare cases"—under which five minutes might not be sufficient, he argued that employees could always grieve any disciplinary action.[313]

Finally, Roberts reached the rule that the union "strongly resented"—the limitation on the number of permissible emergency bathroom visits. The arbitrator was compelled to concede that "[t]here is much truth to the adage 'when you got to go, you got to go.'" In addition, because needs are so variable and unpredictable, "it is most difficult to determine a realistic and fair figure," as witnessed by the company's own increase in the four-week total from an initial eight visits to the 20 it offered at the hearing. Once again, however, he stressed that Rule No. 39 applied only to emergencies—that is, "when an employee unexpectedly has need to use the restroom facilities immediately." He then appeared to exclude from this category toilet needs triggered by medication or the menstrual cycle because they shared elements of predictability. Roberts tried to downplay the significance of such exclusions by pointing to the "special arrangements" in the guidelines for "some of these conditions,"[314] but they were in fact so limited and driven by a spirit of blaming-the-victim that their usefulness would have been marginal at best.

Roberts again confused two distinct issues when he proceeded to mention studies cited by the union indicating that 20 or 24 minutes was adequate personal allowance time for an eight-hour day. The two 10-minute rest breaks approximated this standard, in his view, especially as adjusted by the "extenuating circumstances" of the once-a-day emergency void.[315] However, Roberts overlooked the crucial fact that the total amount of non-working time available is irrelevant if it is so concentrated that the number of breaks is insufficient to enable workers to void when they need to. Ultimately, then, the question boiled down to the specific number of permissible breaks, to which Roberts finally turned.

Before reaching it, however, Roberts stopped long enough to dismiss the union's argument that "any limitation is unreasonable." This first mention of the union's position disclosed that Local 315 had presented five witnesses who had

---

[313]Cagle's Poultry and Egg Co., 73 Labor Arbitration Reports at 45.

[314]Cagle's Poultry and Egg Co., 73 Labor Arbitration Reports at 45.

[315]Cagle's Poultry and Egg Co., 73 Labor Arbitration Reports at 45.

testified about "the trauma and physical difficulties which they encountered when emergency use of the restroom was restricted." In addition, expert testimony had "explained the physical and psychological effect on women of such conditions." The arbitrator's only response was that the original rule, without restrictions, "did not work." Why voiding a total of four times during a workday was unworkable, he did not even try to explain. Instead, he insisted—doubtless without the slightest tinge of sarcasm—that both the employees and the supervisors "needed and deserved help in making it work. The guidelines provided this help." Again, Roberts did not bother to explain how a rule that prevented workers from going to the bathroom when they needed to go helped make a rule on emergency bathroom use work. Instead, he assured the workers that it had been "definitely unreasonable" not to furnish a standard so as to insure "uniform and equitable enforcement," leaving it to each supervisor (and employee) to exercise his or her own judgment. As to why twice a week or once a day was an acceptable number, but twice a day was not, he could offer as an oblique answer only "the importance and justification for disciplining oneself in such matters," citing as authority the opinion of the arbitrator in the Jones Dairy Farm case that "it most certainly is not unreasonable to learn to regulate and train one's self [sic] to work periods of 1½, 2 or 2½ hours. This is expected of us from childhood into school...."[316] Again, without having articulated or cited any independent substantive definition of "abuse," Roberts found additional support for the limitation in its being "merely a way to 'regulate or police' the observance of a rule."[317] Since any other number would have served the same purpose, the arbitrator was no longer able to delay discussion of the reasonableness of the number itself.

The arbitrator's simple answer for upholding the company's once-a-day emergency rule was that, added on to the two breaks, lunch, and bathroom visits before and after work, it fully satisfied the guidelines set by the expert witness. As to why the use of the bathroom on the employee's own time before and after shift should count—a position rejected by Kentucky OSHA in the Jim Beam case—Roberts merely repeated that "[n]ormally it is expected that employees will discipline themselves to use these periods for that purpose in order to avoid the possibility of emergency situations." Ignoring the union's evidence that he had already cited, he held that, absent any suggestion by the union as to a number and any proof of an "unusual or extraordinary hardship," the company's number was substantively reasonable.[318]

Roberts reserved his criticism of Cagle's for its administration of the guide-

---

[316]Cagle's Poultry and Egg Co., 73 Labor Arbitration Reports at 46, citing Jones Dairy Farm, 72-2 Labor Arbitration Awards (CCH) ¶8639 at 5256 (1973).

[317]Cagle's Poultry and Egg Co., 73 Labor Arbitration Reports at 46.

[318]Cagle's Poultry and Egg Co., 73 Labor Arbitration Reports at 47.

lines, which he deemed unreasonable because it lacked flexibility, the company failed to involve a female supervisor or female nurse, "who should have much authority in decision making in this very sensitive area," and in two instances the company let violations from one period carry over to the next instead of wiping the slate clean. He then denied the grievance and Solomonically directed Cagle's to instruct all supervisors to "insure compassion and understanding along with firmness...." [319]

The ruling left workers, who recognized that Roberts had gone "'right along with the company,'" profoundly dissatisfied. In particular, they believed that it was wrong that "[e]verybody should...be made to suffer for what a few people are doing," especially since Cagle's had been treating them "harshly" after Judge Owens' suspension of the rule.[320]

On August 3, noting that the arbitrator had ruled that Rule No. 39 was "reasonable as written,"[321] Judge Owens, with the concurrence of plaintiffs' counsel, dissolved his order preliminarily enjoining the plant rule.[322] In November the union was dismissed from the suit, and in September 1980 the parties entered into a consent decree, which the court approved. In addition to wide-ranging measures for increasing the representation of blacks and women in various positions off the chicken line, the decree abolished Rule No. 39 and did away with the bathroom clock and the requirement of clocking in and out of the bathroom. Instead, the parties—including the union, which agreed that the new rule was "reasonable"and executed the order and decree[323]—created a "new plant rule regarding abuse of the emergency restroom privilege." First, in addition to retaining the mid-morning break, the agreement provided for a 10-minute mid-afternoon break on days when workers worked at least two hours after lunch. The core of the rule then read as follows:

> Two...relief workers shall be provided by Cagles for the purpose of allowing employees to make use of the bathroom on an emergency basis ([a]t times other than before and after work, during lunch break and during rest periods...)[.]
>
> No production line employee shall be permitted to leave his/her position on the production line unless and until his/her place on the line is occupied by a designated relief worker. Any employee who leaves the line without his/her position being properly staffed by a designated relief worker shall be subject to progressive discipline outlined as follows:

---

[319]Cagle's Poultry and Egg Co., 73 Labor Arbitration Reports at 47, 48.

[320]Yvonne Shinhoster, "Cagle's Rule Partly Upheld by Arbitrator," *Macon Telegraph*, July 12, 1979, at 1B, col, 5, at 2B, col. 4.

[321]Bagley v. Cagle's Poultry and Egg Co., Consent Order and Decree, C.A. No. 79-71-MAC, slip op. at 2 (M.D. Ga., Sept. 20, 1980).

[322]Bagley v. Cagle's, Order, C.A. No. 79-71-MAC (M.D. Ga., Aug. 2, 1979).

[323]Bagley v. Cagle's, Consent Order and Decree, slip op. at 5.

(1) FIRST OFFENSE - Verbal warning
(2) SECOND OFFENSE - Written warning
(3) THIRD OFFENSE - Three...day suspension without pay
(4) FOURTH OFFENSE - Immediate discharge

After an employee has had a clear record for twelve...months, their record will be considered as cleared.

Employees will be permitted emergency bathroom visits...in the order of their request.

Those employees who need special use of the bathroom, for medical or personal reasons, may obtain priority on the waiting list...by contacting the company nurse and stating the reason therefore. If the nurse grants priority, the relief workers and the employee's supervisor will be notified. If priority is denied, such priority can only be granted by the presentment of a statement of a medical doctor setting forth the necessity for such a request. ...

Should there be any abuse of the above described bathroom rule, it is agreed that the rule may be reasonably amended by Cagle's in accordance with the terms of the Collective Bargaining Agreement then in existence to enable the company to eliminate the bathroom abuse problem.[324]

To be sure, this new rule was more favorable to the workers than the rule approved by the arbitrator because it eliminated the express maximum of one unscheduled break per day and of five-minute stays and removed the clocks. Nevertheless, the rule was profoundly flawed and would be unlawful under the OSHA toilet standard today. First of all, it retained the employer's focus on "privilege," "abuse," and "emergency," all of which individually and especially taken together are antithetical and inimical to workers' right to be free to void when they need to. Second, the reduction in the number of relief workers to two from 12 constituted a victory for Cagle's, which, before the litigation, had initially demanded a reduction to four and mentioned two as its ultimate goal. Two relief workers for about 150 female workers or 75 workers per reliever created per se a waiting list of completely unworkable proportions, the existence of which was corroborated by Alma Oliver's retrospective comment that sometimes workers had to wait an hour or more, sometimes they never got relief at all before a scheduled break, and sometimes women "messed themselves."[325] The minuscule ratio of relievers to relief-seekers made virtually superfluous an express limitation on the number of daily unscheduled bathroom breaks, while Cagle's was authorized to deal with the length of the individual breaks by virtue of the unilateral power conferred on the company to amend the rule to eliminate "abuse."

But the most important aspect of the decree from Cagle's perspective, as its

---

[324]Bagley v. Cagle's, Consent Order and Decree, slip op. at 3-5.

[325]Telephone interview with Oliver.

then attorney stressed more than 20 years later, was the imposition of the prohibition on leaving the line without permission,[326] thus destroying any at-will custom that had ever existed and restoring to the employer absolute authority over the granting of permission for and the timing and sequence of breaks. By making departure from the line conditional on the availability of a relief worker controlled by management, the new rule eliminated any vestiges of worker collective self-determination under which those workers who did jobs also performed by several other workers standing next to them could be spelled for short periods by their co-workers without the need for a special relief worker.[327]

When asked a quarter-century later why a dispute over going to the bathroom had been so emotional, the company's lawyer observed that standing and working on a continuously moving production line in cold and wet conditions was not pleasant and that employees were therefore looking for "a little excitement."[328] That management's first suspicion takes the form of a not so subtle analogy to powerless school children who know no other way to relieve their boredom than to pretend to have to relieve themselves raises the (rhetorical) question as to why the employer did not instead choose to improve working conditions. The workers themselves offered an entirely different perspective. As Alma Oliver, a shop steward and one of the named plaintiffs, put it, the company's basic position was that the workers could go to the bathroom on their breaks, but "not on the man's time."[329]

Bathroom "abuse" was not the only kind that preoccupied Cagle's at that time: apparently "the man" wanted the women to hold their tongues as well as their urine. Literally just days before *The New York Times* reported on the arbitration decision, it ran a six-column article in its Sunday edition on a then recent trend in labor-management relations for firms to seek enhanced authority to discipline employees for "verbal abuse" of supervisors. Chief among them was Cagle's, which was "[o]ffended by the indelicate speech" of the women in the chicken-dressing department, who "often used...'men's language'—especially toward their supervisors." Cagle's therefore issued a rule in April 1978 "banning 'all profanity including the use of abusive language to supervisors.'" Soon thereafter it fired a woman who "shouted, 'Dammit, I can't!" when her boss ordered her to speed up her work." Cagle's reduced the penalty to a two-week suspension during the grievance procedure, but the union, insisting that "'[t]his is a factory—not a girls' finishing school,'" went to arbitration in order to have

[326]Telephone interview with Victor Cavanaugh (Jan. 30, 2003).

[327]Telephone interview with Oliver.

[328]Telephone interview with Cavanaugh (Jan. 30, 2003).

[329]Telephone interview with Alma Oliver. Oliver worked at the plant for 37 years, stopping only in 2001 when it was closed.

the rule rescinded; the arbitrator, however, upheld the rule and the penalty.[330]

Given the outrageously restrictive character of Cagle's toilet access rule, its obvious race and gender impact, and the workers' perseverance, it is hardly surprising that, as the arbitrator himself noted, "[t]here was much publicity of the dispute in the news media of the entire area"[331]—especially relating to the number of times workers were permitted to void.[332] Although this struggle generated the most intense interest in the Macon area, the reporting was "worldwide,"[333] with interest coming from as far away as Australia.[334] Even the national newspaper of record carried one brief report: bestowing on the Cagle's workers the honor of its attention which it withheld from the Jim Beam dispute 23 years later, *The New York Times* ran a 134-word Associated Press article titled, "Arbitrator Backs Limit on Visits to Restrooms," which presented a very concise account of the course of the events.[335] Although one of the Cagle's lawyers was later surprised to hear of the *Times* piece, mention of it reminded him that at the time the six o'clock national TV-news had also used the event as a closing human interest story. In addition, he, one of the workers' lawyers, and Reverend Ficklin all recalled that the then new NBC television show "Real People" came to Macon to do an episode about the bathroom denial, which also involved the sports stadium mascot, the Famous San Diego Chicken.[336] *U.S. News & World Report* added the dimension of sexism to the media's inability to transcend the jocular approach to the subject of workplace voiding by titling its little filler "Henhouse Spat."[337] Yet in spite of this adverse publicity, Cagle's was not impelled to change its policy.

Both the reasoning and the outcome of the Cagle's case soundly vindicate President Kelley's skepticism of arbitrators. Roberts, who was clearly not blind to the very arduous nature of the work that the employees performed and even summoned up a dose of outrage over the employer's elimination of the afternoon

---

[330]Lawrence Stessin, "Blue-Collar Crime: Chewing Out the Boss," *N.Y. Times*, July 1, 1979, at F3, col. 1-6 (Proquest). Neither the Macon plant nor the RWDSU was involved in this dispute.

[331]Cagle's Poultry and Egg Co., 73 Labor Arbitration Reports at 35.

[332]Cagle's Poultry and Egg Co., 73 Labor Arbitration Reports at 45.

[333]Telephone interview with Oliver.

[334]Telephone interview with Linda Mabry, Atlanta (Jan. 30, 2003).

[335]"Arbitrator Backs Limit on Visits to Restrooms," *N.Y. Times*, July 14, 1979, at 17, col. 1 (Proquest).

[336]Telephone interviews with Cavanaugh (Jan. 23), Mabry (Jan. 30, 2003), and Ficklin. One of the union's lawyers also recalled the extensive reporting. Telephone interview with James Fagan, Atlanta (Jan. 29, 2003). Cavanaugh, though a management-side labor lawyer in a firm specializing in labor law, was also surprised to hear about the OSHA Memorandum.

[337]"Henhouse Spat," *U.S. News & World Report*, Apr. 2, 1979, at 65.

rest period as well as the union's (unexplained) acquiescence in it, nevertheless displayed no understanding of, let alone sympathy for, workers who, not on an exceptional or emergency basis, but on a normal, day-to-day basis, had to void more often than the employer deemed compatible with its production, productivity, and profitability goals. Moreover, the whole notion of "abuse," which underpinned the company's strategy and Roberts's decision, was inappropriate as applied to a factory in which the employer accused virtually the entire workforce of spending too much time in the bathroom. While the charge of "abuse" may make sense in a situation in which some workers who do not need to void stop work, disproportionately burdening their co-workers with performing two employees' jobs, it loses its meaning when all the workers use the bathroom more frequently than management desires; it is more plausible that this latter phenomenon means that supervisors are seeking to suppress necessary voiding and/or that the workers as a group are engaged in a protest over what they view as unacceptable working conditions.[338] Perhaps as a result of his ingrained bias acquired over decades as a personnel manager himself, Roberts was able to perceive in the low-paid black women only malingerers whom a little discipline and self-discipline would not have hurt at all.

The lawsuit and the resolution of the toilet-access dispute reached by the parties (and approved by the court) was also less favorable to the workers than the outcome that rigorous (post-Memorandum) OSHA enforcement could have achieved. In any event, the relief system for which the workers were to able bargain should not have passed muster under today's toilet standard. Although OSHA had been in operation for almost a decade and the dispute had gained widespread publicity, the agency did not intervene. A quarter-century later, the attorneys on both sides stated that the RWDSU did not file an OSHA complaint,[339] but Isach James, the Local 315 representative, insisted that the union had complained to OSHA and to the health department, but that neither had done anything—exactly the same experience that Ficklin of the SCLC reported.[340] Thus, not even a unilaterally imposed rule so restrictive, one-sided, and harsh as to approach self-caricature prompted OSHA to (re-)consider whether its requirement that employers provide toilets included the obligation to make them available to their employees so they could use them when they needed to and not merely when it was profitable for the companies. And even if OSHA had investigated, it would doubtless have taken the same position in 1979 that it maintained all along until the UFCW-*Void Where Prohibited* campaign prompted it to cite Hudson Foods in 1997 and issue the Memorandum in 1998.

---

[338]See below ch. 17.

[339]Telephone interviews with Cavanaugh and Mabry.

[340]Telephone interviews with James and Ficklin.

14

Precocious But Meager Enforcement for Bus Drivers: Washington

It's as if employers believe you can make your kidneys punch a time clock, too.[1]

During the run-up to Federal OSHA's promulgation of its Memorandum in 1998, the spokesman for the Division of WISHA (Washington Industrial Safety and Health Act) Services of the Washington State Department of Labor and Industries told the Associated Press that the agency "mandates that employers make bathrooms available.... And although there is no requirement, the state consistently advises that workers be allowed to use them."[2] Four years later, the Senior Program Manager for Policy and Technical Services at Washington OSHA reported that when his agency had received the April 6, 1998 Memorandum, the staff had all thought that OSHA was "finally coming to its senses," because WISHA had adopted the same position earlier with regard to complaints resolved informally by phone and fax and (he was not certain) perhaps in citations as well.[3]

---

[1]Carrie Mason-Draffen, "Bad News: Bathroom Breaks Can Be Withheld," *Newsday*, Aug. 13, 2000, at F13 (Lexis).

[2]Maggie Jackson, "For Workers with Limited Toilet Breaks, Relief Is on the Way," *Seattle Times*, Mar. 22, 1998, at A13 (Westlaw) (quoting Bill Ripple). Almost five years after the fact, a former compliance officer and now program manager for the statistics section of New Mexico OSHA asserted, without offering any documentation, that that agency even before 1998 had always interpreted the standard requiring that toilets be provided as also requiring that workers be allowed to use them. He sought to buttress this regulatory interpretation by noting that New Mexico OSHA had adopted the same position with regard to other regulations that require employers to provide something, even though the logic of requirement was different when, as with hard hats, workers' use was mandatory rather than discretionary. Telephone interview with George Vigil, Albuquerque (Sept. 20, 2002). Ironically, however, since the promulgation of OSHA's Memorandum on April 6, 1998, New Mexico has failed to issue a single citation to an employer for failure to provide toilets, let alone for denying access to them.

[3]Telephone interview with Michael Wood, Sen. Prog. Mgr. Policy and Tech. Serv.,

And in fact at least once, in 1994, in the midst of a multi-year collective bargaining dispute over bathroom breaks between the Clark County Public Transportation Benefit Area Authority, or C-Tran, in Vancouver, Washington, and Amalgamated Transit Union (ATU) Local 757, which represented all 129 of the system's drivers, WISHA had issued a citation for denial of access.[4] (And again in 1995, WISHA cited Ryder ATE Inc., which provided local passenger transportation, for having "failed to allow mobile employees adequate time for restroom breaks, exposing employees to possible health hazards and injury"; WISHA also imposed a monetary penalty of $1,120, later reduced to $785.)[5] Two years earlier, in 1992, WISHA had issued a citation to the City of Yakima Transportation Department, proposing a penalty of $4,500, because "toilet room facilities for transit drivers were not readily available on a timely basis and employees did not have transportation immediately available to nearby toilet facilities."[6] As a result of this citation, the City of Yakima entered into a settlement agreement with the union (Washington State Council of County & City Employees Local 1122, AFSCME, AFL-CIO) representing the drivers, which had requested an inspection. According to the agreement:

> The City recognizes the need for adequate time for breaks for transit drivers during the course of driving shifts. The transit manager will use his best effort to approve route scheduling that provides adequate time for breaks (including layovers) of approximately five to seven minutes per hour. The Employee Work Force Committee will have the opportunity to be involved in a variety of subjects, which can include route scheduling.... Consideration given to break time will be applicable when routes are extended, so that time points for such routes will also be extended to allow time for breaks. ... The City acknowledges the legitimate concern raised by the issue involved in the above referenced citation. The City will directly communicate its acknowledgment to the transit division drivers and express its regret for the inconvenience and impact they have experienced....[7]

WISHA, Olympia (Oct. 21, 2002).

[4]Public Transportation Benefit Area dba C-Tran, Insp. 115470262 (1994).

[5]Ryder ATE, Inc., Insp. No. 115220816 (Aug. 2, 1995). Because the violation report had been destroyed, further details are no longer available. Fax from Barbara Harris-Williams, Public Disclosure, WISHA, to Marc Linder (Dec. 10, 2002).

[6]City of Yakima - Transit Dept., Insp. No. 111218012 (Oct. 19, 1992) (copy furnished by Wash. Dept. of Labor & Industries). Although the second part of this violation description quotes verbatim the mobile crew exception to the toilet standard, that language is contained in a separate code provision that WISHA did not cite. Wash. Adm. Code sect. 296-24-12007(1)(b) (1995).

[7]Settlement Agreement Between the City of Yakima and Washington State Council of County & City Employees Local 1122, AFSCME, AFL-CIO (June 9, 1993).

After the union and the employer had entered into this agreement, the Department of Labor and Industries and the employer reached their own agreement concerning the latter's appeal from the citation. As a result, the penalty was reduced from $4,500 to $1,350 because the Department determined that "there is no substantial probability that serious physical harm or death could have resulted from the violation."[8]

In contrast, the later citation against C-Tran, in addition to including the failure to provide toilet facilities or "reasonable access" to them, creatively applied the general duty clause of the Washington state occupational health and safety regulations[9]: "The employer did not adopt and use practices which adequately rendered the employment and the place of employment safe in that...schedules did not allow sufficient time for drivers to use the toilets during extended blocks of time...." WISHA also assessed C-Tran a penalty of $1,125 and ordered it to abate the problem.[10] In the language of a contemporaneous local newspaper account, the agency charged that "[n]ot enough time for toilet stops is unsafe for drivers and passengers...." Although C-Tran claimed that drivers could "take a bathroom break whenever they need one, even if it makes the bus late," "no C-Tran runs have time scheduled specifically for driver breaks. ... C-Tran drivers must radio in for permission to stop, tipping off riders about what the unscheduled delay is for," and thus embarrassing the driver.[11]

In response to C-Tran's appeal, a hearing was held in Vancouver, Washington on January 31, 1995, at which a large volume of evidence was presented concerning the lack of bathroom breaks. On some routes drivers went for as long as three, four, five, and even eight hours without urinating.[12] Female drivers were "mortified" to have to "tie their jackets or whatever they could find around them because they haven't been able to get to the bathroom in time."[13] The union

---

[8]Board of Industrial Insurance Appeals, Washington Industrial Safety Health Act, In re City of Yakima - Transit Dept, Safety Report No. 111218012, Docket No. 93W056, Notice to Employees of Proposed Agreement to Settle Appeal (Aug. 24, 1993).

[9]Wash. Adm. Code sect. 296-24-073(2) (1995).

[10]Public Transportation Benefit Area dba C-Tran, Insp. 115470262 (citation issued Nov. 18, 1994), citation item 1, No. 1a; Corrective Notice of Redetermination (Feb. 14, 1995) (issued by the hearing officer). Because the case was archived and more than six years old, WISHA destroyed the paper file. Telephone interview with Deanna Jackson, Public Disclosure, WISHA (Oct. 23, 2002).

[11]Jeanette Steele, "On the Go," *The Columbian* (Vancouver), Mar. 8, 1995, at A1 (Lexis).

[12]In the Matter of Informal Conference on Citations Issued to C-Tran, Employer, of Vancouver, Washington, by the Washington State Department of Labor, Health and Safety Division, Transcript of Proceedings 23 (Jan. 31, 1995) (statement of Mr. Wasela).

[13]In the Matter of Informal Conference on Citations Issued to C-Tran, Transcript of

insisted that drivers who stopped to urinate were either late and were disciplined or had to drive over the speed limit (to make up the time).[14] To be sure, on certain eight-hour runs no breaks at all were scheduled:

> Some operators just forego eating for eight hours. And we're not pleading their case, because they do that because they actually don't want to have an extra half-hour in there. They'd rather get home sooner. But there are other operators who want to eat, and there's a real question of whether it's safe and what the district's liability would be if there's an accident and the operator's reason is, well, I wasn't thinking very clearly, I hadn't eaten in seven hours.
>
> And we believe that the evidence will show that C-Tran's handling of the break issue has created a medical hazard for the employees and safety hazard for the employees and passengers and the general public. [A] driver who is forced to hold it or who has not eaten for eight hours is in no condition to maneuver a bus through urban traffic. If you've been holding it for eight hours and you've got to go, you are distracted and can't concentrate.[15]

Although C-Tran argued that drivers "can use...the restrooms and even be late on their run, and they are not disciplined," it also claimed that it was "pure supposition, speculation, that there are these connections" between urine retention and urinary tract infections.[16] And even if drivers were not sanctioned for lateness, the union argued, "the idea that you've managed to hold it for eight hours, you get overtime because your run is late, the end of your day is late, that isn't proper compensation. That isn't solving the problem, and it isn't removing the safety hazard."[17] Since other bus systems for which the union's members drove did add time to their schedules[18] so that drivers were able to go void, the problem was created and could be solved by C-Tran's management, which "certainly...doesn't have to ask to go to the bathroom."[19] In any event, management failed to respond to the union's suggestion of temporary measures by which part-

---

Proceedings, at 173 (statement of Susan Stoner).

[14]In the Matter of Informal Conference on Citations Issued to C-Tran, Transcript of Proceedings at 31-32 (statement of Susan Stoner).

[15]In the Matter of Informal Conference on Citations Issued to C-Tran, Transcript of Proceedings at 33-34 (statement of Stoner).

[16]In the Matter of Informal Conference on Citations Issued to C-Tran, Transcript of Proceedings at 180, 181-82 (statement of Dennis Duggan, attorney representing C-Tran).

[17]In the Matter of Informal Conference on Citations Issued to C-Tran, Transcript of Proceedings, at 35 (Stoner).

[18]In the Matter of Informal Conference on Citations Issued to C-Tran, Transcript of Proceedings, at 216 (Stoner).

[19]In the Matter of Informal Conference on Citations Issued to C-Tran, Transcript of Proceedings, at 35 (Stoner).

time drivers would go to layover places to perform the work such as boarding passengers and unloading wheelchairs that prevents drivers from getting to go to the bathroom and/or supervisors would watch buses while drivers left to void.[20] In spite of the hearing officer's entreaty to the parties that they settle the dispute amicably and spare him the writing of a decision that neither side would like,[21] the ATU and C-Tran could not,[22] and on February 14, 1995, the hearing officer ruled that the employer's schedules did not allow drivers sufficient time to use toilets and that C-Tran had to allow such time.[23]

During the pendency of C-Tran's appeal to the Board of Industrial Appeals, the ATU's general counsel informed the judge that in the interim only in a small proportion of routes had the employer made changes providing for what appeared to be adequate time for drivers to use the bathroom. Moreover, the changes had been made without surveying drivers and without planners' having ridden the buses. Despite this "abysmally inadequate" response and failure to address the issues raised by the citations,[24] in July 1995—just as WISHA was issuing the aforementioned citation to a local passenger transportation subsidiary of Ryder—as a result of a mediation and settlement conference held on June 28 involving the employer, Local 757, and the Washington Department of Labor and Industries,[25] the citation and penalty were dropped.[26] Although the formal settlement agreement merely vacated the citation without more,[27] a state assistant attorney general stated that the Clark County bus authority "acknowledged that some routes need extra time for breaks. However, the union won only advisory

---

[20]In the Matter of Informal Conference on Citations Issued to C-Tran, Transcript of Proceedings, at 203-204 (Stoner).

[21]In the Matter of Informal Conference on Citations Issued to C-Tran, Transcript of Proceedings, at 261.

[22]In the Matter of Informal Conference on Citations Issued to C-Tran, Transcript of Proceedings, at 257-58 (Heintzman (ATU Local 757 President) and Duggan)..

[23]"C-Tran Loses Potty Break Ruling; Will Rule," *The Reflector* (Battle Ground, WA), Feb. 22, 2995, at A2.

[24]Letter from Susan Stoner to Judge Sally Satwell, Board of Industrial Appeals (June 21, 1995).

[25]In re Public Transportation Benefit Area dba C-Tran, Docket No. 95 W058, Notice of Settlement Conference (Board of Industrial Insurance Appeals State of Washington, June 1, 1995).

[26]In re Public Transportation Benefit Area dba (C-Tran), Docket No. 95 W058, Order on Agreement of Parties (Board of Industrial Insurance Appeals State of Washington, Sept. 20, 1995).

[27]In re Public Transportation Benefit Area dba (C-Tran), Docket No. 95 W058, Agreement of Parties (Board of Industrial Insurance Appeals State of Washington, Sept. 11, 1995).

power over routes...." The state conceded that the "agreement doesn't mean an immediate solution for uncomfortable drivers," but the Department's "'interest was trying to get the drivers an opportunity to go to the bathroom. It's more important that the problem be solved than fight over the penalties.'" The official argued that "[e]verybody anticipates that this will result in some major schedule changes down the line."[28] In fact, however, the problem was not solved: bus drivers there and elsewhere in Washington still have to wait up to six hours to urinate.[29] The union stated that although C-Tran encouraged drivers to take breaks, the system's officials "admitted that the system would 'break down' if they did so."[30]

Susan Stoner, Local 757's general counsel, who argued the case before the hearing officer, reported that regardless of the wording of the citation, the central issue at the hearing was C-Tran's failure to schedule routes so that drivers had enough time to void. She added that even though the citation was rescinded, the union "won" the case and there was eventually some improvement. However, the union's collective bargaining agreements in Washington and Oregon do not provide for formal breaks of any kind (including rest or meal breaks) so that it is a perpetual struggle for drivers to find any time for any of these activities.[31]

The union's account was confirmed in part by C-Tran's Human Resources Director, Arlene Doern, who had held this same position during the 1990s and also attended the hearing. She agreed that, although the citation was for failure to provide a toilet or transportation to a toilet (which prompted her to laugh because the drivers, after all, have transportation), the real dispute at the hearing was over whether drivers had enough time to use the toilet. However, she insisted that drivers could then and can now stop the bus whenever they need to go so long as they inform the dispatcher (in code, if they prefer, to avoid embarrassment in front of the passengers). Moreover, Doern believed that by pointing out that runs average 30 minutes, with the longest being 60 minutes, and asserting that drivers did not need to go to the bathroom at such intervals anyway, she could demonstrate that, despite the fact that drivers had no contractually guaranteed formal breaks, they could still eat and go to the bathroom "in between," and

---

[28]Jeanette Steele, "C-Tran's Drivers Gain a Bit," *The Columbian* (Vancouver), July 9, 1995, F1 (Lexis).

[29]Telephone interview with Susan Stoner, Gen. Counsel, ATU Div. 757, Portland, OR (Oct. 21, 2002).

[30]"C-Tran Loses Potty Break Ruling" (quoting Jason Reynolds).

[31]Telephone interview with Susan Stoner, general counsel, ATU Div. 757, Portland, OR (Oct. 25, 2002); "Agreement Clark County Public Transportation Authority Benefit Area and Amalgamated Transit Union Local No. 757, September 1, 1999 - August 31, 2002, For Fixed Route and Paratransit Operators," on http://atu757.org/contracts/ctran99.pdf.

that, since many worked split shifts, breaks were not a problem.[32] In fact, however, once a driver, as a result of the normal press of driving and dealing with passengers, has fallen behind schedule, "it could," as one driver observed, "easily take the driver 6 to 8 hours to catch up,"[33] thus leaving him or her without a break for that length of time.

Mark Bevington, a veteran full-time C-Tran driver and Local 757 executive board member who was the employee representative at the time of the WISHA investigation, reported that before the citation drivers often had had to go three or four hours without being able to urinate because there was not enough time planned into the schedules for them to take the time to urinate at the end of the line before heading back to the beginning of a run, where the same tightness of schedules might again prevent the driver from urinating. When the union filed a complaint with WISHA, a WISHA inspector actually sat on the bus observing the tight scheduling that made stops to urinate impossible. The outcome of the hearing, in Bevington's view, was WISHA's informing C-Tran of the measures that it had to adopt to comply with the law—building more time into the schedules for stops and providing (for) toilets at stops, both of which C-Tran in part has done. Bevington also noted that as a result of the mediation (which was part of the employer's appeal process), drivers obtained more of a say in scheduling by virtue of sitting on the committee that builds routes. Consequently, drivers did gain somewhat more time to go to the bathroom at the end of a route, although similar problems began cropping up again after a few years. The real and ironic problem is that drivers suffer from both lack of access and stress resulting from the displeasure and anger they perceive or anticipate from passengers who just want to get to work or home as fast as possible and are not focused on the drivers' problems. Bevington, who believed that hardly any drivers stop in the middle of a run to urinate, agreed with Stoner that the drivers' (involuntary) complicity in the suppression of their own voiding would make a successful OSHA case very problematic in the face of the employer's truthful claim that it does not stop drivers from delaying the bus; indeed, he observed that the company would pay overtime to any driver who, by taking the time to urinate, wound up working overtime.[34]

The accompanying cartoon (by Hekate, a long-time C-Tran driver and former

---

[32]Telephone interview with Arlene Doern, Human Resources Director, C-Tran, Vancouver, WA (Oct. 25, 2002); telephone interview with Stoner (Oct. 25, 2002).

[33]Email from Hekate to Marc Linder (Jan. 1, 2003).

[34]Telephone interview with Mark Bevington, C-Tran driver and ATU Local 757 executive board member, Vancouver (Oct. 26, 2002). Interestingly, Bevington did not see himself as in the same position as his colleagues because as a former truck driver he had "learned to hold it."

WHY DON'T THEY JUST CALL IN?!?

union executive board member) nicely captures the multifaceted pressures operating on drivers not to take bathroom breaks. In light of this psycho-socioeconomic situation, the hearing officer's tongue-in-cheek suggestion to C-Tran might be worth considering: "What would you do to take the pressure off the drivers? I mean, would you post big signs on the bus that say, our policy is that the driver gets to stop a bus when they need to go to the restroom, and if you have any complaints about this policy, call Tom?"[35]

The brunt of the burden, according to driver-cartoonist Hekate, is borne by drivers with less seniority, since those, like her, with more seniority can bid for those routes that do have enough time built into them to enable drivers to go to the bathroom at the transit stations at the end of the line.[36] As of late 2001, Local 757 estimated that at least 3,000 of its 4,500 members "do not have immediate access to restroom facilities." To be sure, the union added, drivers often "feel unable to use the restroom when needed...even when there are no institutional barriers to restroom use." As a sample survey revealed, "almost one-half...practices voluntary retention and voluntary dehydration as means of avoiding perceived passenger and supervisory disapproval. Almost one-half also reported experiencing illnesses associated with urine retention and dehydration."[37] Bus drivers' acquiescence in such social pressure constitutes a major obstacle to successful prosecution of an OSHA complaint: employers' (truthful) response that all a driver would have to do is radio in to the dispatcher and explain that he or she is stopping the bus to go to the bathroom would, in Stoner's view, be difficult to overcome.[38] In any event, WISHA has apparently never again issued such a bus-driver-related toilet-access citation to any employer.[39]

Remarkably, bus drivers in Britain are exposed to a similar predicament. According to Martin Mayer, the Transport and General Workers Union's national representative for UK bus drivers:

> Access to toilets is a problem for bus drivers. Although bus stations normally have toilet facilities, and sometimes separate facilities for bus staff, many bus drivers operate routes which do not serve the bus station and so have no access to toilets. The UK has seen a marked decline in public toilet provision over the years, particularly in suburban areas due to vandalism and cost cutting. In Sheffield (my home city) many bus routes are

---

[35]In the Matter of Informal Conference on Citations Issued to C-Tran, Transcript of Proceedings at 236

[36]Telephone interview with Hekate, Vancouver, WA (Dec. 27, 2002).

[37]ATU Div. 757, Application for OR-OSHA Occupational Safety and Health Training and Education Grant Program: The Eight and Two Campaign at 13, 6 (Oct. 26, 2001)

[38]Telephone interview with Susan Stoner (Oct. 25, 2002).

[39]Telephone interview with Mike Rohde, WISHA, Olympia (Oct. 22, 2002), who mistakenly believed that WISHA had not issued any toilet-related citations since 1998.

"cross-city" i.e. travelling back and forth from one outer terminus to another via the city centre but not via the bus station and not timed to spend any time in the city centre. With spells of work of up to 5 hours or more this can be very difficult if no toilets are on the route. The City Council refused to accept it was their responsibility..... The public body in charge of bus stations and bus stops...said it was up to the bus operators. The Bus operators said they would not provide only to find other bus operators' drivers were using the toilets whilst not contributing to the cost. Actually we can only get round the problem by encouraging drivers to call on the radio for a toilet relief, in which case the driver gets permission to drive off route to a toilet facility. Although this is agreed, many drivers will not do so and simply urinate behind bushes at terminus or even use an empty bottle when it's quiet and passengers are off the bus. We are currently in talks with the City Council on public funding for specialised bus shelters at key terminal points with toilet facility attached. In London where special bus drivers' toilets were once common under the public sector, the problem has returned with private bus operators unwilling to foot the bill. Again the impetus is to discuss this with the public sector infrastructure provider, in this case Transport for London. We are concerned at the possible health and safety risks to drivers of "holding it in" on frequent occasions throughout their working career. It is also particularly unattractive to female employees.[40]

In Washington's neighbor, Oregon, which has yet to cite an employer for restricting employees' access to the toilet, the ATU met with the state OSHA on March 2, 1999, to explain bathroom access problems. The union stressed the following obstacles:

> Bus drivers are not allowed timely access to the restrooms (this includes scheduled stops which are not long enough for the drivers to get to, use and return from restrooms).
> Identified bathrooms have no bus parking or hazardous bus parking....
> Identified bathrooms are popular areas (too busy to be able to use in scheduled amount of time).
> Bus drivers are concerned about leaving the bus unattended in poorly lit areas to walk to the bathroom.
> Bus drivers are concerned about walking alone in poorly lit areas during hazardous times...to an identified restroom.
> Bus drivers are often embarrassed to ask permission over a radio/cb unit....
> Bus drivers suffer subtle punishments when their routes are late or if they ask to use the restroom.

In rejecting the union's request that Oregon-OSHA develop a rule to address this situation, the agency replied that section 1910.141(c)(1)(i), as interpreted by the Memorandum of April 6, 1998, addressed the issue and that the agency would

---

[40]Email from Martin Mayer to Marc Linder (Oct. 6, 2002).

enforce the standard if an employee filed a complaint.[41] To be sure, Oregon OSHA (like Washington OSHA) takes the position that a bus driver is a "mobile crew,"[42] subject to a different standard,[43] but, as the Federal OSHA Memorandum itself declares, even such workers are entitled to "equivalent" protection.[44]

Washington OSHA did cite at least one employer for denying workers access to toilets after April 6, 1998. In 1999 WISHA cited Custom Apple Packers, a large commercial packing shed in Quincy (which received a total of 16 citations between 1990 and 2002) because it had not provided toilets. The workers, whose shifts ran from 7 a.m. to 5:30 p.m. with breaks from 9:45 to 10 a.m., 12:30 to 1 p.m., and 3:15 to 3:30 p.m., complained to the agency that the employer

> has restricted employees from using bathroom facilities.... A foreperson...informed Hispanic workers that we could only use bathroom facilities during scheduled breaks. This meant that workers would have to wait two to three hours in order to use bathroom facilities. When we complained about the rule to [ ] (who speaks Spanish), she informed us that this rule was an order from...a supervisor. We were not given anything in writing about this rule. [ ] informed us that workers who violated the rule would be given warnings and then they would be fired. When [ ] asked about changing the rule because she could not wait two to three hours, [ ] told her to wear Pampers. Also, when [ ] informed [ ] that she had problems with her kidneys, [ ] pointed at a bucket on the floor and told her that she could use the bucket. Due to this workplace rule both [ ] and [ ] developed urinary tract infections. [ ] was taken from work to the hospital emergency room....[45]

Omitted from the OSHA report was the part of the complaint that explained that the urinary tract infection had caused one of the women workers "severe and painful abdominal pain during her work shift, resulting in her collapsing to the warehouse floor and wetting herself in front of dozens of co-workers. Along with the physical pain that [ ] has endured as a result of her UTI, the episode has caused her tremendous shame and humiliation."[46]

---

[41]Letter from Barry Jones, Manager of Enforcement, Oregon OSHA, to Susan Stoner, Gen. Counsel (June 9, 1999) (also quoting the points made by the union on Mar. 2).

[42]Telephone interview with Barry Jones, Manager of Enforcement, Oregon OSHA (Oct. 22, 2002); telephone interview with Michael Wood, Sen. Prog, Mgr., WISHA (Oct. 21, 2002).

[43]1910.141(c)(1)(ii).

[44]See below Appendix II.

[45][Washington] Department of Labor & Industries, Alleged Safety or Health Hazards, Custom Apple Packers, Inc., Complaint No. 201176971 (Feb. 24 [1999]). The square brackets in the text represent names that were deleted by WISHA in making the documents available.

[46]Letter from Patrick Pleas, Northwest Justice Project, Wenatchee, to Don Wanamaker, WISHA, Yakima (Mar. 30, 1999).

The inspector's report noted that management had stated that "there had been problems with certain employees" who "would not go during the break, then 5 minutes later they would want to use the bathroom and be gone for up to 20 minutes." The report went on to note that "all the employees interviewed...seemed to be scared, most of them said they had either been denied to [sic] use the bathroom or have seen others denied the use of the bathroom during regular working hours."[47] Without imposing a monetary penalty, the inspector cited the employer: "'Provided means available for use at all times.' Management was not allowing employees the use of the bathrooms during working hours except during break."[48]

Despite the zero-dollar citation, the employer appealed the decision because it was "fearful of being sued by some of the employees and their Attorney." Although the company alleged that some employees "abused the bathroom privileges, by visiting with other employees, using the telephone, or going into the lunch room and heating their lunch early," the hearing officer, ruling that the firm in fact "did deny one or more of their employees...the opportunity to use the bathroom," upheld the citation.[49]

---

[47]WISHA Violation Report, Custom Apple Packers, Inc., Insp. No. 302194816 (Apr. 1, 1999).

[48]Custom Apple Packers, Inc., Insp. No. 302194816 (citation issued May 13, 1999). The sentenced in double quotation marks sounds like some official definition, but, according to Michael Wood, who as the Senior Program Manager is the second highest official at WISHA, it was not taken from any official written source and was probably merely the inspector's way of emphasizing the point. Telephone interview with Michael Wood (Nov. 25, 2002).

[49]WISHA Services Div., Resumption of Jurisdiction Conference Report (June 14, 1999).

15

And Bringing Up the Rear: Cal/OSHA

The best administrative machinery in the world will break down utterly if intrusted to incompetent officials.[1]

California may have the most highly developed and extensive body of labor protective legislation of any jurisdiction in the United States, and its occupational safety and health agency may issue more citations for violations of the general industry (physical) toilet standards (such as number and cleanliness of toilets) than all the other state-plan programs combined, but the state also has the country's only OSHA system that expressly, insistently, and systemically refuses to enforce workers' right to go to the bathroom when they need to do so.

The general industrial sanitation standard enforced by the California Division of Occupational Safety and Health (Cal/OSHA) differs structurally somewhat from the Federal standard, retaining some of the protective language of the American National Standards Institute standard that Federal OSHA deleted in the 1970s, such as the mandatory provision of toilet paper.[2] The Federal OSHA Memorandum was an interpretation of section 1910.141(c)(1)(i), which states that a certain number of toilets "shall be provided" for a certain number of employees. Section 3364(a) of California's General Industry Safety Orders also contains such a provision,[3] but it is followed by section 3364(b), which declares: "Toilet facilities shall be kept clean, maintained in good working order and be accessible to the employees at all times."[4]

When the California Occupational Safety and Health Standards Board issued

[1]"Report of the Standing Committee on Legal Aid Work, Appendix A: First Draft: Of a Model Statute for Facilitating Enforcement of Wage Claims," *Report of the Fiftieth Annual Meeting of the American Bar Association* 323-25 at 332 (1927).

[2]See above ch. 4 and below Appendix I.

[3]Cal. Code of Reg. tit. 8, sect. 3364(a) (2002).

[4]Cal. Code of Reg. tit. 8, sect. 3364(b).

this standard in 1975,[5] it expanded workers' bathroom rights. The previous version, issued by the Division of Industrial Safety in 1972, contained no such language, stating merely that "[e]very place of employment shall be provided with a sufficient number of conveniently located water closets for the use of employees."[6] In addition, as far back as 1889, the California legislature had mandated that all workplaces be "provided, within reasonable access, with a sufficient number of water-closets...for the use of the employees"—a provision formally still in force.[7]

Some light is shed on the meaning of the change in terminology introduced in 1975 by the public hearing that the Occupational Safety and Health Standards Board held on August 23, 1974. The chairman of the Board asked William Steffan, the Supervising Industrial Hygiene Engineer of the Occupational Health Section of the Standards Development Unit, to make a presentation on the sanitation standard and "the rationale for the development of it." Succinctly Steffan stated that the standard then in force in California "is not as effective as the Federal OSHA standard and there was a need to bring California's standard up to 'at least as effective' as the Federal Standard."[8] After several government health officials

---

[5]It was issued as sect. 3260(e)(1)(C); *Register* 75, No. 7 (Feb. 15, 1975); the following year it was renumbered sect. 3364(c); *Register* No. 29, 1976; 8 Code of Cal. Reg. Sect. 3364(c) (1977). The California Occupational Safety and Health Standards Board held a public hearing on the proposed new standard on May 20, 1976; the file at the Board indicates that no comments were made on the renumbered section 3364 at the hearing. In those years, Cal/OSHA was not required to and did not provide the kind of supporting information at the time of publishing its proposed standards that was later required by the Administrative Procedure Act. Notice of Public Hearing and Meeting of the Occupational Safety and Health Standards Board and Notice of Proposed Changes to Title 8, California Administrative Code and Title 24, California Administrative Code (Apr. 13, 1976); Order Adopting, Amending, or Repealing Regulations of the Occupational Safety and Health Standards Board (June 24, 1976); telephone interview with Marley Hart, staff services manager, OSHSB, Sacramento (Nov. 4, 2002).

[6]8 Cal. Code of Reg. sect. 3244(d), *Register* 72, No. 23 (June 3, 1972). See also sect. 3244(d), *Register* 72, No. 6 (Feb. 5, 1972).

[7]Cal. Labor Code sect. 2350 (2002); 1889 Cal. Laws ch. 5, sect. 1 at 3. This provision was incorporated into the state Labor Code by 1937 Cal. Laws ch. 90, at 185, 253. The introductory article in the chapter of the Labor Code dealing with sanitary conditions states: "All employers shall comply with standards relating to sanitary facilities adopted by the Occupational Safety and Health Standards Board...."

[8]Summary, Public Hearing, Occupational Safety and Health Standards Board, San Francisco, August 23, 1974, at 9. Although the chairman of the board referred to Steffan as with the Health Department, a memorandum from Steffan to the Board of Oct. 3, 1974, summarizing changes made in response to the public comments identified him as in the text. Both of these documents from the Board's files were faxed to the author by Andrea

objected that the Board would be adopting regulations that might conflict with other state codes and the Uniform Plumbing Code and Uniform Building Code, an official from the state Attorney General's office gave her informal opinion that "Federal statutes superceded [sic] State statutes and on down the line to the local authority."[9] A representative of the Pacific Telephone and Telegraph Company then raised an objection to the number of toilets required per number of workers, recommending instead that "Federal standards be used throughout."[10] These ratios, which were already present in the 1972 standard and provided for somewhat lower thresholds for the next higher number of toilets, were in fact dropped in favor of the Federal OSHA ratios in the final regulations. As Steffan explained to the Board in October, "it is difficult to substantiate the need for these somewhat more restrictive requirements versus those of the corresponding Federal standard. Consequently, the revised specifications correspond to the Federal standard, in accordance with recommendations from...Pacific Telephone and Telegraph Company,...Caterpillar Tractor Company,...and McDonnell Douglas Corporation."[11]

In a reprise of the Federal OSHA hearings of 1972,[12] Pacific Telephone next objected to the requirement that toilets should be within 200 feet of locations where workers are regularly employed: "They recommend that the 200 foot requirement be deleted and the old language 'every place of employment shall be provided with a sufficient number of conveniently located water closets for the use of employees' be restored, and, if this recommendation is not followed, the Company requests that a statement of exemption for existing buildings be added. Mr. Steffan said the standard is not mandatory, that the standard states 'should' which only gives an idea of what should be done."[13] Although the company's objection seemed to be directed to the 200-foot rule, which was included in the sentence following the sentence providing that "[t]oilet facilities shall...be accessible to the employees at all times," both were contained in proposed section 3260(d)(1)(C)—which, as the Board's draft correctly noted, had no equivalent in

---

Howard on Nov. 19, 2002.

[9]Summary, Public Hearing, Occupational Safety and Health Standards Board, San Francisco, August 23, 1974, at 10-11 (Carol Hunter).

[10]Summary, Public Hearing, Occupational Safety and Health Standards Board, San Francisco, August 23, 1974, at 12 (W. J. Evertz).

[11]Memorandum from William Steffan to California Occupational Safety and Health Standards Board at 2 (Oct. 3, 1974).

[12]See above ch. 4.

[13]Summary, Public Hearing, Occupational Safety and Health Standards Board, San Francisco, August 23, 1974, at 12.

the Federal OSHA standard[14]—and retention of the first sentence would have been quite clumsy if it had been followed by the "conveniently located" language. Presumably, then, this very influential California employer was also seeking to have the new and more protective "accessible...at all times" standard deleted. The only other light shed on the meaning of "accessible" came in the form of the Del Monte Corporation's questioning of the term in a requirement that washing facilities be "readily accessible to all employees" (sect. 3260(f)(1)(A)). Steffan merely responded that "this applies generally to operations where a process requires washing hands." The chairman of the Board then asked whether "readily" could be changed" to "reasonably," and it was.[15]

It is the aforementioned section 3364(b), which more clearly than the Federal standard requires permanent accessibility, that could be used to cite employers for denying access. It currently reads: "Toilet facilities shall be kept clean, maintained in good working order and be accessible to the employees at all times. Where practicable, toilet facilities should be within 200 feet of locations at which workers are regularly employed and should not be more than one floor-to-floor flight of stairs from working areas."[16] Cal/OSHA has cited employers 697 times for violating this standard since July 1990 in addition to issuing 586 citations for failure to provide the required number of toilets.[17] But the only citation that it has issued since the Federal OSHA Memorandum was issued in 1998 that could even have remotely served the purpose of requiring employers to let workers go when they need to go was issued to RPS, Inc., the second largest small-package ground carrier in North America and a subsidiary of Federal Express.[18] At its trucking terminal in Rialto, the Cal/OSHA inspector "observed that all the toilet facilities located in the shipping facilities were kept locked all the times [sic] and keys were in custody of the supervisors and co-ordinators. Therefore, the toilet facilities were not accessible to the employees all the times [sic]."[19] Although

---

[14]Occupational Safety and Health Standards Board, General Industry Safety Order, Standards Presentation (July 16, 1974) at 12 (Aug. 1974). This document is also from the Board's file.

[15]Summary, Public Hearing, Occupational Safety and Health Standards Board, San Francisco, August 23, 1974, at 13; Memorandum from Steffan to Board at 3. The provision is now located at sect. 3266(a).

[16]8 Cal. Code of Regulation sect. 3364(b).

[17]Search ("California and 3364(a)") and ("California and 3364(b)") in Lexis-Nexis OSHAIR file (Dec. 31, 2002).

[18]http://www.fedex.com/us/about/ground/pressreleases/pressrelease040799.html?link=4.

[19]RPS, Inc., Insp. No. 119952893 (Feb. 11, 2000). See also Alpha Recycling, Inc. Insp. No. 119944973 (Oct. 13, 2000), which cited an employer one of whose restrooms "is kept locked and all male employees do not have access to it."

250 employees had been exposed to this hazard, the proposed penalty amounted to only $560 and it was reduced to a mere $100 through informal settlement.[20] To be sure, this act of enforcement was continuous with the position that Federal OSHA itself had taken even before 1998 with respect to the unlawfulness of locking toilets.[21]

Cal/OSHA's total failure and refusal to enforce the right to go to the bathroom set forth in the OSHA Memorandum is underwritten by its personnel from the bottom to the top of its organizational hierarchy. Interviews revealed that several of Cal/OSHA's district offices, which constitute the program's front-line system of inspection, compliance, and enforcement, are managed by officials whose knowledge and attitudes are so strikingly deficient that they augur poorly for the protection of workers' right to a safe and healthful workplace. For example, the manager of the Concord office (located near Oakland), who, when asked whether his office received complaints from workers about lack of toilet access, allowed as how it received many complaints from workers—most of them from employees wanting to harass their employers. Even if some of these complaints had merit, he insisted that Cal/OSHA lacked any power to require employers to give workers access to the bathroom and all he could (and did) do was refer such workers to the state Division of Labor Standards Enforcement (DLSE), although he could not identify any legal provision under which that agency could cite an employer. Immediately after being read the main point of the OSHA Memorandum, he opined that he "could not enforce" such a provision because Cal/OSHA's sanitation standard deals with physical structures and not with access.[22]

Similarly, the manager of the Oakland office at first asserted that OSHA dealt only with the provision of toilets and not with access, and he, too, mentioned that workers could seek relief from the DLSE. When reminded that the Industrial Welfare Orders administered by that agency did not deal with toilet breaks, but only with mandatory 10-minute rest breaks during each four hours of a shift,[23] he noted that employers often did not comply with that mandate and seized on such rest breaks as precisely the solution for workers needing to go to the bathroom (although they are in fact no solution at all for workers needing to go between such breaks and no solution in any respect since the sanction for employers'

---

[20]http://www.osha.gov/cgi-bin/est/est1vd?11995289301003.

[21]Marc Linder and Ingrid Nygaard, *Void Where Prohibited: Rest Breaks and the Right to Urinate on Company Time* 56 (1998).

[22]Telephone interview with Bob McDowell, manager, Cal/OSHA Concord District Office (Oct. 31, 2002).

[23]E.g., Industrial Welfare Commission Order No. 1-2001 Regulating Wages, Hours and Working Conditions in the Manufacturing Industry (Effective July 1, 2002 as amended), 8 Cal. Code of Regulations sect. 11010(12)(A).

failure to provide rest periods is one hour's wage and not an injunction to provide the rest period).[24] Then, when read the aforementioned Cal/OSHA sanitation provision requiring that toilet facilities be "accessible to the employees at all times,"[25] without an audible blink, he immediately said that that phrase meant that employers had to let workers go when they needed to go. Although he had no recollection of ever having seen the Federal OSHA Memorandum, he commented that that was the way Cal/OSHA had always viewed the matter and it did not really need Federal OSHA's advice in this area. Nevertheless, he related that he was unaware of any complaints to his office from workers whose employers had been "metaphorically chaining" them to their work stations: employers had moved far enough into the twentieth century not to do that, though he admitted that failure to provide enough—or, in construction and agriculture, any—toilets was a problem.[26]

The manager of the Foster City office in the Bay Area observed that although he did get complaints about bathroom access, he had never cited an employer because, when an inspector goes to a workplace, it is "easier" to count the number of toilets and to cite an employer for failing to have an adequate number than to cite for restricting access, which requires sorting out contradictory claims of employees and management.[27] The Senior Industrial Hygienist in the Research and Standards Development Unit of the Division of Occupational Safety and Health, while not condoning this approach, sought to explain that since considerably greater amounts of inspectors' time would have to be devoted to resolving issues of credibility in such disputes than with respect to objectively quantifiable violations, it would not be surprising if few employers were cited for restricting workers' access.[28]

The manager of the Santa Rosa office reported that in 13 years he had never issued a citation for denial of access; the only telephone calls he received on the matter came from employers asking whether they had to let workers go to the bathroom, and his only response to them was to read them the provision.[29] An inspector in the Modesto office stated that when she received telephone complaints from manufacturing workers, she resolved them by faxing employers

---

[24]8 Cal. Code of Regulations sect. 11010(12)(B).

[25]Cal. Code of Regulations, tit. 8, sect. 3364(b).

[26]Telephone interview with Michael Horowitz, manager, Cal/OSHA Oakland District Office (Nov. 1, 2002).

[27]Telephone interview with Michael Frye, manager, Cal/OSHA Foster City District Office (Oct. 30, 2002).

[28]Telephone interview with Bob Barish, San Francisco (Nov. 5, 2002).

[29]Telephone interview with Jimmie Jones, manager, Cal/OSHA Santa Rosa District Office (Oct. 30, 2002).

information without follow-up investigations.[30]

Nor was this policy and attitude confined to the district offices. The manager of Cal/OSHA's Anaheim regional office in charge of four district offices unequivocally stated on a speaker telephone, together with an official of the agency's Research and Standards Development Unit (who was also a former district office manager), that according to enforcement history, "accessible" in section 3364(b) merely means "not locked." Any other aspect of access, such as denial of permission, would, they both agreed, be a matter of labor-management relations or for referral to the DLSE, although they too were at a loss to cite a relevant legal provision that the DLSE could enforce in this regard. Neither had ever heard of the OSHA Memorandum, while the regional director commented that she had never heard of a worker who had said that his employer would not let him go to the bathroom.[31]

This cavalier approach becomes somewhat more understandable against the background of this pithy response by Cal/OSHA's Professional Development and Training Unit Coordinator to a question about enforcement of the OSHA Memorandum: "We don't enforce Federal interpretations."[32] The only serious, albeit inadequate, response the author was able to secure came from Cal/OSHA's chief legal counsel, who at least suggested that the author write a letter to all the highest officials of the agency explaining that inspectors and district managers denied that Cal/OSHA had any authority to require employers to let workers go to the bathroom and another letter to Federal OSHA explaining that Cal/OSHA was failing to enforce the standard as interpreted by the Memorandum.[33]

When asked, the DLSE stated that it often receives complaints from workers (especially in fruit and vegetable packing houses) who are not allowed to go to the bathroom between their statutory rest and meal breaks; it informs such workers that under the law they must be allowed to go. Unsurprisingly, the law the DLSE has in mind is not the labor standards regulations that it enforces, but

---

[30]Telephone interview with Karen Helton-Jorgensen, inspector, Cal/OSHA Modesto District Office (Oct. 30, 2002).

[31]Telephone interview with M. Kenady and Joel Foss, Cal/OSHA, Anaheim (Nov. 18, 2002).

[32]Email from Jack Oudiz to Marc Linder (Oct. 31, 2002). When the author replied that California, like all state-plan states, was required both by the Memorandum itself and as a condition of being approved by Federal OSHA to enforce OSHA standards, including this interpretation, at least as effective as Federal OSHA's, and that if Cal/OSHA was openly refusing to enforce the interpretation, he would like to get that statement officially confirmed by the director of Cal/OSHA for presentation to Federal OSHA, the official responded: "Thanks for the lecture. This conversation is over. Take your business elsewhere." Email from Jack Oudiz to Marc Linder (Oct. 31, 2002).

[33]Telephone interview with Michael Mason, San Francisco (Nov. 27, 2002).

those under the jurisdiction of Cal/OSHA, to which it also refers complaining workers for help. Informed of all the high- and low-ranking Cal/OSHA officials who had denied having such jurisdiction and had declared that they referred all such complainants to the DLSE, one of the Division's field investigators laughed and offered to get to the bottom of the confusion. Although he failed to do so, he did refer the author to the DLSE legal staff.[34] A lawyer in the DLSE San Francisco office, who initially stated unequivocally that under the law workers did have a right to go to the bathroom, cited the following sections of the agency's *Interpretations Manual* as the relevant legal authority:

Rest Period Is Not Limited To Toilet Breaks. The intent of the Industrial Welfare Commission regarding rest periods is clear: the rest period is not to be confused with or limited to breaks taken by employees to use toilet facilities. This conclusion is required by a reading of the provisions of IWC Orders, Section 12, Rest Periods, in conjunction with the provisions of Section 13(B), Change Rooms and Resting Facilities, which requires that "Suitable resting facilities shall be provided in an area separate from the toilet rooms and shall be available to employees during work hours."

Allowing employees to use toilet facilities during working hours does not meet the employer's obligation to provide rest periods as required by the IWC Orders. This is not to say, of course, that employers do not have the right to reasonably limit the amount of time an employee may be absent from his or her work station; and, it does not indicate that an employee who chooses to use the toilet facilities while on an authorized break may extend the break time by doing so. DLSE policy simply prohibits an employer from requiring that employees count any separate use of toilet facilities as a rest period.[35]

When the author pointed out to the DLSE attorney that these provisions not only in no way vindicated an independent right to go to the bathroom apart from and outside of the rest breaks mandated by the DLSE, but in fact stated that no such right existed even as an extension of a legally required break, the official made a complete about-face, stating abruptly: "This is all the law there is—we have no jurisdiction. All we can do is collect wages."[36]

The acting chief counsel of the DLSE, confirming, after reading the text aloud, that the aforementioned interpretation did not confer any such right to go to the bathroom, added that the question of the right to go to the bathroom "hasn't

---

[34]Telephone interview with Ruben Navarrette, field investigator, California DLSE, Fresno (Nov. 19, 2002).

[35]Division of Labor Standards Enforcement, *The 2002 Update of the DLSE Enforcement Policies and Interpretations Manual* sect. 45.3.4-45.3.4.1 at 45-8 (June 2002), on http://www.dir.ca.gov/dlse/DLSEManual/dlse_enfcmanual.pdf.

[36]Telephone interview with Ramon Garcia, attorney, DLSE, San Francisco (Nov. 20, 2002).

come up in our enforcement." Moreover, whatever enforcement the agency engaged in with regard to employers' failure to give their employees rest periods focused on securing workers an hour's wages, not on enjoining employers to comply with the law. When prompted, she promised to speak to someone at Cal/OSHA about the fact that the two agencies have been pointing at each other.[37]

Finally, the author contacted Federal OSHA's San Francisco Regional Office, which has oversight over the Cal/OSHA program. The official in charge of state plans, the Director of Analysis and Evaluation, had never heard of the April 6, 1998 Memorandum and wondered why his interlocutor knew so much about the general industry sanitation standard. Although at first he strenuously contended that Cal/OSHA had been correct in claiming that it had no obligation to insure that its standard and interpretation be at least as effective as the Federal OSHA standard, after heated discussion he was finally constrained to agree that, if Cal/OSHA officials did in fact make the statements and formulate the policy reported in the preceding paragraphs, the agency was not in compliance with its legal obligations. He stated that if the author presented these facts in writing, he would investigate and the author would be pleased with the result, although it might take six months for the policy to be changed and the agreement to do so would not be in writing.[38]

---

[37]Telephone interview with Anne Stevason, acting chief counsel of the DLSE (November 27, 2002). The DLSE's web-based Info Line invites the "general public" to email inquiries to receive "information...concerning the rights and responsibilities of employees and employers in the State of California." http://www.dir.ca.gov/dlse/dlse infoline.htm. An email inquiry concerning bathroom breaks generated this instantaneous response: "Due to staffing constraints, DLSE is unable to answer the high volume of emails sent to this mailbox." DLSEInfo@dir.ca.gov (Nov. 23, 2002).

[38]Telephone interview with Alan Traenkner (Dec. 5, 2002).

Part VI

Human Waste and Capitalist Efficiency

"The current agreement...provides for spacing of rest periods and lunch periods so that 2 to 2½ hours is the maximum work period without a break during an 8-hour work day. ...

"It most certainly is not unreasonable to expect one to learn to regulate and train one's self [sic] to work periods of 1½, 2, or 2½ hours. This is expected of us from childhood on into school, traveling, meetings, social events, and finally work. Without this minimum discipline, it would be chaos trying to schedule most organizations."[1]

---

[1]Jones Dairy Farm v. Amalgamated Meat Cutters and Butcher Workmen of North America, Local P-1236, 72-2 Labor Arbitration Awards (CCH) ¶ 8639 at 5256 (1973) (company response to union grievance over three-day suspension of worker for making nine emergency-basis bathroom visits over 27 months).

16

A Golden Age for the Golden Stream?

> [I]t wouldn't cause problems to let workers go when they have to go. "When workers are adults...they can be trusted to go to the bathroom."[1]

This key question posed by this book, as formulated in its opening sentence, is whether OSHA's issuance of its toilet-standard Memorandum in 1998 has in fact meant that millions of workers whose employers had previously restricted their access to the bathroom can now void when they need to. In lieu of conducting a large-scale national workplace voiding interview survey—which remains a desideratum of empirical research—it is necessary to rely on an array of other information and informants in order to gain a sense of the Memorandum's impact.

The data presented in Part V demonstrate that after the promulgation of the Memorandum, Federal and state OSHA began to cite a few firms whose employees had complained to the agency that their employers were not letting them go to the bathroom. Such enforcement marks a programmatic step forward compared with OSHA's pre-1998 pat response to complainants that they had no such right. Today, as far as can be determined, if a worker—outside of the most populous state, California—who is a current employee files such a complaint, OSHA will at least promptly investigate and, once in a while, even issue a citation, though only Iowa OSHA, galvanized and motivated by an openly pro-labor commissioner, has thus far imposed a monetary penalty that can even arguably be considered a deterrent to a large employer. Based on fragmentary data, Chapter 11 also offered a very tentative estimate of the much larger volume of complaints about restricted toilet access that OSHA might receive annually, but the number of workers who acquiesce in such violations is as unknown as the proportion of the covered workforce that is even aware of the existence of a right to void at work.

---

[1]Scott Whisnant, "Hog Plant Union Talks Include Call for Restroom Breaks,"*Morning Star* (Wilmington, NC), Aug. 21, 1997, at 1B, 3B (Lexis) (quoting UFCW spokesman Greg Denier).

However, inspections, citations, fines and other formal aspects of enforcement do not exhaust the indicators of the potential dismantlement of employers' autocratic control over working time in general and the duration of and intervals between workers' withdrawal from work in order to empty their bladders and colons. After all, even though much more egregious violations of workers' toilet-break rights remained totally hidden to public view, and even though the Jim Beam citation involved no monetary penalty at all, widespread reporting about that case not only brought the existence of this employer obligation and correlative employee right to the attention of many companies that had either never heard of it or failed to take it seriously, but may even have persuaded them that such colossally bad publicity and corresponding ill will had made voiding rights a distinctly suboptimal field in which to confront labor and labor-protective government regimes. Likewise, awareness of the legal backing furnished by OSHA's new interpretation may have mobilized unions and workers to press employers more aggressively to abandon rigidly held restraints.

Two possible sources of information about this process are labor union and OSHA officials.

Unions' Initiatives and Experiences

An appropriate place to begin an inquiry into the impact of the Memorandum on voiding rights in unionized workplaces is the birthplace of the resistance movement that eventually prompted OSHA to declare at-will bathroom breaks an enforcible right. At the Hudson (now Tyson) poultry slaughter plant in Noel, Missouri, where the UFCW workers took up the struggle against autocratic bans on voiding that led to the first OSHA citation in 1997,[2] Karen Smith, the business agent of Local 2008, reported that bathroom access has not been a problem since. A clear understanding prevails that management has 15 minutes to get a worker relief and if the worker is not promptly relieved, he or she just walks away from the line; when, as sometimes happens, workers do have to exercise that discretion, no disciplinary consequences follow.[3]

A picture-perfect example of how vigorous national and local union action backed up by the OSHA Memorandum can force overreaching employers to drop their plans is offered by Local 2 of the UFCW, which has organized workers in animal slaughter plants in Oklahoma and Kansas. In 2002, Seaboard Farms decided to restrict toilet access at its Guymon, Oklahoma hog slaughterhouse[4] in

---

[2]See above ch. 2.

[3]Telephone interview with Karen Smith (Dec. 16, 2002).

[4]On the plant and company, see http://www.sierraclub.org/factoryfarms/rapsheets/

order to deal with what it regarded as abuse of restroom breaks. The new rule it devised ordained that, in the absence of a medical excuse, workers would be permitted to go to the bathroom, outside of their 15-minute mid-morning break and 30-minute lunch break, only a couple of times a week. When asked what exactly this latter phrase meant, the union's feisty secretary-treasurer, Martin Rosas, replied, "like once on Monday and Thursday." Rosas informed management that even if it were true that some among the plant's 2,300 workers went to smoke or make phone calls instead of going to the bathroom, the company could not punish all the workers merely to discipline the few abusers.[5] He then asked the national union's occupational safety and health office to fax him a copy of the Memorandum—two years earlier the office had discussed the Memorandum with stewards of the local, at which time Rosas became familiar with it[6]—which he then showed to management. After familiarizing themselves with it, the managers decided that very same day that it was not worth trying to implement the new rule, which they rescinded before it ever went into effect. Unlike the Jim Beam management, Seaboard's did not complain that the OSHA standard interpretation had deprived it of its power to discipline workers; on the contrary, it intended to continue monitoring workers and documenting those who went elsewhere than to the toilet on their bathroom breaks. Rosas also attributed the union's successful counter-attack to the number of relief workers or lead-men (about one per 20 or 40 workers depending on the area of the plant), for the hiring of whom it had recently been able to negotiate. Rosas reported that a successful outcome to a similar confrontation had also taken place at the Excel-Cargill cattle slaughterhouse in Dodge City, Kansas.[7]

At three poultry processing plants organized by UFCW Local 1996, the union was able to turn supervisors' interference with workers' right to go to the bathroom into durable vindications of that right. According to area director Curtis Williams, the union, having learned about the OSHA Memorandum from meetings with the UFCW's occupational safety and health office, filed grievances at Gold Kist in Athens, Georgia, Crider's in Stillmore, Georgia, and Columbia Farms in Columbia, South Carolina, all of which were resolved at the first level with the human resources directors. The upshot has been that workers in those plants just tell the line leader and go; they do not have to wait for the line leader to find a "pee boy" to replace them because that procedure is precisely what led to the problem in the first place. After the successful resolution of those griev-

---

oklahoma/seaboard_guymon.asp.

[5]Telephone interview with Martin Rosas (Dec. 13, 2002).

[6]Email from Jackie Nowell, Director, Occupational Safety and Health Office., UFCW, Washington, D.C., to Marc Linder (Dec. 15, 2002).

[7]Telephone interview with Rosas.

ances, workers in those plants have not reported any further problems. Significantly, Williams observed that the union keeps the workers informed of their right to go to the bathroom because it is "one of the few positive things we have in those plants."[8]

UFCW Local 1149, which represents the 950 workers at the IBP-Tyson hog slaughter plant in Perry, Iowa (all of whose output is exported to one company in Japan), achieved a similar success. In 2001, according to Jim Olesen, the president of the local, some supervisors initiated "goofy shit" such as threatening to limit unscheduled bathroom breaks to once a week or to demand that workers go to a doctor at their own expense to determine what was allegedly wrong with them medically.[9] After a worker made a complaint to Iowa OSHA, the agency conducted an inspection of the plant, but failed to issue a citation to IBP-Tyson.[10] Nevertheless, according to Olesen, who attended the closing conference,[11] the inspector put the company on notice that future violations would be cited—a warning "put the clamp on it right away." Since then the company has agreed that supervisors will provide relief within 15 minutes of being asked for it; the union has told its members to wait 15 minutes and then to go, and if there are any disciplinary consequences, the union will back them up. As an example of the company's compliance following the OSHA warning, the union president observed that when a supervisor on his own began monitoring the number and duration of workers' bathroom breaks, the union complained to upper manage-

---

[8]Telephone interview with Curtis Williams, Area Director, UFCW Local 1996, Sewanee, GA (Dec. 26, 2002).

[9]Telephone interview with Jim Olesen, President, UFCW Local 1149, Perry, IA (Dec. 16, 2002).

[10]IBP-Tyson, Insp. No. 304790884 (complaint filed Apr. 22, 2002; case closed May 9, 2002). http://www.osha.gov/cgi-bin/est/est1xp?i=304790884. The case file was subject to an open records request, but because of an "extensive" backlog, Iowa OSHA was unable to redact and make the records available in time for publication. Email from Mary Bryant, Administrator, Iowa OSHA, to Marc Linder (Jan. 8, 2003).

[11]Olesen stated that the inspector did not explain at the conference why Iowa OSHA decided not to issue a citation. Telephone interview with Jim Olesen (Jan. 8, 2003). In response to an inquiry as to why Iowa OSHA was giving IBP-Tyson two bites at the enforcement apple (email from Marc Linder to Byron Orton, Jan. 8, 2003), Byron Orton, using what may have been bureaucratic understatement to express the same warning mentioned by Olesen, explained: "A citation was not issued in this instance because the CSHO did not have sufficient evidence showing a violation of the sanitation standard. During the closing conference the CSHO made some observations intended to assist the employer in continued compliance with the standard and my interpretation issued in Jan. '98." Email from Byron Orton to Marc Linder (Jan. 14, 2003).

ment, which ordered the supervisor to stop monitoring.[12]

At the Swift slaughter plant in Marshalltown, Iowa, to which OSHA had issued two citations for denial of access in 2000 and 2001,[13] the secretary-treasurer of UFCW Local 1149, Ross Boyer, emphasized that the April 6, 1998 Memorandum "really helped us" because it "gave us clout to force the employer to give breaks." Boyer, who began working at the plant in the 1960s, noted that during the time from then until the 1980s, when the union was much stronger, workers would simply shut down the line in mass and go to the toilet. However, as the strength of the union declined, workers lost not only that informal power, but also their scheduled mid-afternoon break; consequently, they had (and have) no scheduled break during the three hours following their 30-minute lunch, although about three-fourths of the workers have to use the bathroom during that time. The advent of the OSHA Memorandum marked a turning point, enabling the union to force Swift to take corrective action before the union had to file a formal complaint with OSHA. Nevertheless, Swift has not totally abandoned its hard-line position: at the end of September 2002, it began to station an employee in the hall in front of the bathrooms near the kill floor all day long whose sole task was to ask workers their names and to write down the time they entered and left the bathroom. The company maintained this monitoring for a couple of months, presumably, in Boyer's view, to let workers know that it was keeping an eye on them. To be sure, Boyer conceded that a small proportion (perhaps one-tenth) of the workers "abuse" breaks (for example by staying in the bathroom 25-30 minutes), "screwing things up" for the other workers, who not only get to go to the toilet less often, because supervisors say there are not enough relief workers to go around, but must also in many cases do those workers' jobs for them until they return.[14]

At the John Morrell hog slaughter plant in Sioux City, Iowa, which employs 1,200 workers, of whom about 80 percent are Hispanic, Bill Buckholtz, the secretary-treasurer of UFCW Local 1142, who has worked there for 22 years, observed that the root cause of the denials of access that had led to the OSHA citation in 2001[15] was the foremen's totally ignoring workers' requests to use the toilet—to the point that workers urinated in their pants. This illegality, in his view, was and is fostered by upper management's having given too much power to supervisors. Interestingly, at the time, he had not even been aware of the OSHA Memorandum: he filed a complaint with OSHA simply because he did not know what

---

[12]Telephone interview with Olesen (Dec. 16, 2002).

[13]See above ch. 12.

[14]Telephone interview with Ross Boyer, secretary-treasurer, UFCW Local 1149, Marshalltown, IA (Oct. 3 and Dec. 30, 2002).

[15]See above ch. 12.

else to do. Buckholtz noted that at the closing conference it was made clear to management that in the future when workers on the chain asked to go to the toilet, if the company could not provide relief within 10-15 minutes, the workers could just walk off the line. In order to facilitate relief, the union has told management that it will waive the collective bargaining agreement's strict ban on management's performing bargaining unit work; thus with the need to go to the bathroom treated as an emergency, supervisors can replace workers when no relief worker is promptly available. These bathroom breaks are vital because workers have only a 15-minute break in the morning and 30 minutes for lunch, with no afternoon break except on days when they work overtime. Since some workers have to spend as much as 10 minutes doffing and donning protective equipment and getting to and from the toilets, and the bathrooms at break times are so crowded that they also often have to wait in line, rest is one thing that they do not get during their 15-minute mid-morning rest breaks.[16]

In spite of these understandings, the year 2002 witnessed yet another violation when a (female) supervisor told a worker who said she needed to go to the bathroom to deal with menstrual bleeding that she did not need to go; after the worker bloodied her clothing, Buckholtz told the plant manager that he was going to file another complaint with OSHA; the manager pleaded with Buckholtz not to do so, offering to remove the offending supervisor, who Buckholtz assumed was transferred to quality control, if she still works there at all. Buckholtz believed that OSHA's citation—backed up by extensive and embarrassing local media attention—has "carried a lot of weight" in keeping the company compliant. However, he said that he would be very surprised if toilet access problems did not flare up again because management was constantly testing and feeling out the union on all issues to see how far it could push its powers. Although the local did inform workers of their rights after OSHA issued the citation—Buckholtz had apparently been unaware of Orton's memorandum of January 21, 1998 until the author mentioned and faxed it to him—with 20 to 30 new workers hired weekly and an annual turnover approaching 100 percent, the union did not and could inform workers every day; consequently many workers, especially those from Latin America, may be unaware of their right to void when they need to and may not complain when supervisors prohibit them from leaving the line.[17]

A more differentiated picture of developments at the John Morrell plant in Sioux City emerged from the remarks of Ron Derochie, president of UFCW Local 1142. Derochie, who has worked there since 1959, observed that until the late 1980s, the custom had always been that someone working on the chain who

---

[16]Telephone interview with Bill Buckholtz, secretary-treasurer, UFCW Local 1142, Sioux City, IA (Oct. 16, 2002).

[17]Telephone interview with Buckholtz.

needed to go to the toilet would shout to the foreman that he needed relief and the foreman would send a spell-out man; if the spell-out man did not come soon, the worker would shout again that he needed relief and would just go. Since the late 1980s, however, as the union's strength waned, it has been a constant struggle because young foremen, trying to make a name for themselves, periodically refuse to let workers go, some of whom have urinated on themselves. Although the OSHA Memorandum has helped, Derochie noted that the union had to "jump on" upper management every once in a while to rein in these foremen. To be sure, Derochie could not say for certain whether the foremen were operating on their own when engaging in such unlawful practices.[18]

In spite of the $36,000 penalty that Iowa OSHA assessed against the Excel plant in Ottumwa,[19] the prospects of future compliance by the employer seemed dim in light of the policy remarks made by the corporate spokesperson. Denying that there had been any further problems, Mark Klein stated that employees with medical notes could go to the bathroom whenever they needed to; asked about toilet access for those without medical notes, he commented that they could go during breaks, which he incorrectly asserted were three in number—mid-morning, lunch, and mid-afternoon. In fact, unless there is overtime, workers have no afternoon break and must work three hours straight after lunch. In response to a question as to whether such workers could go to the bathroom between breaks, Klein stated that the supervisor would let them go "if they really have to." When the author expressed mystification as to how anyone (including a supervisor) could possibly know when another human being "really had to" void, Excel's spokesperson pooh-poohed this skepticism, asserting that perhaps it would be difficult to know in the abstract, but in actual circumstances "common sense tells you" whether someone "really" has to go. The crucial point here is not that Klein was unwilling (and presumably unable) to specify those circumstances, but that the corporate media representative of the huge Excel-Cargill food processing empire felt no compunction about telling a university researcher who purported to be an expert on the subject that the corporation had implemented a policy that Klein should have known (if he did not know) was blatantly unlawful under OSHA, which accords the employer no power to await the results of the deployment of a urinary divining rod or lie-detector test before letting employees go to the bathroom.[20]

---

[18]Telephone interview with Ron Derochie, president, UFCW Local 1142, John Morrell plant, Sioux City (Hinton, IA, Oct. 12, 2002).

[19]See above ch. 12.

[20]Telephone interview with Mark Klein, Minneapolis (Oct. 15, 2002). When the author called the Excel plant, Ken Larson, the assistant director of human resources, referred him to "corporate" and specifically Klein, whose title he characterized as "media." Tele-

That Excel may in fact use precisely such methods of harassment to deter workers from exercising their right to answer nature's call when they need to was made plausible by an interview with Ron Brown, the president of UFCW Local 230 in Ottumwa, who remarked that it was "funny" that the author had called to interview him about bathroom breaks at Excel because at that very moment he was dealing with an incident that had just occurred the previous week. Brown related that workers' access to the toilet had improved since Iowa OSHA had cited Excel and that the union had discussed workers' right to go to the bathroom in its newsletter. As a result of the resolution of the citation, management and the union had reached an understanding that there would be a three-step process in the future when a worker needed to go: he or she would tell a supervisor, who would have five to ten minutes to find a replacement; if no supervisor were available, the employee would wait five to ten minutes for a utility person; if no utility were available, the worker would just tell the person next to him or her and go. Although this system generally functioned, Brown was aware of cases where workers had to wait as long as 90 minutes. During the previous week, a woman on the evening shift had gone to the bathroom to deal with heavy menstruation flow; on discovering that she lacked a quarter to buy a hygienic pad in the machine in the bathroom, she returned to the line to borrow the coin, but the supervisor, doubting her statement, told her that she had already spent enough time away from the line and ordered her to get back to work. The worker, going over the head of this supervisor—who was a relatively new assistant supervisor, who, as is often the case with novice straw bosses, was trying to establish his authority—repeated her request to the next higher supervisor, who at first asked her whether she was perhaps talking about "hog" blood she might have picked up on the line or her own blood. Since by this time she had already bloodied her clothes, he relented, letting her go home to change her clothes.[21] Brown reported that toward the end of 2002 another worker had filed a bathroom access complaint with OSHA, which then did an inspection, but because he had not yet had the time to read the letter that OSHA had sent the union about it, he did not know the details of the complaint or the outcome.[22] In fact, OSHA had been "unable to establish that any employees had been denied a restroom break or punished as the result of using the restroom."[23]

---

phone interview with Ken Larson (Oct. 15, 2002).

[21]Telephone interview with Ron Brown, president, UFCW Local 230, Ottumwa, IA (Oct. 14, 2002). Byron Orton, who was aware of the incident, expressed hope that the case could be resolved amicably. Telephone interview with Byron Orton (Oct. 18, 2002).

[22]Telephone interview with Ron Brown (Dec. 30, 2002).

[23]Fax letter from Mary Bryant, IOSH Administrator, to Marc Linder (Dec. 31, 2002). Iowa OSHA did however cite Excel for several violations of the standard governing

UFCW Local 222, headquartered in Sioux City, Iowa—and reputed to be the strongest at any IBP plant[24]—reported that at the Tyson-IBP cattle slaughter plant across the Missouri River in Dakota City, Nebraska, it had taken a while after 1998 to deal with bathroom access problems, but that union officials had pushed hard. The company, especially after Tyson acquired IBP, has been well aware of the import of the OSHA ruling. The informal rule at the plant—whose 3,800 employees have the lowest annual turnover rate (4 percent) in the IBP system—according to local union president, Marv Harrington, is that workers have to ask a supervisor for permission, but if they do not receive permission within five minutes or no supervisor is available for five minutes, they are free to leave the line. Since the company employs no utility or relief workers, a foreman or extra gang member takes the worker's place.[25]

But, as Chuck Clayton, Local 222's business agent, conceded, it is easy for full-time union officials like himself to tell workers that if they have to go to the bathroom, they should tell their supervisors and, if the latter do not provide relief, go straight to the bathroom and come straight back without any detours (a trip which, including doffing and donning protective clothing, takes about five minutes); many line workers, however, justifiably fear that supervisors will hassle them later for this act of independence.[26] This fear accounted for the fact that sometimes foremen made workers wait as long as an hour. Harrington explained that the residual root problem is a "power thing" on the part of individual foremen; nevertheless, although upper management does reprimand foremen who violate the foregoing procedure, the president believed that the company was in fact complicit. When asked whether management could really regard such harassment as a means of increasing worker productivity, Harrington replied that, with a large pool of job applicants at its disposal, the company probably did not think about that issue one way or the other. During 2002 several workers who had been made to wait defecated in their pants, one of whom was too embarrassed to let the union file a complaint with OSHA. At the very end of 2002 two more workers whose supervisors had made them wait an hour defecated on themselves; after going to the bathroom to clean themselves off, they went home

---

controls of hazardous energy. Excel Corp., Insp. No. 305712721 (Dec. 12, 2002). Because the complaint file was not yet closed, Iowa OSHA was unable to provide additional information as to why it had been unable to confirm a violation. Email from Mary Bryant to Marc Linder (Jan. 8, 2003).

[24]Telephone interview with Jackie Nowell, UFCW, Washington, D.C. (Dec. 13, 2002).

[25]Telephone interview with Marv Harrington, President, UFCW Local 222, Sioux City, IA (Dec. 17, 2002).

[26]Telephone interview with Chuck Clayton, Business Agent, UFCW Local 222, Sioux City, IA (Dec. 13, 2002).

because they were so upset. The company then terminated them for leaving the plant, and the cases went through the grievance process to pre-arbitration. In none of these instances, however, did the union file a complaint with OSHA.[27] If a succession of such complaints had been filed and if OSHA had issued a citation, the basis would have been laid for imposing much costlier monetary penalties for willful and/or repeated violations[28] in addition to reinstatement and back pay for having discharged employees for exercising their right to go to the bathroom when they needed to under OSHA.[29] One possible reason for the union's reluctance to file complaints may be its belief that, because the company tells workers that they are free to go to the bathroom with relief, it might be complicated to prove violations to OSHA.[30]

The president of UFCW Local 431, which has organized the IBP plant in Waterloo, the Oscar Mayer plant in Davenport, and the Hormel plant in Knoxville, Iowa, explained that it has not had any toilet access problems. It printed the new OSHA rule in its newsletter once, and has informed the workers and management.[31]

Some sense of the difference that the OSHA Memorandum has meant can also be gleaned from an account given by Gary Best, the very articulate secretary-treasurer of UFCW Local 227 in Kentucky, which has a total membership of 20,000, many of them employed in chicken and hog slaughter plants. Best recounted that the significant toilet access problems that had prevailed at these plants (such as Tyson, Cagle's, and Keystone) before April 6, 1998, were greatly diminished by a combination of union aggressiveness, which included distributing handbills with the text of the Memorandum, and increased compliance by employers, which had learned of their new obligation independently. But he emphasized that this positive impact was confined to unionized plants; at unorganized plants, such as Perdue, the freedom to void remains a problem, and workers who complain about it are simply fired.[32] Unsurprisingly, toilet access has also played a role in organizing drives. During Local 227's campaign to organize the Cagle's-Keystone Foods poultry plant in Albany, Kentucky in 2000, one of the huge number of unfair labor practice charges contained in the complaint

---

[27]Telephone interview with Harrington (Dec. 17, 2002 and Jan. 8, 2003). In one instance the company admitted its responsibility and paid for the worker to go home and change her pants. Telephone interview with Clayton.

[28]29 USC sect. 666(a).

[29]29 USC sect. 660(c).

[30]Telephone interview with Clayton.

[31]Telephone interview with John Honeycutt, President, UFCW Local 431, Davenport, IA (Dec. 16, 2002).

[32]Telephone interview with Gary Best, Louisville, KY (Oct. 2, 2002).

that the National Labor Relations Board filed against the company was threatening employees that they would get only two bathroom passes per week if they voted for the union.[33] However, later that year the workers voted overwhelmingly in favor of union representation[34] and bathroom access has not been a problem since.[35]

From a broader perspective, Best frankly admitted that members sometimes are just "goofy," abusing bathroom breaks by making telephone calls, smoking, and even leaving the plant. In turn, however, employers are also at times "goofy": although they clearly have the right to discipline workers for such violations, instead of doing so, they have issued blanket bans on unscheduled toilet breaks for everyone, as Jim Beam did. The OSHA memo was especially important in putting an end to such employer tactics, and if the ultimate outcome in the Jim Beam case had been to vindicate such measures, the UFCW would, in his view, have been in real trouble. Best also noted that during boom times, a unionized employer could not get away with banning toilet access—workers would just say "fuck you" and leave for another job. Moreover, he insisted that for himself personally, if a boss in a unionized workplace told him to wait 30 minutes to use the toilet, he would either systematically ask permission 30 minutes early or find one way or another to take back the 30 minutes. Best doubted very much that employers would prefer a rule requiring them to let workers use the toilet every x minutes: employers can live with the reasonableness standard and adjust production. To buttress this claim of flexibility, he pointed to plants organized by Local 227 in which workers finish their work after six hours and go home and get paid for eight hours. If employers tried to give them additional work to fill up eight hours, they would soon discover that it took workers nine hours to do.[36]

The Communications Workers of America, which has organized about 10 percent of call-center workers—one of the fastest growing occupational groups with an estimated 3.5 to seven million workers[37]—on whom even union employers have imposed very tight toilet access restrictions, quickly informed employers of the OSHA Memorandum. Dave LeGrande, the union's director of oc-

---

[33]Joseph Gerth, "Labor Board Lists 153 Charges Against Cagle's," *Courier-Journal* (Louisville), June 28, 2000, on http://www.courier-journal.com/localnews/2000/0006/28/000628chik.html.

[34]"Cagle-Keystone Workers Stand Up for a Voice on the Job," U.S. Newswire, Dec. 8, 2000 (Lexis).

[35]Telephone interview with Gary Best (Jan. 22, 2003).

[36]Telephone interview with Gary Best (Oct. 2, 2002).

[37]"Telephone Call Centers: New Factory Floors," on http://www.workindex.com/editorial//whar/whar0207-01.asp;http://www.cwa-union.org/workers/customer/call_center.asp; Stephanie Luce, "Stress in the Call Center" (Oct. 16, 2002), on http://www.fguide.org/Bulletin/callcenter.htm.

cupational safety and health, praised it for having given the union greater "leverage," as a result of which the newly interpreted standard has been "really great in bringing employers up to speed."[38]

Nevertheless, when informed later that OSHA had cited the Convergys call center for having restricted toilet access,[39] LeGrande was surprised. Having assumed at the time the Memorandum was issued that OSHA would apply its new interpretation only to assembly-line-type settings, he had not imagined that the agency would enforce the standard if it meant interfering with the kind of employer-imposed workplace structures that, for example, dictated work routines in call centers. To be sure, LeGrande's skepticism had in part been shaped by the union's experience with a complaint that it had filed with Maryland OSHA in the period before the Memorandum was issued. When that state-plan program refused to cite an employer in Baltimore based on the position—which OSHA in general took at that time—that section 1910.141(c)(1)(i) did not confer power on OSHA to deal with the intervals at which employers permitted workers to go to the bathroom, LeGrande spoke to the then (Clinton-administration) administrator of Federal OSHA, who stated that state-plan states were free to interpret such provisions totally independently of Federal OSHA. After having been apprised of the OSHA action against Convergys, however, LeGrande was persuaded that that citation could play a part in future organizing campaigns of call-center workers at Convergys or elsewhere.[40]

That the presence of a union can radically improve access was shown in the case of Faneuil, a 300-employee call center, where before the advent of the UFCW, "you went into break mode to go to the bathroom." After the union was elected as the workers' representative, they gained "the freedom to go to the bathroom anytime."[41]

Nevertheless, suppression of voiding rights remains an acute problem even for unionized call center workers. A survey of stress among call center workers in Detroit, Boston, New Jersey, and Ohio, conducted on behalf of the Utility Workers Union in the summer of 2002, elicited these open-ended comments:

> The supervisors that have no work to do will see you leave your desk to go break or bathroom + they'll make a point to let you see them look at their watch noting the time you left (like we don't watch our break times). Supervisors...will follow you into the bathroom to check on you.

---

[38]Telephone interview with Dave LeGrande, director of occupational safety and health, Communications Workers of America, Washington, D.C. (Oct. 3, 2002).

[39]See above ch. 11.

[40]Telephone interview with Dave LeGrande (Nov. 14, 2002).

[41]"Little Things Lead to Big Change," *Union*, September 2002, at 15.

You are timed to the second for everything you do. You can't even go to the bathroom without being timed. There is entirely too much pressure and something needs to be done.

What upsets me is the fact that we have to be on the phones 7 hours and 15 minutes out of a 7 ½ hour day. We also have to sign off the phones with a special code for the bathroom. They call it a bio break. I personally have to go to the bathroom 3-4 times a day. This kills about 15-20 minutes a day. So you see this already puts me below the time I'm supposed to be on the phone.

More, more, more…faster, faster, faster…. and whatever you do, don't drink a lot or you will have supervisors coming in the bathroom after you.

We may as well be chained to our desks for any kind of freedom to move around & even go to the bathroom.

Only have 14 minutes to go to restroom - what if I am ill?[42]

The managerial mindset facing workers and unions seeking less stressful working conditions in this industry is illustrated by letters from two companies to OSHA in response to complaints of restriction of toilet access. Amusingly, in spite of the seamlessly oppressive control that call centers impose in large part by virtue of the very computers that the employees use, one company sought to disarm OSHA by explaining: "The Omaha workplace is that of a typical large open office space. No manufacturing operations occur at this office location. No chemicals are used in this office setting. Employees sit at desks and use telephones and computers, much like the activities that occur in OSHA's own offices."[43] The other took the opposite approach, virtually boasting to OSHA of the all-encompassing technological surveillance it was creating in response to a complaint of access:

[W]e are implementing another AUX Code in the Customer Service Center that would allow the employee to hit a button on their phone and it would log them off the system and identify that they were using the restroom. ... This offers a second AUX Code from

---

[42]These comments were collected in connection with Stephanie Luce and Tom Juravich, "Stress in the Call Center: A Report on the Worklife of Call Center Representatives in the Utility Industry," Report submitted to the Utility Workers Union of America (August 29, 2002). The comments were emailed to the author by Stephanie Luce, who also gave permission to use them (Dec. 20, 2002). The study itself can be read at http://www.uwua.org/callcenter.htm.

[43]Letter from Denise Serrett, Human Resources Manager, APAC Customer Services, Inc., Omaha, NE, to Duty Officer, U.S. Department of Labor, Omaha (June 17, 2002), in OSHA, APAC Customer Service [sic], Complaint No. 203761291 (May 31, 2002).

the current AUX Code 1 that only identifies when you are on a break. In doing so, this will allow us to track usage of the restroom and create a bell curve for average usage and abnormal usage and lengths of bathroom breaks needed for employees. With this statistical information we will have a better handle on employees going to the restroom or on their assigned breaks as agreed to, through their labor contract.[44]

Further confirmation of the effectiveness of the Memorandum as a threat came from Mike Wright, the director of Occupational Safety, Health, and Environment at the United Steelworkers, who observed that frequently during organizing drives, especially at smaller firms in little-unionized industries in which employers do deny access to the toilet, the union has been able to give workers a sense of entitlement by telling them about the OSHA Memorandum. Even where the union eventually did not succeed in winning a representation election, Wright stated that the workers nevertheless felt more empowered to demand and gain access.[45]

Interestingly, local officials of the Steelworkers reported that bathroom access has not been a significant problem in unionized steel plants because many of the jobs are "overstaffed"; consequently, individual workers are not indispensable and if a worker has to void, a co-worker can perform his job.[46] Terry Davidek, the hourly safety coordinator (a new full-time position bargained for by the union) at an Allegheny Ludlum steel plant went so far as to say that, if, for example, the operator of a line suddenly had diarrhea and had to go to the toilet immediately, he could just shut down his line and if there were others working on it, they could do clean up or some other work—"there's always something to do"—for the few minutes until the operator returned. Because the unionized workers more or less "run the plant," they do not need to ask a supervisor's permission to go to the bathroom. In 29 years at the plant, Davidek had never heard of a worker's not being allowed to go to the bathroom.[47]

The American Postal Workers Union, whose members obtained OSHA coverage on September 28, 1998, when, thanks largely to lobbying by UPS and Federal Express—of which OSHA has to date conducted a total of 2,735 and 443

---

[44]Letter from Aquila Networks, Lincoln NE, to OSHA, Omaha (Aug. 12, 2002), in OSHA, Aquila, Complaint No. 203761739 (July 17, 2002).

[45]Telephone interview with Mike Wright, Director of Occupational Safety, Health, and Environment, United Steelworkers, Pittsburgh (Oct. 10, 2002).

[46]Telephone interview with Jim Blainiar, Hourly Safety Coordinator, Allegheny Ludlum, West Leechburg, PA USW Local 1138 (Oct. 30, 2002).

[47]Telephone interview with Terry Davidek, hourly safety coordinator, Allegheny Ludlum, Brackenridge, PA (Nov. 1, 2002). To be sure, this description contrasts sharply with that given by the president of the Steelworkers local at Steel of West Virginia, which OSHA cited for not permitting an employee to go to the bathroom. See above ch. 11.

inspections, respectively[48]—Congress enacted the Postal Employees Safety Enhancement Act, which amended OSHA,[49] reported that before that time, when federal workers were formally covered by 29 CFR section 1960, which conferred no enforcement powers, inside workers were often not permitted to go to the toilet by individual supervisors. After postal workers became covered by OSHA, the union's occupational safety and health director, Corey Thompson, using the April 6, 1998 Memorandum, pressed the U.S. Postal Service to comply, and he reported that the problem has largely been resolved.[50] In fact, although OSHA has issued at least two citations to the USPS in recent years for failing to provide the proper number of toilets,[51] none has been issued for denying access to toilets. However, OSHA's Austin Area Office noted that it has received several toilet-access complaints from postal workers that were resolved short of an inspection,[52] and the Omaha Area Office received one in 2002,[53] while a safety engineer with the OSHA regional office in Seattle reported that in the period after April 6, 1998, while he was an area director in Bellevue, Washington, he received a complaint from a worker at a USPS sorting facility about a supervisor who clocked workers' bathroom visits and followed them to the bathroom. As a result of OSHA's intervention requesting that the USPS deal with the union on this issue, the Postal Service issued an apology to the workers and a written statement declaring that the workers had the right to go to the bathroom. The OSHA official, who was in frequent contact with the union, never heard of problems again.[54]

Local 100 of the Transport Workers Union in New York City reported that bus drivers often have two-hour routes during which they cannot leave the bus; and even if they could, there is often no bathroom they can use.[55] Alternatively, they may have a five-minute break at the end of a run, but may have to walk a half-mile to find a toilet. Some drivers have been reduced to urinating into a bottle, a stratagem not readily available to female drivers, many of whom suffer

---

[48]http://www.osha.gov/cgi-bin/est/est1.

[49]Postal Employees Safety Enhancement Act, Pub. L. 105-241, 112 Stat. 1572 (1998), codified at 29 USC sect. 652(5). For a condensed legislative history, see *Occupational Safety and Health Law* 53-55 (2d ed.; Randy Rabinowitz ed. 2002).

[50]Telephone interview with Corey Thompson, Washington, D.C. (Oct. 7, 2002).

[51]U.S. Postal Service, Insp. No. 301505343, Big Rock, TN (2002) (toilet out of service for 10 days); U.S. Postal Service, Insp. No. 300675501, Albuquerque, NM (2000) (insufficient number of toilets).

[52]Telephone interview with Elizabeth Slatten, Asst. Area Dir., OSHA Austin Area Office (Nov. 7, 2002).

[53]See above ch. 11.

[54]Telephone interview with Dale Cavanaugh, OSHA Regional Office, Seattle (Dec. 2, 2002).

[55]Rob Ortiz, Safety Dept., TWU Local 100, New York (Oct. 22, 2002).

from urinary tract infections and related illnesses. Although the union had been unaware of the OSHA Memorandum, a vice chairperson of its Transit Authority Surface division was intrigued by the suggestion that the union file complaints with OSHA, especially in conjunction with the fall 2002 round of collective bargaining.[56] Local 100 also reported a further practice clearly at odds with employers' obligation to let workers urinate when they need to: employees working in subway token booths may be stuck there for four or even eight hours if their relief person is sick and no one comes to relieve them, because they are forbidden to leave the booth where they are required to guard $10-12,000.[57]

At the Retail, Wholesale, and Department Store Union, which is affiliated with the UFCW, the Health and Safety Department has created a fact sheet explaining that reasonable access is a right and not merely a privilege, and that employers "cannot dictate how often employees can use bathroom facilities during a work day."[58] (To be sure, instead of underscoring the OSHA Memorandum's injunction that "employers...make toilet facilities available so that employees can use them when they need to do so,"[59] it overcautiously admonishes RWDSU members that "OSHA regulations do not authorize employees to go to the bathroom whenever they please.")[60] In addition to posting an excellent explanation of the Memorandum on its website,[61] the union has recently begun vigorously propagandizing its locals on the matter.[62]

The RWDSU has organized three chicken processing plants in Georgia, owned by Tyson, Cagle's, and Equity-Keystone. According to the union's energetic and articulate International Representative in Atlanta, Edgar Fields, bathroom access became a problem in 2001-2002, as employers made workers wait for as long as 30 minute to leave the line, in large part because the companies had not hired enough relief workers. The situation in chicken processing plants is of particular interest because, according to Fields, a worker's leaving the line before a relief worker arrives generally does not inflict any economic or material damage on the employer: it is not the case that any machinery or product would be damaged let alone destroyed; and because three or four workers generally per-

---

[56]Telephone interview with Lloyd Archer, vice chairperson, TA Surface, TWU Local 100, New York (Oct. 23, 2002).

[57]Telephone interview with George McDonald, TWU Local 100, New York (Oct. 21, 2002).

[58]RWDSU, "Gimme a Break: Access to Bathroom Facilities at Work" (n.d.).

[59]See below Appendix II.

[60]RWDSU, "Gimme a Break: Access to Bathroom Facilities at Work" (n.d.).

[61]http://www.rwdsu.org/health_safety.html#bathroom_breaks

[62]Telephone interview with Steve Mooser, director of safety and health, RWDSU, New York City (Oct. 7, 2002).

form the same work as the (would-be) urinator, they can do his or her share for the few minutes he or she is away.[63] Since no material detriment accrues to the employer and the crux of the dispute is merely the employer's desire to retain unilateral control over workers' time, the employer cannot adduce any objective disruptive impact on its operations that would reasonably justify delaying the onset of the worker's bathroom visit.[64] At the Equity plant in Camilla, Georgia, the employer fired a worker who had left the line before the employer had granted him relief because he urgently had to void. The union filed a grievance over the dismissal (and the worker was reinstated after two weeks with back pay), but then, after Fields spoke to the author, it decided to file its first OSHA complaint as well.[65] Instead, however, choosing to rely on its own power and resources, the RWDSU embarked on its enforcement campaign at the three Georgia poultry plants. It distributed the aforementioned factsheet, "Gimme a Break: Access to Bathroom Facilities at Work," to the members and informed the managements that in the future workers would be adhering to this procedure taken from the factsheet, which in turn is in conformity with the OSHA Memorandum:

Going to the bathroom when you need to is not a privilege. IT IS A RIGHT! Follow this procedure for using the bathroom facilities:

*Ask your supervisor for permission to go to the bathroom.
*If you are told to wait, ask when you will be able to go.
*If that waiting time is unreasonable for you, tell your supervisor you must go sooner.
*If your supervisor still says no, go ahead and use the bathroom.
*Your union will protect you in any disciplinary action.[66]

As of the beginning of 2003, management was still complying with the RWDSU's straightforward interpretation of the Memorandum. This success, Fields stressed, has been due in part to the union's firm and aggressive stance, but above all to the concerted day-to-day enforcement by well-informed shop stewards against supervisory recidivism, for which self-defense there is no substitute. Fields astutely pointed out that although supervisors' petty interference with workers' biological needs sometimes appears to be driven by ego trips, it should also never be forgotten that these lower-level bosses—who of course have no union to protect them—are under constant pressure to produce more with what they regard as understaffed production lines; when their bosses tell them that they will simply have to make do with the workers they have, they react by trying to

---

[63]Telephone interviews with Edgar Fields, Atlanta (Oct. 7, 2002 and Jan. 7, 2003).
[64]For an explanation of this interpretation, see above ch. 6.
[65]Telephone interviews with Edgar Fields.
[66]RWDSU, "Gimme a Break: Access to Bathroom Facilities at Work" (n.d.).

cover themselves by extracting every second of working time they can from every employee present. One of the consequences of these production campaigns is that it becomes very easy for harried supervisors to presume, if they did not already believe it, that workers who request permission to stop working in order to go to the bathroom do not "really" need to void.[67]

This account of the RWDSU's successes in the southern chicken-slaughter plants should be tempered by the attitude of and position taken by the chairman and chief executive officer of one of these companies. The CEO of Cagle's, the eighth biggest producer in the United States,[68] who instantly recalled the details of the dispute over bathroom breaks at the Macon plant a quarter-century earlier,[69] stated in 2003 that he had never heard of the OSHA Memorandum. The practice in his firm's plants was that workers can go to the bathroom outside of scheduled breaks (mid-morning, lunch, and mid-afternoon if the line operates more than 2.5 hours in the afternoon) as long as their relief person is not relieving someone else. Estimating that one relief worker was employed for about 25 production workers, he insisted that even if four workers performed the same task next to one another, the pace was such that three could not fill in for one who went to the toilet even for just a couple of minutes.[70] If any employee sought relief more often than the others, the relief worker would note this fact and report it. If employees are sick, they are allowed to go to the bathroom more often—but only until they can "clear up" their health problem.[71]

To be sure, not all unions have aggressively pushed for bathroom rights, and the members of the weak ones that have not may have benefited as little as unorganized workers.[72] The United Auto Workers is an interesting example of a strong union with a remarkable system of collectively bargained-for regulation of hours that nevertheless appears unaffected by the OSHA Memorandum. In the summer of 1997—on precisely the same day that Federal OSHA was issuing its first toilet access citation to Hudson Foods—the union had struck the General Motors parts plant in Warren, Michigan, where the company's failure to replace

---

[67]Telephone interview with Edgar Fields (Jan. 6 and 7, 2003).

[68]Gary Thornton, "Acquisitions, Restraint Key Factors in 2001," Watt Poultry USA, Jan. 2002, at 26A-F, on http://www.wattnet.com/Archives/Docs/0102wpA.pdf.

[69]See above ch. 13.

[70]Alma Oliver, who worked at the Cagle's Macon plant for 37 years, observed that for most jobs co-workers performing the same job could also do the job of someone who leaves the line for about five minutes; a relief person is necessary only for those jobs that are done by one person or when someone needs to use the bathroom for more than five minutes. Telephone interview with Alma Oliver, Macon (Jan. 29, 2003).

[71]Telephone interview with Doug Cagle, Atlanta (Jan. 28, 2003).

[72]Telephone interview with Robyn Robbins, Asst. Dir., Occupational Safety and Health, UFCW, Washington, D.C. (Oct. 3, 2002).

retiring workers for two years had led to a shortfall of hundreds of workers. One result was that assembly line workers were "'waiting an average of 30 to 45 minutes to urinate....'"[73]

What was an abuse provoking a successful strike—G.M. quickly agreed to hire 420 additional workers[74]—is the bureaucratically sanctioned norm at the Ford assembly plant in Louisville, Kentucky, where the five-day week/ten-hour day has been in effect for about 25 years and workers are entitled to six minutes of break time per hour; this time, however, must be taken in two blocks of 40 minutes in the morning and 20 minutes in the afternoon. These two rest periods are taken, beginning one hour after the start of the workday and one hour after the lunch break, in (tag) rotation as one worker is relieved by a relief worker and then returns and the next worker leaves. If a worker needs to void between breaks, a "quality upgrader" can relieve him or her, but Ford has up to 40 minutes to provide this relief. The author's suggestion to the president of UAW Local 862 that a 40-minute wait would be unlawful under OSHA's sanitation standard and that the union might want to consider filing a complaint failed to elicit a direct response.[75]

---

[73]Robyn Meredith, "Strike Closes Parts Factory in Job Dispute," *N.Y. Times*, July 24, 1997, at A12, col. 6 (Lexis) (quoting Ronald L. Campbell, Jr., a paint inspector). The version of this article now available on Lexis lacks this part, as does the edition of the *Times* on microfilm and Proquest, in all of which the article appears in section D.

[74]"Union Approves Deal with G.M. to End Strike," *N.Y. Times*, July 28, 1997, A9, at col. 1 (Lexis).

[75]Telephone interview with Rocky Comito, president, UAW Local No. 862, Louisville, KY (Oct. 7, 2002). The local contract includes a letter of understanding pertaining to a "relief ratio" varying according to job classification; converted into minutes, these ratios provide for as much as a 10-minute per hour break for paint sprayer to 8 minutes for welder arc, acetylene and gas. *Local Agreements, Letters of Understanding and Rates Between Local 862 UAW and the Ford Motor Company Louisville Assembly Plant Agreement Dated October 29, 1999* at 152. The maximum 40-minute waiting time for bathroom relief is not set out in the national or local agreements, but Comito explained that it has been in effect for as long as he can remember and derives from the 40 minutes that the relief person is occupied giving scheduled relief in the tag rotation system. Telephone interview with Comito (Oct. 22, 2002). According to the latest collective bargaining agreement between the Ford Motor Co. and the UAW: "On line operations relief men will be designated to make relief available at all times and in a ratio to provide each employee with at least 24 minutes of actual personal relief per 8 hour shift." The agreement permits details to be worked out at the local level. In addition, by agreement of Sept. 18, 1964, "an additional relief allowance of 12 minutes per eight-hour shift, for a total of 36 minutes, will be provided to those employees on operations where their manual operations are continuous and cannot be left unattended and for which tag relief is furnished...." By agreement of Dec. 7, 1970, workers received an additional relief al-

In spite of such OSHA violations, the staff of the UAW's Health and Safety Department reported that they were unaware of any toilet access problems in auto plants under contract.[76] The director of the department stated that bathroom access for auto workers "is not perceived by anybody as a problem." Moreover, although it was unclear who at the UAW would even have a national overview of the state of bathroom access in the industry, "anybody who'd know wouldn't talk to a law professor" because the UAW "isn't going to admit" that members have problems getting to go. Nevertheless, he did stress that the issue of staffing levels, which do have a crucial impact on a broad variety of working conditions including breaks, overtime, and vacations, is a significant problem.[77]

In contrast, the Canadian Auto Workers (CAW) have achieved a break system that its director of Work Organization and Training was willing to discuss. At Daimler-Chrysler, for example, where workers work 7.5 hours (for 8 hours' pay) and get 45 minutes of mass breaks (11 minutes during first half of shift, 24 minutes of a meal break, and 10 minutes in the second half):

> In addition to this negotiated relief there is also emergency relief. An example: In the chassis operations where there are 240 workers 3 emergency relief workers are available to cover them for first aid, pass and pee. Usually a worker is readily accommodated. There are no time requirements in our agreement. As the plant chairperson puts it: "When you have to pee you have to pee."
>
> There are times—maybe 10-12 times a year with a plant workforce of about 5200—when a worker is not relieved in time and has to walk of[f] the job to go to the bathroom. In these situations management always starts to discipline the worker but the union is successful in most cases of getting the discipline buried.[78]

Evidence for the vitality of the OSHA Memorandum can also be drawn from the negative example of UNITE (Union of Needletrades and Industrial, and Textile Employees). Asked whether OSHA's action had had any impact on the

---

lowance of eight minutes per eight-hour shift, for a total of 44 minutes. *Agreements Between UAW and the Ford Motor Company* 1:18, 362, 365 (Oct. 9, 1999). The author originally heard about the 40-minute wait at the Ford plant from Jo Anne Kelley, the president of UFCW Local 111-D, who, at the height of the Jim Beam publicity, received many calls from workers about bathroom-break practices, including this one.

[76]Telephone interview with Peter Dooley, Occupational Safety and Health, UAW, Detroit (Oct. 21, 2002); telephone interviews with Sylvia Johnson, Occupational Safety and Health, UAW, Detroit (Oct. 2002).

[77]Telephone interview with Dr. Franklin Mirer, Director, Health and Safety Dept., UAW, Detroit (Oct. 22, 2002).

[78]Email from David Robertson, Director, Work Organization and Training, CAW, to Marc Linder (Oct. 17, 2002).

membership's access to toilets, its director of occupational safety and health explained that complaints about bans on voiding would rarely come to his attention because most workers do not regard the problem as one of safety and health, but of human dignity. Just as his interlocutor was about to ask him whether it was not the union's obligation to bring to its members' attention—as the UFCW and CWA had done in systematic campaigns—the fact that the issue was both one of dignity and occupational health and that since 1998 the government agreed that workers have a right to void at work when they have to, the director self-critically observed that perhaps he should have done so earlier and that perhaps he would do so now.[79]

Surprisingly, the director of safety and health at the Service Employees International Union, which has organized tens of thousands of low-paid unskilled workers, knew of no problems with bathroom access; he suggested (partly in jest) that the workers who might be most exposed to such treatment, janitors, worked in bathrooms and therefore presumably had adequate access.[80] Similarly, the director of occupational safety and health at the American Nurses Association observed: "I know that many nurses don't get to the bathroom as often as they need or like—sometimes throughout their whole shift. However, ANA hasn't specifically asked this question on any of our recent on-line surveys. To be honest, I was not actually aware of this memorandum from OSHA."[81] Despite the fact that Teamsters locals have filed numerous OSHA complaints about denial of access, the international's director of safety and health stated that he was unaware of any problems.[82] The safety and health director at the American Federation of Teachers, who was very well aware of anecdotal information about teachers' access problems, was, after she had been informed of the survey that the authors of *Void Where Prohibited* had conducted of teachers' toilet access problems, eager to replicate the study, on a larger scale, with the AFT's membership.[83]

---

[79]Telephone interview with Eric Frumin, director of occupational safety and health, UNITE, New York City (Oct. 10, 2002).

[80]Voice mail message from Bill Borwegen, director of occupational safety and health, SEIU, Washington, D.C. (Oct. 3, 2002).

[81]Email from Karen Worthington, director of occupational safety and health, American Nurses Association, to Marc Linder (Oct. 16, 2002).

[82]Telephone interview with Chris Madar, director of occupational safety and health, International Brotherhood of Teamsters, Washington, D.C. (Oct. 8, 2002). On the OSHA complaints, see above ch. 11.

[83]Telephone interview with Daryl Alexander, director of occupational safety and health, American Federation of Teachers, Washington, D.C. (Oct. 15, 2002). According to the industrial hygienist of the United Federation of Teachers in New York City: "While this is not one of the top ten problems our staff face it is one that comes up." Email from Ellie Engler to Marc Linder (Oct. 25, 2002).

Opinions of the Authors of the Federal and Iowa OSHA Memorandums

The views of John Miles, who as Director of Compliance Programs issued the Memorandum in 1998 and since then has had the opportunity to observe its enforcement as Regional Administrator of OSHA's Dallas Region, merit special attention. Miles ventured his "gut feeling" almost five years later that, even if the agency has not issued many citations for restriction of access, the Memorandum's greatest impact has probably been raising employers' awareness. That OSHA presumably did not anticipate that that impact would be very great emerges from Miles's astounding remark, concerning frequency of voiding, that "what we had in mind back then was every 2-3 hours—that's what's normal." However, in effect advising workers to engage in guerrilla warfare against his interpretation of his own interpretation, he added that, knowing that he has to void every two hours, if he were a worker whose employer made him wait 30 minutes to go to the bathroom, he would simply signal an hour and a half before he had to go.[84]

As author of the January 21, 1998 Iowa OSHA Memorandum that anticipated Federal OSHA's action ten weeks later, Iowa Labor Commissioner Byron Orton has, as already noted, actively intervened in the state's enforcement of workplace voiding rights.[85] Orton reported that overall he was "pleased" with the results of the aforementioned enforcement actions against slaughter plants: there had been no repeat complaints, the situations had become "much more workable," and "management is making a good-faith effort" to comply.[86] Orton stressed that, since "news travels fast in the employer community," as part of his overall enforcement and compliance strategy, he had made sure at the time that the Excel case, and especially the "painful" monetary penalty, got publicized.[87] While recognizing that employer intimidation does deter workers in unorganized workplaces from filing complaints with OSHA, Orton conjectured that, as a result of nonunion employers' fear of being inspected—though programmed inspections would be very unlikely—even in such settings the situation was "not totally hope-

---

[84]Telephone interview with John Miles (Nov. 12, 2002). The same questions posed to Miles's successor elicited a response three weeks later that answers would be drafted by staff and possibly reviewed by the Solicitor's Office and would take about 60 days. Email from Marc Linder to Richard Fairfax (Dec. 10 and 25, 2002); email from Helen Rogers to Marc Linder (Dec. 30, 2002); letter from Richard Fairfax to Marc Linder (Dec. 31, 2002). No such answer had been received as of Mar. 6, 2003 when the page proofs were returned to the printer.

[85]See above ch. 12.

[86]Telephone interview with Byron Orton, Iowa Commissioner of Labor, Des Moines (Oct. 17, 2002).

[87]Telephone interview with Byron Orton (Oct. 18, 2002).

less."[88] The fact, however, that Orton spoke prematurely—that even the unionized companies that Iowa OSHA had fined have proved to be recidivist and continue to test workers' determination to defend their right to void—strongly suggests that Orton may well have been overly optimistic about the nonunion sector.

---

[88]Telephone interview with Orton (Oct. 18, 2002).

17

Is At-Will Voiding Now "The Law"?

Needing permission to use the bathroom is a hassle that may not pass after you get a job.[1]

When asked by the UFCW's lawyer at the OSHA hearing in August 2002 whether the practice of tag rotation or mini-breaks in between scheduled breaks had been ended by the side-agreement letter attached to the collective bargaining agreement of a year earlier, Jim Beam's plant manager inadvertently alluded to one of the most important questions concerning labor standards legislation: "That practice had been in effect for a long time, and it did not...end just because we said that it would. I mean...it ended when we implemented the policy and controlled it. ... How does a practice end with a signed piece of paper? No it didn't: it did not end at that time."[2]

What is true of a policy in force at one relatively small workplace and enforcible by managers and/or owners with a pecuniary interest in compliance, is a fortiori true of a regulation theoretically in force at millions of places of employment, but policed by a small number of inspectors alerted to violations by complaints submitted by those (largely union) workers who happen not to be intimidated by the threat of dismissal or other reprisals. To be sure, the mere existence of a piece of paper (or a website), announcing a government agency's interpretation of its own regulation, stating that workers have a right should not give rise to a presumption that they actually do enjoy that right; but, conversely, the existence of a number of violations should also not per se give rise to the opposite presumption. Until more is known about the real world of workplace toilet access, the status of that right must remain indeterminate.

---

[1]*React*, July 6-12, 1998, at 5.

[2]Commonwealth of Kentucky, Occupational Safety and Health Review Commission, Administrative Action No. 02-KOSH-0015 KOSH 3681-01, Secretary of Labor, Commonwealth of Kentucky v. Jim Beam Brands Co. at 202 (Aug. 29, 2002) (Jeff Conder, plant manager, Jim Beam Clermont plant).

The mere fact, however, that no state or federal court has as yet upheld OSHA's interpretation of its toilet standard in no way undermines its validity or legitimacy. The reason that no judge has been called on yet to uphold or invalidate the Memorandum—Jim Beam Brands Company came closest to prosecuting such an appeal—is ambiguous: employers may take its validity for granted, not find compliance burdensome, or regard OSHA's enforcement activity as so feckless and sporadic that they can continue to violate the standard with impunity. Alternatively, OSHA may be choosing its citations very carefully because it would rather not litigate the provision.

Despite the evidence of continuing widespread violations of employers' obligation to make toilets available so that workers can use them when they need to, it is unclear whether noncompliance is so systematic, rampant, and massive, OSHA so deficient in its enforcement initiatives and so unresponsive to workers' requests for intervention, and monetary penalties so inadequate that it would be plausible to argue that the legal rule/right is insufficiently embedded in general public and employers' and workers' consciousness and conscience to have acquired the status of a moral right. In contrast, for example, under the National Labor Relations Act, the "level of employer lawlessness"[3] has become so high and deterrence so meager—for every ten votes for unions in elections in 1985 one illegally discharged worker was reinstated by the National Labor Relations Board[4]—that it is plausible to characterize the right to self-organization as having become largely illusory for the workers who most urgently need its support.[5]

If the argument were accepted that because, for example, Jim Beam, Convergys, and Liz Claiborne violated the law and interfered with workers' right to void, what was purportedly a right for workers had been converted into a mere privilege to be dispensed at employers' whim, then by the same logic, it would follow that because many employers violate the Fair Labor Standards Act by failing to pay workers overtime premiums, the right to be paid time and a half had also been turned into a mere privilege. Such a position would create a very high threshold for labor standards; indeed, it would come perilously close to arguing that since all laws are violated, no rights exist. To apply that approach it would be necessary to calculate empirically just how widespread and massive the violations are and how tough and intense the enforcement.

If, to take a hypothetically extreme example, 95 percent of employers employing 99 percent of workers violated the law, paying none of their covered

---

[3]Paul Weiler, *Governing the Workplace: The Future of Labor and Employment Law* 239 (1990).

[4]Weiler, *Governing the Workplace* at 112.

[5]See, e.g., Craig Becker, "Elections Without Democracy: Reconstructing the Right to Organize," *New Labor Forum*, No. 3, at 97-109 (Fall/Winter 1998).

workers time and a half, and the enforcement agency prosecuted only very few, then it would be meaningful to characterize the right as illusory; but to reach the same conclusion merely because a few employers violated the law (and the workers involved filed complaints with OSHA, which actually cited the employers and brought about, aided by publicity, restoration of the status quo ante) would be premature.

An analysis of almost 75,000 cases brought by the U.S. Department of Labor (DOL) between 1991 and 1995 revealed that at 2.11 "victims" per 100 workers, construction firms recorded the highest rate of failure to pay premium overtime wages.[6] Yet as dismayingly high as one violation for every 50 workers is, it still does not seem rampant enough to warrant regarding the legal right as having been downgraded to a de facto privilege. Perhaps the DOL's finding of overtime payment violations among 83.9 percent of security guard services, 80.9 percent of janitorial services, 69.4 percent of hotels and motels, and 60.6 percent of restaurant employers that it investigated in 1997[7] and of unpaid hours of work violations in 100 percent of 51 poultry processing plants surveyed in 2000[8] might justify such a conclusion, but even these levels of illegality might be insufficient if, despite such violations, even these chiselers nevertheless paid most of their workers overtime premiums most of the time.

The prerequisites for meaningful enforcement of labor protective norms include free and cooperative vigilance by well-informed workers and unions and a government agency staffed by motivated officials who can bring to bear adequate deterrence: "The agency should have a set of thumbscrews so assorted as to fit every unfairly grasping hand."[9] If OSHA rarely if ever looks for toilet-access problems in connection with scheduled inspections, but the vast majority of current employees are too fearful of retaliation to file complaints with OSHA and the complaints of ex-employees will rarely enable OSHA to conduct an on-site inspection, the efficacy of complaint-driven enforcement of workers' right to void when necessary is seriously compromised. Moreover, if OSHA has only once imposed a monetary penalty large enough to make an employer think twice about risking even larger fines for future repeated or willful violations of the

---

[6]G. Zachary, "Shortchanged: Many Firms Refuse to Pay for Overtime, Employees Complain," *Wall Street Journal*, June 24, 1996, at A1 (Lexis).

[7]Brian Tumulty, "Many Low-Wage Workers Getting Shortchanged on the Job," Gannett News Service, Dec. 20, 1997 (Lexis).

[8]U.S. Department of Labor, "Poultry Processing Compliance Survey Fact Sheet" (Jan. 2001), on http://www.ufcw.org/pdf/Usdept~1.pdf.

[9]"Report of the Standing Committee on Legal Aid Work, Appendix A: First Draft: Of a Model Statute for Facilitating Enforcement of Wage Claims," *Report of the Fiftieth Annual Meeting of the American Bar Association* 323-25 at 325 (1972).

toilet-access standard,[10] firms that systematically restrict access in the belief that they are increasing working time, production, productivity, and profitability, may easily make a cost-benefit calculation indicating that even in the improbable case that one of their (especially nonunion) employees filed a complaint and OSHA actually issued a citation accompanied by a fine, the cumulative financial impact of the unlawful practice would far exceed the cost.

Nor is it likely that these severe agency failings would be overcome by amending OSHA to permit a private right of action so that workers would not be totally reliant on the agency to enforce their rights. Apart from the political implausibility of enactment of such a congressional amendment for the foreseeable future, the prolonged litigation associated with such individual enforcement would, for dealing with the daily problem of getting to go to the bathroom, be a poor substitute for an administrative process that offers workers—provided that they are not under the jurisdiction of Cal/OSHA—a same-day telephone call from OSHA to the employer requiring it to respond within five days and an on-site inspection by OSHA within five days of receiving a rebuttal of the employer's response to the complaint from a union representative or a complaining worker who is a current employee.[11] Introduction of a private right of action would, moreover, be ill designed to eliminate the problem of retaliation: workers who are afraid of jeopardizing their livelihood by filing anonymous complaints with OSHA are not likely candidates to become plaintiffs in federal court, although it might strengthen the enforcibility of the complaints of ex-employees, for whom OSHA can do little. And, finally, as the Cagle's dispute demonstrated,[12] there is no reason to assume that federal judges would be more knowledgeable and insightful (let alone pro-worker) than OSHA inspectors concerning the conflict between the perceived needs of workers to empty their bladders and employers to fill their coffers.

Wide swaths of the working world in the United States are dominated by managements that bully and infantalize workers in the apparent belief that fear gives productivity a fillip.[13] Perversely, but all too predictably, it is precisely workers in such firms who are least likely to feel self-confident and secure enough to risk filing the complaints that are realistically the only method for inducing OSHA to investigate low-priority toilet-access violations. Even in or-

---

[10]On the $36,000 penalty that Excel had to pay in addition to a $25,000 gift to public agencies, see above ch. 12.

[11]Email from Elizabeth Slatten, OSHA Assistant Area Director Austin Area Office, to Marc Linder (Dec. 24, 2002): OSHA, Complaint Policies and Procedures, CPL 2.115 (June 14, 1996).

[12]See above ch. 13.

[13]E.g., New River Industries, Inc. v. NLRB, 945 F.2d 1290 (4th Cir. 1991).

ganized settings, some unions report that management periodically or constantly tests workers to see how far they can be pushed. In this context, control over their bladders is only one among many battlegrounds. Some unions have been much more activist and innovative in asserting workers' voiding rights than others. The UFCW, without whose intervention OSHA would never have issued the Memorandum in the first place, has also been the leader, especially through its national occupational safety and health office, in informing its locals' officials and members of their rights.

Whatever positive results the combined efforts of aggressive union monitoring, resistance, and demands and forceful OSHA inspections may have achieved, the great majority of workers in nonunion workplaces do not directly benefit from them. Relatively few nonunion workers are intrepid enough to file complaints with OSHA against their current employer about any health or safety violation, let alone something as intimate as eliminating bodily waste, and those who do may well find themselves disciplined or harassed, if not fired. Nevertheless, even for unorganized workers the potential of government intervention exists and some have availed themselves of the protection: of all citations issued by Federal OSHA for violations of section 1910.141(c)(1)(i) in 2000, 2001, and 2002, 20 percent were triggered by complaints filed by workers in nonunion workplaces, while the corresponding figure for citations issued specifically for restriction of access by Federal and state OSHA programs after April 6, 1998, was 35 percent.[14]

Whether actual experience with OSHA would encourage nonunion workers to be persistent is another matter. Consider, for example, the view of a 28-year OSHA veteran in the Denver Regional Office: A complaint by a worker about toilet access would be considered "too minor" to merit anything but streamlined phone/fax treatment, and only if the worker later called to state that the employer had not abated the problem would OSHA ever do an on-site inspection, and then only as a low-priority matter and much later. In any programmed inspection, the inspector on seeing toilets would assume that workers had access to them and would not interview workers about the matter; perhaps if a union were present and complained, the inspector would look into the matter.[15] Such attitudes may explain why some union occupational safety and health officials who view with skepticism OSHA's (and the Department of Labor's) record of insuring, and capacity and desire to insure, safer and more healthful workplaces, have concluded that OSHA should be more of a back-up for union action than a primary tool for safety and health campaigns.[16]

---

[14]See above ch. 11.

[15]Telephone interview with Cindi Cross, Duty Officer, Denver Regional OSHA Office (Oct. 23, 2002).

[16]Telephone interview with Jackie Nowell, director of occupational safety and health,

Indeed, under the unabashedly pro-employer George W. Bush administration, OSHA might be more likely to withdraw the Memorandum altogether than to secure enforcement of workers' right to void at will. The chief constraint on such withdrawal may be recent rulings by the federal appellate courts (especially the D.C. Circuit) that once an agency has interpreted a regulation, it can substantially change its interpretation only in the same manner that it can change the regulation itself—by notice and comment rulemaking.[17] But just as the Clinton administration avoided (or, perhaps, evaded) notice and comment in order to deprive employers of a forum in which to attack the newly conferred right, so, too, the Bush administration would presumably have little interest in giving workers and unions a forum in which to attack the termination of such a right. As the Jim Beam episode demonstrated, it is imaginable that even employers themselves might prefer continuation of OSHA's sporadic and low-intensity enforcement of their obligation not to interfere with workers' right to void to a highly publicized and embarrassing national debate over their demand for the restoration of their power to discipline workers for being human beings.

Short of such outright withdrawal or revocation of the toilet-access standard interpretation, what some employers (like Jim Beam) demand is guidelines setting out what kinds of counter-measures they may lawfully take to deal with workers who "abuse" their new right to void when they need to. This issue is of special significance in nonunion workplaces in which OSHA-imposed at-will bathroom breaks (supported by a statutory prohibition of retaliation and provision for reinstatement with back pay)[18] represent a salient inroad against employer-imposed at-will employment regimes. Thus even if it is unlawful for an employer to eliminate unscheduled breaks for everyone just to punish the small number of "abusers," a question still remains as to what extent an employer retains its pre-statutory power to discipline such workers, whose existence even union officials, both on and off the record, concede—that small proportion of workers who "abuse" breaks and make workplace life harder for everyone else.[19]

UFCW, Washington, D.C. (Oct. 18, 2002).

[17]Paralyzed Veterans of America v. D.C. Arena, 117 F.3d 579, 586 (D.C. Cir. 1997); Alaska Professional Hunters Ass'n v. Federal Aviation Adm., 177 F.3d 1030, 1034 (D.C. Cir. 1999); Shell Offshore, Inc. v. Babbitt, 238 F.3d 622, 629 (5th Cir. 2001). For the view that because the Administrative Procedure Act exempts from notice-and-comment rulemaking a document that actually does interpret a regulation, "the Supreme Court may eventually overrule *Alaska Hunters*, but the case will impact OSHA in the meantime," see *Occupational Safety and Health Law* 504 (2d ed. Randy Rabinowitz ed. 2002).

[18]29 USC sect. 660(c) (2000).

[19]See above ch. 16 (telephone interviews with Boyer, Best, LeGrande, Rosas, and Olesen and one UFCW local president who stated "off the record" that worker abuse once in a while was the only bathroom problem).

It is noteworthy that employers' claims about abuse are generalized and invariant across labor standards protections. At the same time, for example, that Jim Beam was complaining that it was powerless to defend itself against abuse of toilet breaks, employers reacted similarly to legislation enacted in California in September 2002 providing workers with six weeks of paid leave annually funded by an increase in employees' contributions to the State Disability Insurance Fund. Contrasting this pioneering regime with the federal Family and Medical Leave Act, which offers no compensation, Randel Johnson, the vice president for labor policy at the U.S. Chamber of Commerce, lamented: "Paid family leave presents more room for employee misuse of the time off than the unpaid option.... With the leave prescribed under the FMLA...employees are not paid so they are less likely to take the leave under false pretense. 'When it's paid leave, there is going to be more incentive for an employee to abuse the leave....' Johnson also complained that the California leave law does not give employers enough control over when and why workers can take family leave. He contends that if the leave is being subsidized, employers should have more say in how it is used."[20] This same logic underlay the aforementioned fee-to-pee regimes imposed by several Canadian employers designed to make workers less likely to use toilet breaks and more likely to abuse their bladders.[21]

In the initial analysis of the Memorandum in Chapter 6, the issue of abuse was raised, but not resolved, in connection with identifying the model created by OSHA as an "at-will voiding regime" that is subject to a substantively and temporally limited employer veto in those cases in which (1) an employee's immediate departure for the bathroom would disrupt operations unacceptably and (2) requiring the employer to employ sufficient relief workers to enable workers to leave immediately would be economically too burdensome for the firm. Acknowledgment of such a limited veto power had also appeared in union settings before OSHA issued its Memorandum. For example, the Steelworkers, in a dispute over an employer's failure to give workers "tied to" an assembly line adequate relief, did "not argue that power belt employees should have a right to go to the bathroom whenever they choose. It recognizes the necessity of a relief system."[22]

The issue of burdensomeness might then arise at the intersection of OSHA and the Americans with Disabilities Act. As the Director of Legal Support at Virginia OSHA, who astutely pointed out that employers' polices of limiting bathroom access to specified times will make some people so anxious about having

---

[20]Simon Nadel, "New California Law, Other Proposals Intensify Argument Over Paid Family Leave," *U.S. Law Week* 71(15):2259-61 at 2259-60 (Oct. 22, 2002).

[21]See above ch.9.

[22]Mor-Flo Industries, Inc., 62 Labor Arbitration Reports (BNA) 398, 400 (1974).

to void at other times that they will actually have to void more often, observed, at the extremes, workers who need to go to the bathroom at much shorter intervals than other workers may well raise an issue under the ADA that would require a decision as to whether the employer must accommodate the worker's disability or whether the burden to the employer in terms of lost output is excessive.[23]

To be sure, OSHA's at-will rule is definitionally coupled with the need to void; consequently, if workers, under the guise of exercising their right to void, in fact engage in other activities not covered by the Memorandum, employers presumably retain whatever powers they possessed before April 6, 1998 to discipline them for "stealing time" (the expression of choice at United Parcel Service, Wal-Mart, and other firms, for taking unauthorized breaks).[24] With regard to bathroom access in the unionized setting, labor arbitrators, even after the advent of OSHA (though before the issuance of the Memorandum), did not rule that employers lacked the means under collective bargaining agreements—which overall confer greater rights on employees than OSHA does—to detect abuse; on the contrary, they declared that, corroborated by written records of relief requests (and the time taken) that "would have the effect of discouraging any tendency to abuse," "individual abuse can be dealt with individually, if necessary, through disciplinary procedures."[25] Elsewhere, too, workers and their union, far from "blam[ing] the Company for wanting to curb the wasteful practices of a small number of loiterers" in the bathrooms, "demonstrated that they [would] cooperate in proven cases of loitering."[26]

The Memorandum would be of no help to an employee who, instead of going to the bathroom, went to the cafeteria to make a telephone call or outside to smoke. Employers could, in these situations, detect such detours and frolics without engaging in non-traditional surveillance, and would, therefore, have no basis for accusing OSHA of making it impossible for them to punish abuse. A related, but somewhat more complicated situation would arise if employees did in fact go to the bathroom, but, instead of voiding, smoked, read, talked to others, or (in the age of cellphones) made telephone calls.

Can employers lawfully emulate Henry Ford's "Ford Service," his plant po-

---

[23]Telephone interview with Jay Withrow, Director, Office of Legal Support, Virginia OSHA (Oct. 31, 2002).

[24]E.g., United Parcel Service v. Administrative Review Bd., 1998 U.S. App. Lexis 24979 (6th Cir.) (Lexis); Cline v. Wal-Mart, 144 F.3d 294 (4th Cir. 1998). See generally, Laureen Snider, "Crimes Against Capital: Discovering Theft of Time," *Social Justice* 28(3):105 (Sept. 2001) (Lexis).

[25]Mor-Flo Industries, Inc., 62 Labor Arbitration Reports (BNA) at 400-401. See also United-Carr Tennessee, 59 Labor Arbitration Reports (BNA) 883, 888 (1972).

[26]Schmidt Cabinet Co., 75 Labor Arbitration Reports (BNA) 397, 400 (1980).

lice? "Servicemen were used to check on the men constantly.... How thorough a job they did is indicated by the fact that employees were even routinely followed to the toilets."[27] To the extent that supervisors can detect such OSHA-unauthorized activities without engaging in what would otherwise be an actionable invasion of privacy, there can be no valid complaint that the Memorandum has interfered with the exercise of any pre-existing managerial powers. A complication might arise if an employee simultaneously engaged in OSHA-authorized and -unauthorized activity. However, unless the latter prolonged the former—a risk factor that could be reduced, for example, by prohibiting smoking—an employer would, again, have no valid basis to complain about OSHA's interference with management's disciplinary powers.

Employers' real complaint relates not to detecting abuse among workers who go somewhere other than the bathroom or who do go there but engage in easily detectable OSHA-unauthorized activities. Rather, employers such as Jim Beam demand that OSHA tell them what lawful means are available to them to determine whether workers who go to the bathroom more frequently (or, secondarily, spend more time there) than management deems 'normal,' but who do not go so frequently that they need a doctor's note, are really voiding.

The Iowa Labor Commissioner, Byron Orton, as already noted, declared early on that since OSHA does not deprive employers of their normal powers of discipline, if workers use their right to void for other purposes, employers are free to discipline them.[28] Asked, however, whether an employer could station a monitor in the bathroom or position a camera to determine whether certain employees who exercise their right to go to the bathroom more often than others, without purporting to have a medical condition or doctor's explanation, are in fact voiding, Orton observed that those methods were probably illegal.[29] Asked what methods would then be available to employers, he replied that that was employers' problem—OSHA lacks the power to tell employers how to run their businesses and they "just gotta deal with it."[30] Voicing similar skepticism, one state labor standards official suspected that it is very unlikely that workers who abuse bathroom breaks are otherwise stellar employees: since this kind of behavior is probably correlated with other deficiencies, employers would be in a position to

---

[27]Harry Bennett, *Ford: We Never Called Him Henry* 58 (1987 [1951]).

[28]See above ch. 3.

[29]Telephone interview with Byron Orton (Oct. 17, 2002).

[30]Telephone interview with Byron Orton (Oct. 17 and 18, 2002). Nevertheless, Orton did state, for example, when asked whether Iowa OSHA would cite school districts if elementary school teachers were unable to go to the bathroom for three hours, that he viewed a school like an assembly line and that a relief worker system (if necessary involving the administrative staff) would be an appropriate method of abatement.

discipline such workers with poor overall records.[31]

Orton may or may not be right that a camera in a panopticonic bathroom would constitute an unlawful invasion of privacy,[32] but, as the questioning of UFCW Local 111-D President Jo Anne Kelley at the Jim Beam hearing revealed, posting a human attendant or monitor in the bathroom to determine whether workers are engaged in OSHA-unauthorized activities may well be lawful.[33] It is difficult to discern the possible basis of an employee's alleged reasonable expectation of privacy exclusively vis-à-vis the employer's attendant in a bathroom used by many other co-workers simultaneously, all of whom can see and hear exactly the same acts of the employee that the attendant can.[34] It seems implausible that courts would accept the argument that the employee has a reasonable expectation that—despite the fact that all of his co-workers can watch him—the employer's attendant not see, for example, that in addition to urinating for two minutes, the employee also used his OSHA-mandated voiding break to comb his hair, chat, make cell-phone calls, read, or smoke for five minutes, for which time the employer is legally obligated to compensate him. For the same reasons, however, the reasonable expectation of privacy would apply to any time that the employee spent within a toilet stall—which OSHA requires to be surrounded by "a door and walls or partitions between fixtures sufficiently high to assure privacy."[35]

The upshot of these considerations is that, Jim Beam's complaints to the contrary notwithstanding, it may be legally permissible for employers to mount

---

[31]Telephone interview with Richard Ervin, Employment Standards Manager, Washington Dept. of Labor and Industry, Olympia, WA (Oct. 21 and Nov. 1, 2002).

[32]"It is unlawful for any employer or the agent or representative of an employer, whether public or private, to operate any electronic surveillance device or system, including, but not limited to, the use of a closed circuit television system, a video-recording device, or any combination of those or other electronic devices for the purpose of recording or monitoring the activities of the employees in areas designed for the health or personal comfort of the employees or for safeguarding of their possessions, such as rest rooms, shower rooms, locker rooms, dressing rooms and employee lounges." W. Va. Code sect. 21-3-20(a) (2002). See also Conn. Gen. Stat. sect. 31-48b(b) (2001). But the NLRB has ruled merely that installing a surveillance camera in a restroom is a mandatory bargaining subject, not that it is per se unlawful. Colgate-Palmolive Co., 323 NLRB 515 (1997).

[33]See above ch. 13.

[34]As an arbitrator noted in a case involving an employer-imposed rule establishing maximum permissible personal relief time and requiring workers to register with their foremen: "Employees of either sex who...avail themselves of the prerogatives of personal relief to an unusual degree or with unprecedented frequency may be observed by their fellow workers and may even become the target for scatological comment." Detroit Gasket & Mfg. Co., 27 Labor Arbitration Reports (BNA) 717, 721 (1956).

[35]29 CFR sect. 1910.141(c)(2)(i).

effective searches for evidence of employees' fraudulent use of OSHA-mandated toilet breaks. In fact, employers' real concern may not be the lawfulness of such monitoring at all, but rather its counter-productive impact on workers' morale: the blatant lack of trust symbolized by the bathroom monitor could hardly be expected to engender or sustain a spirit of cooperativeness in a workforce resentful of being treated like wayward elementary school pupils.[36] And even if employers could ignore such problems with impunity, they may not be able to disregard other costs. As an arbitrator noted with regard to a pre-OSHA employer-imposed monitoring program:

> The prospect of employees and their supervisors spending substantial amounts of time in rehearsing the details of an employee's washroom habits is not to be faced with equanimity or indifference. Indeed, the more diligently the foremen seek to administer the program, the more searching will be their questions, the more time may be taken from productive activities and the more likely will personal irritations be provoked.[37]

If for no other reason, the question of the legality of employers' efforts to detect abuse should be clarified because it will eventually be raised in litigation, as it would have been had Jim Beam not dropped its appeal. If an employer could show that it was bereft of any plausible means of determining whether its employees were in fact voiding during OSHA-mandated compensable bathroom breaks, it might be able to persuade the Occupational Safety and Health Review Commission or a federal appeals court (or their state counterparts) that a regulation that left it no way to protect itself against fraudulent use of the right to void was arbitrary and capricious and thus invalid.

Like Orton, John Miles, the former OSHA Director of Compliance, rejected such complaints of employer defenselessness, but his reasoning bodes ill for advocates of urinary freedom: Not only do employers, in his opinion, retain control of the situation, but since someone going to the bathroom every 30 minutes or every hour is not normal—Miles added that OSHA has doctors on staff whom he spoke to about the issue while preparing the Memorandum—if such em-

---

[36]At a unionized plant where workers "enjoyed the right to leave their work area (at times other than their rest periods) to avail themselves of the restrooms," an arbitrator upheld a rule that the employer imposed requiring workers to check out and in, but nevertheless added that "it does not appeal to the Arbitrator as being a particularly sound method for correcting abuses while retaining employee morale...." Elgin Instrument Co., 37 Labor Arbitration Reports (BNA) 1064, 1066 (1961). For numerous examples of bathroom pass systems that doubtless do prepare children for life at workplaces unconstrained by the OSHA Memorandum while helping teachers identify "bogus potty breaks," see http://www.teachnet.com/how-to/manage/cantwait 011399.html.

[37]Detroit Gasket & Mfg. Co., 27 Labor Arbitration Reports at 720.

ployees did not have a medical excuse, an employer could lawfully exercise its disciplinary powers in such a case.[38]

Since employers are concerned primarily with abuse by individual workers rather than with some concerted class-struggle time-war by many or all the workers in a plant using bathroom breaks to resist what they regard as an unacceptable length or structure of working hours and/or intensity of labor, the problem may be less intractable than employers believe with a partial solution coming from unexpected quarters.[39] Workers at most workplaces know what the custom is and who is taking more than his or her fair share of breaks. Jo Anne Kelley, the president of the union at Jim Beam, offered this perspective:

> I do believe that everyone knew who was taking more than their "fair share." The question was not so much "fair share," but sometimes what they were doing when they took a break. It would have been a rare instance if someone took a break more than 1 time between scheduled breaks. Some may have stayed away from the line longer than they should, but the majority of the time that was handled "in house." Example, the person relieving them might say, "If you want me to relieve you again, you'd better not stay so long unless you've got something wrong with you." That usually took care of the problem.[40]

To be sure, Kelley herself recognized that certain employers might still demand a quantifiable precision and uniformity that human variability makes unattainable: "I also believe that the company was looking for the union, OSHA or someone to tell them exactly how many times is reasonable. Unfortunately we were unable to do that. Everyone is different. That's what makes us human beings. We are not machines."[41]

An even more striking illustration of "in-house" handling of abuse ratified by OSHA that undercuts employers' objection of defenselessness comes from the

---

[38]Telephone interview with John Miles (Nov. 12, 2002).

[39]In proposing "an at-will voiding regime," *Void Where Prohibited* conceded that its adoption "might prompt some workers to dissimulate." However, the book went on to argue that if workers engaged in "pseudo-excretory guerrilla warfare" as resistance against what they perceived as employers' illegitimate appropriation of their time and energy, a resolution of the conflict had to be sought in new norms of work and non-work rather than in sharper disciplinary intervention. Marc Linder and Ingrid Nygaard, *Void Where Prohibited: Rest Breaks and the Right to Urinate on Company Time* 160 (1998). Although this macrosocietal analysis remains valid, it manifestly transcends the current and foreseeable political-economic horizon and would not be applied by OSHA or the courts. The discussion in the text is an attempt to provide an analysis that could plausibly be adopted by agency administrators and judges to bolster at-will voiding breaks.

[40]Email from Jo Anne Kelley to Marc Linder (Oct. 6, 2002).

[41]Email from Jo Anne Kelley to Marc Linder (Oct. 6, 2002).

Bridgestone/Firestone Tire plant in Des Moines, Iowa, where workers work a 12-hour day (during alternating three- and four-day weeks) with breaks every two hours of 10, 10, 20, 10, and 10 minutes. According to the vice president of Local 310 of the Steelworkers (with which the Rubber Workers merged) and the safety chairman, Dennis Green, the production process there is sufficiently different from an automobile assembly line that workers can go to the bathroom when they need to. One worker, who frequently and publicly verbalized her position that the workers there worked too hard and too much and that she was going to take more breaks, went to the bathroom every odd hour for 10 to 15 minutes all day long for two to three years. Since her co-workers, who did not take these unscheduled breaks, had to fill in for her, they did not appreciate her absences. In 2002 the company began disciplining and ultimately fired her for taking excessive breaks. The union did file an (apparently halfhearted) grievance on her behalf, which turned out to be indefensible in light of the worker's documented statements of her real purpose in taking unscheduled breaks. The employee herself filed a complaint with OSHA,[42] which carried out an inspection, but decided not to issue a citation.[43] The inspector's narrative report explained that documentation furnished by the employer

> shows that the employee that states she was not able to take bathroom breaks was asking other employees to take multiple bathroom breaks during off times. I spoke to the employee that had been disciplined for excessive breaks. She stated that she takes breaks to sit down. That if she has to go to the bathroom, she feels she shouldn't have to do it on break time. She further states that it takes her 8 minutes to walk to the bathroom that she likes to use. Her breaks are only 10 minutes. Not much time to "rest" afterwards. The employee also stated to...(IOSH's clerical staff) that she was not going to be forced to take bathroom breaks on her sit breaks. This is not an isolated incident that she needed to use the restroom facility and was denied. There is [sic] several signed statements from co-workers stating that she has asked them to slow their work down.[44]

One conclusion to be drawn from such a case is that where co-workers' interests are impaired by a worker the frequency and/or length of whose bathroom breaks systematically exceed the workplace norm by a wide margin and are unsupported by any medical basis, "abuse" appears to be an apt characterization of that worker's behavior. To be sure, any employer systems for ferreting out abuse would, to pass muster, have to be created in good faith in order to detect abuse

---

[42]Telephone interview with Dennis Green, vice president, USWA Local 310, Des Moines (Oct. 14, 2002).

[43]Bridgestone/Firestone Tire, Insp. No. 304792542 (Mar. 1, 2002).

[44]Excerpt from Bridgestone/Firestone Tire, Insp. No. 304792542 (furnished by fax by Mary Bryant, IOSH Administrator, Oct. 30, 2002).

and not to create burdens on—or obstacles for the worker to overcome that might have the effect of substantially or unnecessarily burdening—the worker's right to void when the worker needs to go. And, as the real world of arbitration demonstrates, the evidentiary barriers facing employers are hardly insurmountable. For example, in the pre-OSHA era at one meat processing plant, where the United Packinghouse Workers had negotiated a 10-minute break in the morning and afternoon in addition to an hour for lunch, the company, after warnings and "as an example to all workers that this abuse could not be countenanced any longer," fired two workers who, to a greater extent than the other workers, had persistently and regularly taken four 12- to 15-minute mini-breaks in the "ladies room." The arbitrator had no difficulty finding the bathroom visits too regular to "be justified as...necessary, natural, and unavoidable"; on the contrary, "betoken[ing] a determination to disregard one's duty to the job and the legitimate need for co-operation with management," they constituted proper cause for discharge.[45]

Employers' animus against OSHA's loose-fitting performance standard may be rooted in the recognition that the 'you-gotta-let-'em-go-when-they-gotta-go' rule creates an unwelcome forum for contests over control of working time. For reasons that OSHA presumably never contemplated, let alone intended, a performance standard is much more conducive to such struggles than a fixed or quantitative standard—which some employers, such as Jim Beam, seem, at least rhetorically, to prefer to a reasonableness standard—framed in terms of breaks at set intervals. Since fixed breaks might create a windfall for workers who do not need to void that frequently—for them the additional breaks would constitute rest periods, which OSHA could mandate, albeit not under a sanitation standard—employers' disenchantment with the flexible performance standard that OSHA created hardly means that they would welcome a rigid system requiring a fixed number of breaks of fixed length.[46]

If employers nevertheless insisted on some quantification of OSHA's reasonableness standard, then OSHA's Memorandum could be amended to require management and workers/unions to form a committee to formulate a specific workplace rule. One possible rule might empower workers to use the toilet once an hour on an ongoing basis, no questions asked. If on more than an occasional

---

[45]Nevertheless, he recommended that the employer "give earnest consideration" to rehiring them if they gave assurances that they would "not abuse the privileges to which workers are entitled and which the company is cheerfully willing to grant." Boston Sausage & Provision Co., 2 Labor Arbitration Reports (BNA) 128, 129 (1946). Despite certain similarities, these mini-breaks differed from those at Jim Beam because they were not an employer-acknowledged universal practice of tag-relief.

[46]As an Iowa OSHA official put it, whichever method OSHA chose, at least half of employers would complain. Telephone interview with Rich Calonkey, Senior Industrial Hygienist, Iowa OSHA, Des Moines (Oct. 7, 2002).

basis (triggered, for example, by diarrhea or heavy menstruation), a worker went to the bathroom more than once an hour, she or he would either have some explaining to do or be required to submit a doctor's note.[47] The committee would also have to establish a framework for determining the maximum period of time that the employer was permitted to require a worker to wait before stopping work when immediate departure would be too disruptive. Once the committee defined a presumptively unreasonably long waiting period, it could determine how many relief workers would be necessary. Employers might find that their complaints about the cost of additional relief workers would be met with complaints about the cost, in terms of health and comfort, to the workers of not hiring them. Such a democratic process could easily make an employer appreciate the virtues of the reasonableness standard.

A fixed standard would, moreover, also have to specify how long people 'normally' stay in the bathroom (excluding the walking time to and from the toilet). At such decision points, however, the problems inherent in a numerical standard become so glaring that even advocates could be tempted to retreat to a reasonableness standard. President Kelley, for example, guessed that five minutes would, barring any problems, represent the average duration of a bathroom visit, but she added perspicaciously: "How do you regulate that without invading someone's privacy? I have personally been through enough problems that I don't care to know the details about why someone might need extra time. It would certainly be a challenge to come up with a rule that did not require an employee to report all the details."[48]

One low-tech tactic that might suit certain employers as a way of deterring "abusers" (as well as rightful users) from going to the bathroom is some version of the Canadian fee-to-pee system.[49] John Miles himself conjectured, when asked, that it might be lawful for an employer to charge workers to use the toilet[50]—a practice that OSHA expressly prohibits in agriculture.[51] He regretfully felt constrained to mention this possibility as the potential consequence of an adverse decision of the Occupational Safety and Health Review Commission, which held unreasonable OSHA's interpretation of a standard stating that certain per-

---

[47]Calling the once-an-hour regime a "very liberal rule," Jo Anne Kelley commented: "Even the 6 times a day which the Co. claimed to provide would have been quite adequate if you were allowed to go when you needed to go and not when the Co. said you could go, at least for the majority of employees." Email from Jo Anne Kelley to Marc Linder (Oct. 6, 2002).

[48]Email from Jo Anne Kelley to Marc Linder (Oct. 6, 2002).

[49]See above ch. 9.

[50]Telephone interview with Miles.

[51]29 CFR sect. 1928 110(c) (2002).

sonal protective equipment "shall be provided" as requiring that the employer must also pay for it.[52]

A related tactic that employers might try to adopt to discipline "abusers" (as well as others) would be to require assembly-line-type workers to go off the clock while exercising their right to void. Off the top of his head, John Miles felt that this measure was "probably legal," though he hastened to add that he did not know enough about wage and hour law to know whether taking employees off the clock was unlawful, which he thought it might be.[53] However, even if OSHA refused to declare such a practice unlawful on the grounds that it unduly deterred workers from exercising their right to void, the Federal Wage and Hour Administrator has adopted a position clearly inconsistent with such pay-docking. According to a long-standing interpretive regulation issued under the Fair Labor Standards Act: "Rest Periods of short duration, running from 5 to about 20 minutes,...must be counted as hours worked. Compensable time of rest periods may not be offset against other working time...."[54] If there had been any doubts as to whether a toilet break is included under the category of rest breaks, a series of opinion letters dealing with smoking breaks issued by the Wage and Hour Division of the U.S. Department of Labor during the 1990s dispelled them by concluding that "it is immaterial with respect to compensability of such breaks whether the employee drinks coffee, smokes, goes to the restroom, etc." Moreover, "if the employer allows affected employees to work beyond their shifts to make up the time, the employer is incurring additional liability for such makeup time under the FLSA."[55] In one place of employment workers were allowed to leave their work stations to smoke for three to four minutes at a time for up to 15 minutes per day; this break time exceeded that allowed other employees. Even where the four smoking breaks totaled 40 minutes daily, the agency declared them compensable.[56] More systematically the Wage and Hour Administrator opined in response to a query from another employer:

> Employees have always taken short work breaks, with pay, for a myriad of non-work

---

[52]Union Tank Car Co., 18 OSHC (BNA) 1067 (1997) (Lexis). Instead of litigating the interpretation, OSHA chose to engage in rulemaking. "OSHA to Change Rule, Not Appeal Protective Equipment Payment Decision," *Daily Labor Report*, Dec. 17, 1997, 1997 DLR 242 d12 (Lexis); *Federal Register* 62:15402 (1999).

[53]Telephone interview with Miles.

[54]29 CFR sect. 785.18 (2001).

[55]Wage and Hour Div., U.S. Dept. of Labor, Opinion Letter 1995 WL 1032460 (Westlaw) (Jan. 25, 1995).

[56]Wage and Hour Div., U.S. Dept. of Labor, Opinion Letter 1994 WL 1004839 (Westlaw) (June 16, 1994). See also Wage and Hour Div., U.S. Dept. of Labor, Opinion Letter 1996 WL 1004840 (Westlaw) (June 16, 1994).

purposes—a visit to the bathroom, a drink of coffee, a call to check the children, attending to a medical necessity, a cigarette break, etc. The Department has consistently held for over 46 years that such breaks are hours worked under the FLSA, without evaluating the relative merits of an employee's activities. This position, found at 29 C.F.R. 785.18, is based squarely in the premise that short breaks are common in industry, promote the efficiency of employees and are customarily treated as work time by employers.

The compensability of short breaks by workers has seldom, if ever, been questioned. Any modification of the Department's long held position to accommodate your request would require a series of tests to evaluate the relative benefit provided to employee and employer and the impact on employee efficiency of each and every small work break ever taken by any employee.

We believe that such tests would be an undesirable regulatory intrusion in the workplace with the potential to seriously disrupt many employer-employee relationships. Further, it would be difficult, if not impossible, to design practical tests applicable to all workplace circumstances.

While we fully appreciate the extraordinary difficulties presented to employers by smoking in the workplace, we believe that the government should not be in the business of determining what employees do on short work breaks, much less attempting to evaluate which short breaks merit or do not merit compensation. We strongly believe that employers and employees are best served by the bright line time test currently provided in Section 785.18.

We are unwilling, for these reasons cited above, to modify the existing position. The FLSA does not require an employer to provide its employees with rest periods or breaks. If the employer decides to permit short breaks, however, the time is compensable hours worked. Even if the employees agree to forego compensation for the break time, the time is still compensable, because employees may not waive their statutory rights through an agreement with the employer. If the employer permits its employees to take a series of short smoke breaks, the employees must be compensated for their time.[57]

The cri de coeur that Jim Beam's management uttered over the alleged disappearance of its power to police and punish fraudulent urinators might also galvanize employers' interest in hi-tech solutions. It is, as its author, sociologist Gary Marx, noted, "an interesting commentary on our society," that when he published the following satirical piece in the *Los Angeles Times* on April Fools Day 1987, "many readers thought it was real and some even wrote and asked where the system could be purchased." Though in part blatantly in violation of OSHA since April 6, 1998, the proposal would presumably trigger the same responses today:

---

[57]Wage and Hour Div., U.S. Dept. of Labor, Opinion Letter 1996 WL 1005233 (Westlaw) (Dec. 2, 1996).

TO: ALL EMPLOYEES
FROM: EMPLOYEE RELATIONS DEPARTMENT
SUBJECT: RESTROOM TRIP POLICY (RTP)

An internal audit of employee restroom time (ERT) has found that this company significantly exceeds the national ERT standard recommended by the President's Commission on Productivity and Waste. At the same time, some employees complained about being unfairly singled out for ERT monitoring. Technical Division (TD) has developed an accounting and control system that will solve both problems.

Effective 1 April 1987, a Restroom Trip Policy (RTP) is established.

A Restroom Trip Bank (RTB) will be created for each employee. On the first day of each month employees will receive a Restroom Trip Credit (RTC) of 40. The previous policy of unlimited trips is abolished. Restroom access will be controlled by a computer-linked voice-print recognition system. Within the next two weeks, each employee must provide two voice prints (one normal, one under stress) to Personnel. To facilitate familiarity with the system, voice-print recognition stations will be operational but not restrictive during the month of April.

Should an employee's RTB balance reach zero, restroom doors will not unlock for his/her voice until the first working day of the following month.

Restroom stalls have been equipped with timed tissue-roll retraction and automatic flushing and door-opening capability. To help employees maximize their time, a simulated voice will announce elapsed ERT up to 3 minutes. A 30-second warning buzzer will then sound. At the end of the 30 seconds the roll of tissue will retract, the toilet will flush and the stall door will open. ...

To prevent unauthorized access (e.g., sneaking in behind someone with an RTB surplus, or use of a tape-recorded voice), video cameras in the corridor will record those seeking access to the restroom. However, consistent with the company's policy of respecting the privacy of its employees, cameras will not be operative within the restroom itself.

...

Management recognizes that from time to time employees may have a legitimate need to use the restroom. But employees must also recognize that their jobs depend on this company's staying competitive in a global economy. These conflicting interests should be weighed, but certainly not balanced. The company remains strongly committed to finding technical solutions to management problems. We continue to believe that machines are fairer and more reliable than managers. We also believe that our trusted employees will do the right thing when given no other choice.[58]

An April Fools joke—and yet the endless possibilities for electronic surveillance and control at the workplace have made such great technological advances in recent years that (Gary) Marx's nightmarish scenario has become only all too

---

[58]Gary Marx, "Raising Your Hand Just Won't Do," on http://web.mit.edu/gtmarx/www/raising.html. Marx's website privacy policy states that his "interest is in giving away ideas and in encouraging/provoking thought...."

possible.[59] For example, employers (or at least those unfettered by the aforementioned interpretation of the FLSA) can take solace that, while brutally intrusive managerial methods survive for verifying employees' claims of being humans with bodily needs,[60] at least one meat factory in the United Kingdom, where employees have to go through a turnstile to get to the toilet, has issued them "smart cards which deduct their pay for the time they're away from the factory floor."[61]

Having survived *Das Kapital*, World War I, the Bolshevik Revolution, the Great Depression, World War II, and the Chinese Communist Revolution, capitalism seems unlikely to be toppled by at-will bathroom breaks. After all, many employers in the United States do not interfere with the frequency and duration of their employees' acts of waste elimination, and there are even rumored to be whole capitalist countries in Europe in which workplace voiding freedom is taken for granted. And yet, firms' accumulation- and profit-driven tendency to secure and retain as much unilateral control over working time as possible will presumably continue to manifest itself in myriad ways that conflict with workers' own conceptions of individual and collective autonomy, self-development, and co-determination. Consequently, even reformist struggles over the need to stop work in order to attend to other needs—including voiding—will continue to test capitalism's compatibility with the most elementary dictates of humanity.

---

[59]"A new high-tech watchdog may soon monitor the personal hygiene habits of health care, food service and other workers every time they use the bathroom at work. The first system of its kind, dubbed Hygiene Guard, was installed at the Tropicana Casino and Resort in Atlantic City yesterday to track whether 20 chefs, dishwashers and waiters use soap dispensers and wash their hands after using the toilet. Under the system, employees will be required to wear a battery-powered 'smart badge.' The badge communicates with sensors in the bathroom that are connected to a computer in a manager's office. It also beeps periodically to remind employees to wash their hands. Unless an employee uses the soap dispenser and stands for a required amount of time in front of a sink with running water, an infraction will be recorded on the computer." Robert O'Harrow Jr.,"Big Brother in Workplace Bathrooms?" *Washington Post*, Aug. 30, 1997, at A1 (Lexis).

[60]Thus, at one of Levi Strauss's "hired factories" in Indonesia a "contractor...was found strip-searching female workers to determine whether they, as they claimed, were menstruating—and thus entitled to a day off with pay, according to the law of that Muslim country...." G. Zachary, "Exporting Rights: Levi Tries to Make Sure Contract Plants in Asia Treat Workers Well," *Wall Street Journal*, July 28, 1994, at A1 (Westlaw).

[61]Billy Paterson, "Have a Break and a Quick C***: New Staff Motto After Bosses Dock Them £1000 a Week for Toilet Time," *Sunday Mail*, Jan. 5, 2003, at 21 (Lexis) (discussing Brown Brothers Manufacturing Ltd. in Kirkconnell, Scotland). Inspired by the campaign launched by *Void Where Prohibited*, Rory O'Neill, the editor of the British occupational safety and health magazine *Hazards*, prompted the TUC to initiate a campaign on Feb. 21, 2003 to outlaw such practices. "Give Us a Break!" *Hazards*, No. 81 at 18-19 (Jan. - Mar. 2003).

Appendix I

ANSI and OSHA Workplace Toilet Standards from 1935 to the Present

On April 27, 1971, pursuant to sections 3(9) and 6(a) of OSHA, Secretary of Labor Hodgson designated as "national consensus standards" certain sanitation standards of the American National Standards Institute.[1] The toilet standard promulgated by OSHA was virtually identical with section 6 of the United States of America Standards Institute, *USA Standard Requirements for Sanitation in Places of Employment*.[2] The principal differences were the exclusion of the advisory ("should" or "may") provisions and recommendations. For example, OSHA did not adopt this provision (sect. 6.3.3) on the non-split toilet seat: "Integral water-closet seats may be used where specifically permitted by the health authorities having jurisdiction." Despite the Labor Secretary's statement that he had adopted only mandatory provisions, the toilet standard did adopt the rule that "[a]s far as practicable, toilet facilities should be located within 200 feet of all locations at which workers are regularly employed" (sect. 6.1.2; 1910.141(c)(1)(ii)). The 1968 standard was, in turn, identical with section 6 of the 1955 American Standards Association, *American Standard Minimum Requirements for Sanitation in Places of Employment*.[3] In turn, the 1955 version is very closely patterned after sections 3.11-3.14 of the 1935 American Standards Association *American Standard Safety Code for Industrial Sanitation in Manufacturing Establishments*.[4] Many provisions were identical. Others differed because the earlier version had still been committed to sex and/or sexist differences as well as to a more and less hygienic standard.[5]

---

[1]*Federal Register* 36:10466 (May 29, 1971).

[2]United States of America Standards Institute, *USA Standard Requirements for Sanitation in Places of Employment* (USA Z4.1-1968, Feb. 26, 1968).

[3]American Standards Association, *American Standard Minimum Requirements for Sanitation in Places of Employment* (ASA Z4.1-1955, Mar. 28, 1955).

[4]American Standards Association *American Standard Safety Code for Industrial Sanitation in Manufacturing Establishments* (ASA Z4.1-1935, Apr. 1, 1935)

[5]See Marc Linder and Ingrid Nygaard, *Void Where Prohibited: Rest Breaks and the Right to Urinate on Company Time* 56-57 (1998).

Set out below are the various standards and amendments discussed above in Chapter 4.

American Standards Association *American Standard Safety Code for Industrial Sanitation in Manufacturing Establishments* (ASA Z4.1-1935, Apr. 1, 1935):

3.11 Toilet Facilities in General

(a) Every place of employment shall be provided with adequate water closets, chemical closets, or privies, separate for each sex. ...

(b) Covered receptacles shall be kept in all toilet rooms used by females.

(c) An adequate supply of toilet paper shall be provided for every toilet facility.

(d) Unless the general washing facilities are on the same floor and in close proximity to the toilet rooms, adequate washing facilities shall be provided in every toilet room or adjacent thereto.

(e) Toilet rooms shall be readily accessible to employees using them. No toilets shall be more than one floor above or below the regular place of work of the persons using them. This rule shall not apply when passenger elevators are available for employees' use in going to and from toilet rooms.

(f) Toilet facilities (closets) shall be provided for each sex according to the following table. The number to be provided for each sex shall in every case be based on the maximum number of persons of that sex employed at any one time at work on the premises for which the facilities are furnished. When persons other than employees are permitted the use of toilet facilities (closets) on the premises, a reasonable allowance shall be made for such other persons in estimating the minimum number of toilet facilities (closets) required.

Number of Persons	Minimum number of facilities
1 to 9	1.
10 to 24	2.
25 to 49	3.
50 to 74	4.
75 to 100	5.
Over 100	1 for each additional 30 persons

Whenever urinals are provided, one facility less than the above specified number of facilities may be provided, for males, for each urinal, except that the number of facilities in such cases may not be reduced to less than two-thirds of the number specified above. Two feet of acid-resisting porcelain enamel urinal shall be considered as equivalent to one urinal.

3.12 Construction of Toilet Rooms

(a) In toilets for women, each toilet facility (closet) shall occupy a separate compartment. In toilets for use of men, each toilet facility (closet) shall occupy a separate compartment, or fixtures shall be separated by simple partitions extending at least 15 inches in front of the fixture, with such arrangement that the toilets do not face one another.

(b) The walls of compartments or partitions between fixtures may be less than the height of the room walls, but the top shall not be less than 6 feet from the floor and the bottom not more than 1 foot from the floor.

(c) The door to every toilet room shall be fitted with an effective self-closing device and screened so as not to be visible from the workroom. All "compartment" doors shall be supplied with latch.

(d) In all toilet rooms hereinafter installed, the floors and side walls, to a height of 6 inches, shall be water-tight and impervious to moisture, including the angle formed by the floor and side walls.

(e) The floors, walls, and ceilings of all toilet rooms shall be of a finish that can be easily cleaned.

(f) The walls of every toilet room shall be of solid construction and shall extend to the ceiling, or the area shall be independently ceiled over. Above the level of six feet the wall may be provided with glass that is translucent but not transparent. This rule shall not apply to foundries, rolling mills, furnaces, and other places where females are not allowed to enter; in such cases the enforcing authority may specify less rigid rules when requested to do so.

(g) In new installations the minimum floor space allotted for toilet facilities (closets), lavatories (wash basins), and urinals shall be as follows:

	Minimum Width	Minimum Depth	Minimum total floor space
Facilities	32 in.	3.6 ft	16 sq ft
Lavatories	24 in.	3.6 ft	12 sq ft
Urinals	24 in.	3.6 ft	12 sq ft

(h) Natural or artificial means shall be provided to insure a comfortable and healthful atmosphere throughout the premises.

3.13 Construction and Installation of Toilet Facilities

(a) The construction and maintenance of toilet fixtures shall comply with the State building and plumbing codes where such codes exist. In other cases and until such time as national specifications are developed, it is suggested that the requirements of Bulletin B. H. 13, of the U.S. Department of Commerce, be followed.

(b) Every water closet bowl shall be set entirely free and open from all enclosing woodwork, and shall be so installed that the space around the fixture can be easily cleaned.

(c) Every water closet except those of integral seat type shall have a hinged open-front seat made of substantial material having a nonabsorbent finish. If absorbent material is used the seat shall be finished to make it impervious to moisture, and shall be preferably light in color.

3.14 Chemical Closets and Privy Vaults

(a) Chemical closets and privies shall not be permitted except where there is no sewer accessible, and no privy shall be permitted where it is impossible to construct and maintain the same without danger of contaminating any source of drinking water.

(b) When chemical closets are permitted they shall be of a type approved by the health authorities having jurisdiction and shall be maintained in a sanitary condition.

(c) The container shall be changed frequently enough that it will not be allowed to become more than two-thirds full, the contents at all times to be disposed of in accordance with the regulations of the health authorities having jurisdiction.

(d) In every establishment employing more than 25 persons, the use of privies is prohibited. In such instances where they are permitted they shall be constructed and maintained in accordance with Specifications for the Sanitary Privy—Z1.3-1935....

(e) No privy shall be located within 100 feet of any room where foodstuffs are stored or handled.

OSHA National Consensus Standard (effective Aug. 27, 1971)[6]

1910.141 Sanitation

(c)—(1) *General.* (i) Every place of employment shall be provided with adequate toilet facilities which are separate for each sex. The sewage disposal method shall comply with requirements of the health or other authorities having jurisdiction.

(ii) Toilet facilities shall be provided so as to be readily accessible to all employees. Toilet facilities so located that employees must use more than one floor-to-floor flight of stairs to or from them are not considered as readily accessible. As far as practicable, toilet facilities should be located within 200 feet of all locations at which workers are regularly employed.

(iii) Water closets shall be provided for each sex according to the following table. The number to be provided for each sex shall in every case be based on the maximum number of persons of that sex employed at any one time at work on the premises for which the facilities are furnished. When persons other than employees are permitted the use of toilet facilities on the premises, a reasonable allowance shall be made for such other persons in estimating the minimum number of toilet facilities required.

---

[6]*Federal Register* 36:10593-94 (May 29, 1971); 29 Code of Federal Regulations (1972).

Number of Persons	*Minimum number of facilities*
1 to 9	1.
10 to 24	2.
25 to 49	3.
50 to 74	4.
75 to 100	5.
Over 100	1 for each additional 30 persons

Where 10 or more are employed, urinals may be provided. One water closet less than the number specified in the foregoing may be provided for each urinal, except that the number of water closets in such cases may not be reduced to less than two-thirds of the number specified in the foregoing. Two feet of trough urinal shall be considered as equivalent to one individual urinal.

(iv) An adequate supply of toilet paper with paper shall be provided for every water closet.

(v) Covered receptacles shall be provided in every toilet room used by women.

(vi) Adequate washing facilities shall be provided in every toilet room or be adjacent thereto.

(2) *Construction of toilet rooms.* (i) Each toilet facility (closet) shall occupy a separate compartment, which should be equipped with a door, latch, and clothes hanger.

(ii) The walls of compartments or partitions between fixtures may be less than the height of room walls, but the top shall not be less than 6 feet from the floor and the bottom not more than 1 foot from the floor.

(iii) The door to every toilet room shall be fitted with an effective self-closing device, and the entrance to the toilet room shall be so screened that the interior of the toilet room is not visible from the workroom.

(iv) In all toilet rooms installed after July 1, 1971, the floors and side walls, to a height of at least 6 inches, including the angle formed by the floor and side walls, shall be of watertight construction.

(v) The floors, walls, ceilings, partitions, and doors of all toilet rooms shall be of a finish that can be easily cleaned. In installations made after July 1, 1971, cove bases shall be provided to facilitate cleaning.

(vi) Toilet rooms, except those in work places accessible to men only, shall be completely enclosed with solid materials that is nontransparent from the outside.

(3) *Construction and installation of toilet facilities.* (i) Every water closet bowl shall be set entirely free and open from all enclosing structures and shall be so installed that the space around the fixture can be easily cleaned. This provision does not prohibit the use of wall-hung type water closets.

(ii) Every water closet shall have a hinged openfront seat made of substantial material having a nonabsorbent finish.

(4) Chemical closets and privies. Chemical closets and privies shall be constructed and maintained in accordance with § 1910.143.

Proposed Amendment (July 10, 1972)[7]

1910.141

(c) *Toilet facilities*—(1) *General.* (i) Toilet facilities shall be provided in accordance with Table J-1 of this section.

(ii) The sewage disposal method shall not endanger the health of employees.

TABLE J-1

Type of employment	Water closets		Urinals	Lavatories	
	No. of employees	No. of fixtures		No. of employees	No. of fixtures
Office	1-9	1	Wherever urinals are provided one water closet less than specified may be provided for each urinal installed except the number of water closets in such cases shall not be reduced to less than 2/3 of the minimum specified	1-9	1.
	10-20	2		10-24	2.
	25-49	3		25-49	3.
	50-74	4		50-74	4.
	75-100	5		75-100	5.
	Over 100	One fixture for each 30 employees		Over 100	One fixture for each 30 employees
All other	Same schedule as for offices		Same schedule as for offices	1-100 employees: 1 fixture for each 10 employees. Over 100 employees: 1 fixture for each additional 15 employees.	

(iii) Toilet facilities shall be readily accessible to all employees, and in no case more than one floor-to-floor flight of stairs, or no more than 200 feet from any place where employees regularly work.

(iv) When persons other than employees are permitted the use of toilet facilities on the premises, the number of such facilities shall be appropriately increased in accordance with Table J-1 of this section in determining the minimum number of toilet facilities required.

(v) Toilet paper with holder shall be provided for every water closet.

(vi) Covered receptacles shall be kept in all toilet rooms used by women.

---

[7] *Federal Register* 37:13997-98 (July 15, 1972).

(vii) For each required toilet facility at least one lavatory shall be provided either in the toilet room or adjacent thereto.

(2) *Construction of toilet rooms.* (i) Each water closet shall occupy a separate compartment with walls or partitions between fixtures sufficiently high to assure privacy.

(ii) In all toilet rooms installed on or after August 31, 1971, the floors and sidewalls, to a height of at least 6 inches, including the angle formed by the floor and sidewalls, shall be of watertight construction.

(iii) The floors, walls, ceilings, partitions, and doors of all toilet rooms shall be of a finish that can be easily cleaned. In installations made after August 31, 1971, cove bases shall be provided to facilitate cleaning.

(3) *Construction and installation of toilet facilities.* (i) Every water carriage toilet facility shall be set entirely free and open from all enclosing structures and shall be so installed that the space around the fixture can be easily cleaned. This provision does not prohibit the use of wall-hung type water closets or urinals.

(ii) Every water closet shall have a hinged seat made of substantial material having a nonabsorbent finish. Seats installed or replaced after (effective date of this provision) shall be of the open-front type.

(iii) Nonwater carriage toilet facilities and disposal systems shall be in accordance with § 1910.143.

Amended Standard (effective June 4, 1973)[8]

1910.141

(c) *Toilet facilities.*—(1) *General.*—(i) Except as otherwise indicated in this subdivision (i), toilet facilities, in toilet rooms separate for each sex, shall be provided in all places of employment in accordance with table J-1 of this section. The number of facilities to be provided for each sex shall be based on the number of employees of that sex for whom the facilities are furnished. Where toilet rooms will be occupied by no more than one person at a time, can be locked from the inside, and contain at least one water closet, separate toilet rooms for each sex need not be provided. Where such single-occupancy rooms have more than one toilet facility, only one such facility in each toilet room shall be counted for the purpose of table J-1.

---

[8]*Federal Register* 38:10933 (May 3, 1973); 29 Code of Federal Regulations (1973).

TABLES [sic] J-1

Number of employees:	*Minimum number of water closets*[1]
1-15	1.
16-35	2.
36-55	3.
56-80	4.
81-110	5.
111-150	6.
Over 150	1 additional fixture for each additional 40 employees

[1]Where toilet facilities will not be used by women, urinals may be provided instead of water closets, except that the number of water closets in such cases shall not be reduced to less than 2/3 of the minimum specified.

(ii) The requirements of subdivision (i) of this subparagraph do not apply to mobile crews or to normally unattended work locations so long as employees working at these locations have transportation immediately available to nearby toilet facilities which meet the other requirements of this subparagraph.

(iii) The sewage disposal method shall not endanger the health of employees.

(iv) When persons other than employees are permitted the use of toilet facilities on the premise [sic], the number of such facilities shall be appropriately increased in accordance with table J-1 of this section in determining the minimum number of toilet facilities required.

(v) Toilet paper with holder shall be provided for every water closet.

(vi) Covered receptacles shall be kept in all toilet rooms used by women.

(vii) For each three required toilet facilities at least one lavatory shall be located either in the toilet room or adjacent thereto. Where only one or two toilet facilities are provided at least one lavatory so located shall be provided.

(2) *Construction of toilet rooms.*— (i) Each water closet shall occupy a separate compartment with walls or partitions between fixtures sufficiently high to assure privacy.

(ii) In all toilet rooms installed on or after August 31, 1971, the floors and sidewalls, including the angle formed by the floor and sidewalls, and excluding doorways and entrances, shall be watertight. The sidewalls shall be watertight to a height of at least 5 inches.

(iii) The floors, walls, ceilings, partitions, and doors of all toilet rooms shall be of a finish that can be easily cleaned. In installations made after August 31, 1971, cove bases shall be provided to facilitate cleaning.

(3) *Construction and installation of toilet facilities.* (i) Every water carriage toilet facility shall be set entirely free and open from all enclosing structures and shall be so installed that the space around the fixture can be easily cleaned. This provision does not prohibit the use of wall-hung type water closets or urinals.

(ii) Every water closet shall have a hinged seat made of substantial material having a nonabsorbent finish. Seats installed or replaced after June 4, 1973, shall be of the open-front type.

(iii) Nonwater carriage toilet facilities and disposal systems shall be in accordance with § 1910.143.

Proposed Revocation/Revision (1977)[9]

(c) *Toilet Facilities.*— [T]oilet facilities shall be provided in all places of employment.

The sewage disposal method shall not endanger the health of employees.

Revised Standard (1978)[10]

(c) *Toilet facilities.*—(1)*General.*—(i) Except as otherwise indicated in this subdivision (i), toilet facilities, in toilet rooms separate for each sex, shall be provided in all places of employment in accordance with table J-1 of this section. The number of facilities to be provided for each sex shall be based on the number of employees of that sex for whom the facilities are furnished. Where toilet rooms will be occupied by no more than one person at a time, can be locked from the inside, and contain at least one water closet, separate toilet rooms for each sex need not be provided. Where such single-occupancy rooms have more than one toilet facility, only one such facility in each toilet room shall be counted for the purpose of table J-1.

TABLE J-1

Number of employees	*Minimum number of water closets*[1]
1-15	1.
16-35	2.
36-55	3.
56-80	4.
81-110	5.
111-150	6.
Over 150	1 additional fixture for each additional 40 employees

[1]Where toilet facilities will not be used by women, urinals may be provided instead of water closets, except that the number of water closets in such cases shall not be reduced to less than 2/3 of the minimum specified.

(ii) The requirements of paragraph (c)(1)(i) of this section do not apply to mobile crews or to normally unattended work locations so long as employees working at these locations have transportation immediately available to nearby toilet facilities which meet the other requirements of this subparagraph.

(iii) The sewage disposal method shall not endanger the health of employees.

(2) *Construction of toilet rooms.*— (i) Each water closet shall occupy a separate compartment with walls or partitions between fixtures sufficiently high to assure privacy.

---

[9]*Federal Register* 42:62801 (Dec. 13, 1977).

[10]*Federal Register* 43:49736-37 (Oct. 24, 1978); 29 Code of Federal Regulations 1910.141 (1979).

Since the major revocation project of 1978, section 1910.141(c) has remained unchanged and thus still reads:

29 CFR 1910 (2002)

1910.141 Sanitation.

(a) *General*—(1) *Scope*. This section applies to permanent places of employment.
(2) *Definitions applicable to this section.*

...

Number of employees means, unless otherwise specified, the maximum number of employees present at any one time on a regular shift.

...

Toilet facility, means a fixture maintained within a toilet room for the purpose of defecation or urination, or both.

Toilet room, means a room maintained within or on the premises of any place of employment, containing toilet facilities for use by employees.

...

Urinal means a toilet facility maintained within a toilet room for the sole purpose of urination.

Water closet means a toilet facility maintained within a toilet room for the purpose of both defecation and urination and which is flushed with water.

...

(c) *Toilet facilities*—(1) *General.* (i) Except as otherwise indicated in this paragraph (c)(1)(i), toilet facilities, in toilet rooms separate for each sex, shall be provided in all places of employment in accordance with table J-1 of this section. The number of facilities to be provided for each sex shall be based on the number of employees of that sex for whom the facilities are furnished. Where toilet rooms will be occupied by no more than one person at a time, can be locked from the inside, and contain at least one water closet, separate toilet rooms for each sex need not be provided. Where such single-occupancy rooms have more than one toilet facility, only one such facility in each toilet room shall be counted for the purpose of table J-1.

Table J-1

Number of employees	Minimum number of water closets[1]
1 to 15	1
16 to 35	2
36 to 55	3
56 to 80	4
81 to 110	5.
111 to 150	6
Over 150	([2])

[1]Where toilet facilities will not be used by women, urinals may be provided in-stead of water closets, except that the number of water closets in such cases shall not be reduced to less than 2/3 of the minimum specified.

[2]1 additional fixture for each additional 40 employees.

(ii) The requirements of paragraph (c)(1)(i) of this section do not apply to mobile crews or to normally unattended work locations so long as employees working at these locations have transportation immediately available to nearby toilet facilities which meet the other requirements of this subparagraph.

(iii) The sewage disposal method shall not endanger the health of employees.

(2) *Construction of toilet rooms*. (i) Each water closet shall occupy a separate compartment with a door and walls or partitions between fixtures sufficiently high to assure privacy.

(ii)[Reserved]

Despite the revocations that OSHA has undertaken, ANSI, in its 1995 revised *American National Standard for Sanitation in Places of Employment—Minimum Requirements*, has not only, like Cal/OSHA,[11] retained several of the provisions deleted by OSHA, but also added others. It has, for example, retained the 200-foot rule, the mandate that toilets "shall be provided so as to be readily accessible to all employees," and the requirement of receptacles for sanitary napkins. It has also introduced new rules that the adequate supply of toilet paper "shall be maintained at all times" and that grab bars be installed.[12]

---

[11]Title 8 Cal. Code of Reg. sect 3364 (2002).

[12]American National Standards Institute, *American National Standard for Sanitation in Places of Employment—Minimum Requirements* sect. 6.1.2, 6.1.4, 6.1.5, 6.2.3 (Z4.1-1995, Aug. 8, 1995). These provisions were also all present in American National Standards Institute, *American National Standard for Sanitation in Places of Employment—Minimum Requirements* sect. 6.1.2, 6.1.4, 6.1.5, 6.2.3 (Z4.1-1986, July 31, 1986), and American National Standards Institute, *American National Standard Minimum Requirements for Sanitation in Places of Employment* sect. 6.1.2, 6.1.4, 6.1.5, 6.2.3 (Z4.1-1979, Jan. 15, 1979).

Appendix II

OSHA's Interpretive Memorandum of April 6, 1998[1]

April 6, 1998

MEMORANDUM FOR:	REGIONAL ADMINISTRATORS STATE DESIGNEES
FROM:	JOHN B. MILES, JR., Director Directorate of Compliance Programs
SUBJECT:	Interpretation of 29 CFR 1910.141(c)(1)(i): Toilet Facilities

OSHA's sanitation standard for general industry, 29 CFR 1910.141(c)(l)(i), requires employers to provide their employees with toilet facilities:

Except as otherwise indicated in this paragraph (c)(l)(i), toliet [sic] facilities, in toilet rooms separate for each sex shall be provided in all places of employment in accordance with Table J-1 of this section .... [emphasis added]

This memorandum explains OSHA's interpretation that this standard requires employers to make toilet facilities available so that employees can use them when they need to do so. The employer may not impose unreasonable restrictions on employee use of the facilities. OSHA believes this requirement is implicit in the language of the standard and has not previously seen a need to address it more explicitly. Recently, however, OSHA has received requests for clarification of this point and has decided to issue this memorandum to explain its position clearly.

Background

The sanitation standard is intended to ensure that employers provide employees with sanitary and available toilet facilities, so that employees will not suffer the adverse health effects that can result if toilets are not available when employees need them. Individuals

---

[1] http://www.osha.gov/pls/oshaweb/owadisp.show_document?p_table=INTERPRETATIONS&p_id=22932&p_text_version=FALSE

vary significantly in the frequency with which they need to urinate and defecate, with pregnant women, women with stress incontinence, and men with prostatic hypertrophy needing to urinate more frequently. Increased frequency of voiding may also be caused by various medications, by environmental factors such as cold, and by high fluid intake, which may be necessary for individuals working in a hot environment. Diet, medication use, and medical condition are among the factors that can affect the frequency of defecation.

Medical studies show the importance of regular urination, with women generally needing to void more frequently than men. Adverse health effects that may result from voluntary urinary retention include increased frequency of urinary tract infections (UTIs), which can lead to more serious infections and, in rare situations, renal damage (see, e.g., Nielsen, A. Waite, W. [sic; should be Walter, S.], "Epidemiology of Infrequent Voiding and Associated Symptoms," Scand J Urol Nephrol Supplement 157). UTIs during pregnancy have been associated with low birthweight babies, who are at risk for additional health problems compared to normal weight infants (see, Naeye, R.L., "Causes of the Excess [sic; should be Excessive] Rates of Perinatal Mortality and the [sic] Prematurity in Pregnancies Complicated by Maternity [sic; should be Maternal] Urinary[-]Tract Infections," New England J. Medicine 1979; 300(15); 819-823). Medical evidence also shows that health problems, including constipation, abdominal pain, diverticuli, and hemorrhoids, can result if individuals delay defecation (see National Institutes of Health (NJH) Publication No. 95-2754, July 1995).

OSHA's field sanitation standard for Agriculture, 29 CFR 1928.110, based its requirement that toilets for farmworkers be located no more than a quarter mile from the location where employees are working on similar findings. This is particularly significant because the field sanitation standard arose out of the only OSHA rulemaking to address explicitly the question of worker need for prompt access to toilet facilities.

The Sanitation Standard

The language and structure of the general industry sanitation standard reflect the Agency's intent that employees be able to use toilet facilities promptly. The standard requires that toilet facilities be "provided" in every workplace. The most basic meaning of "provide" is "make available." See **Webster's New World Dictionary, Third College Edition**, 1988, defining "provide" as "to make available; to supply (someone with something);" Borton Inc. V. OSHRC, 734 F.2d 508, 510 (l0th Cir. 1984) (usual meaning of provide is "to furnish, supply, or make available"); Usery v. Kennecott Copper Corp., 577 F.2d 1113, 1119 (10th Cir, 1978) (same); Secretary v. Baker Concrete Constr. Co., 17 OSH Cas. (BNA) 1236, 1239 (concurring opinion; collecting cases); Contractors Welding of Western New York, Inc., 15 OSH Cas. (BNA) 1249, 1250 (same).[1] Toilets that employees are not allowed to use for extended periods cannot be said to be "available" to those employees. Similarly, a clear intent of the requirement in Table J-1 that adequate numbers of toilets be provided for the size of the workforce is to assure that employees will not have to wait in long lines to use those facilities. Timely access is the goal of the standard.

The quoted provision of the standard is followed immediately by a paragraph stating that the toilet provision does not apply to mobile work crews or to locations that are normally unattended, "provided the employees working at these locations have transportation immediately available to nearby toilet facilities which meet the other requirements" of the standard (29 CFR 1910.141(c)(1)(ii)[)] (emphasis supplied). Thus employees who are members of mobile crews, or who work at normally unattended locations must be able to leave their work location "immediately" for a "nearby" toilet facility. This provision was obviously intended to provide these employees with protection equivalent to that the general provision provides to to [sic] employees at fixed worksites. Read together, the two provisions make clear that all employees must have prompt access to toilet facilities.

OSHA has also made this point clear in a number of letters it has issued since the standard was promulgated. For example, in March 1976, OSHA explained to Aeroil Products Company that it would not necessarily violate the standard by having a small single-story building with no toilet facilities separated by 90 feet of pavement from a building that had the required facilities, so long as the employees in the smaller building had "unobstructed free access to the toilet facilities." Later that year, it explained again, in response to a question about toilet facilities at a U-Haul site, "reasonableness in evaluating the availability of sanitary facilities will be the rule." Again in 1983, OSHA responded to a request for a clarification of the standard by stating, "([i]f [sic] an employer provides the required toilet facilities ... and provides unobstructed free access to them, it appears the intent of the standard would be met."

In light of the standard's purpose of protecting employees from the hazards created when toilets are not available, it is clear that the standard requires employers to allow employees prompt access to sanitary facilities. Restrictions on access must be reasonable, and may not cause extended delays. For example, a number of employers have instituted signal or relief worker systems for employees working on assembly lines or in other jobs where any employee's absence, even for the brief time it takes to go to the bathroom, would be disruptive. Under these systems, an employee who needs to use the bathroom gives some sort of a signal so that another employee may provide relief while the first employee is away from the work station. As long as there are sufficient relief workers to assure that employees need not wait an unreasonably long time to use the bathroom, OSHA believes that these systems comply with the standard.

Citation Policy

Employee complaints of restrictions on toilet facility use should be evaluated on a case-by-case basis to determine whether the restrictions are reasonable. Careful consideration must be given to the nature of the restriction, including the length of time that employees are required to delay bathroom use, and the employer's explanation for the restriction. In addition, the investigation should examine whether restrictions are general policy or arise only in particular circumstances or with particular supervisors, whether the employer policy recognizes individual medical needs, whether employees have reported adverse health effects, and the frequency with which employees are denied permission to use the toilet facilities. Knowledge of these factors is important not only to determine whether a citation will be issued, but also to decide how any violation will be characterized.

It is important that a uniform approach be taken by all OSHA offices with respect to the interpretation of OSHA's general industry sanitation standard, specifically with regard to the issue of employee use of toilet facilities. Proposed citations for violations of this standard must be forwarded to the Directorate of Compliance Programs (DCP) for review and approval. DCP will consult with the Office of Occupational Medicine. DCP will approve citations if the employer's restrictions are clearly unreasonable, or otherwise not in compliance with the standard. **(NOTE: See 08/11/00 Memorandum to RAs attached below.)---Added this note**

State Plan States are not required to issue their own interpretation in response to this policy, however they must ensure that State standards and their interpretations remain "at least as effective" as the Federal standard. Regional Administrators shall offer assistance to the States on this issue, including consultation with the Directorate of Compliance programs, at the State's request.

If you have any questions, contact Helen Rogers in the Office of General Industry Compliance at (202) 219-8031/41 x106.

Footnote(1) This decision was later vacated pursuant to a settlement, but the Commission has continued to cite it. See Secretary v. Baker Concrete Constr. Co., supra. The issue in Contractors Welding and the other cited cases has been whether the meaning of the term "provide," in various standards requiring employers to provide certain equipment or other materials, is not limited to making something available, but may also mean that the employer must pay for what it provides and must require it to be used. Those broader meanings are not relevant to this issue, however, where the sanitary facilities the employer is required to provide are a physical part of its workplace, and the question is not whether employees must be required to use those facilities, but whether they will be allowed to do so.

August 11, 2000----Added this memo

MEMORANDUM FOR: REGIONAL ADMINISTRATORS

FROM: RICHARD E. FAIRFAX, Director
Directorate of Compliance Programs

SUBJECT: Interpretation of 29 CFR 1910.141(c)(1)(i): Toilet Facilities

On April 6, 1998 we issued an interpretation of 1910.141(c)(1)(i), which requires employers to make toilet facilities available so that employees can use them when they need to do so. A copy of that memorandum is attached.

The 1998 memorandum states that proposed citations for violations of this standard are to be forwarded to the Directorate of Compliance Programs (DCP) for review and approval. Shortly after the interpretation was issued, it was decided that the review and approval was to be at the Regional Office level, but that copies of any citations issued based on the April 6, 1998 interpretation should still be sent to DCP.

This topic continues to generate interest from the public. Early this year we had a Freedom of Information Act (FOIA) request for copies of citations issued. Therefore, please continue to send copies of any citations issued pursuant to the 1998 interpretation to the National Office. If you have any questions, please contact Helen Rogers at (202) 693-1850. The copies should be sent to the following address:

Richard E. Fairfax, Director
Directorate of Compliance Programs
U.S. Department of Labor - OSHA
200 Constitution Avenue, NW Room N-3603
Washington, DC 20210

Appendix III

OSHA's Agricultural Sanitation Standard[1]

(a) Scope. This section shall apply to any agricultural establishment where eleven (11) or more employees are engaged on any given day in hand-labor operations in the field.

(b) Definitions. Agricultural employer means any person, corporation, association, or other legal entity that:

(i) Owns or operates an agricultural establishment;

(ii) Contracts with the owner or operator of an agricultural establishment in advance of production for the purchase of a crop and exercises substantial control over production; or

(iii) Recruits and supervises employees or is responsible for the management and condition of an agricultural establishment.

Agricultural establishment is a business operation that uses paid employees in the production of food, fiber, or other materials such as seed, seedlings, plants, or parts of plants.

Hand-labor operations means agricultural activities or agricultural operations performed by hand or with hand tools. Except for purposes of paragraph (c)(2)(iii) of this section, hand-labor operations also include other activities or operations performed in conjunction with hand labor in the field. Some examples of hand-labor operations are the hand-cultivation, hand-weeding, hand-planting and hand-harvesting of vegetables, nuts, fruits, seedlings or other crops, including mushrooms, and the hand packing of produce into containers, whether done on the ground, on a moving machine or in a temporary packing shed located in the field. Hand-labor does not include such activities as logging operations, the care or feeding of livestock, or hand-labor operations in permanent structures (e.g., canning facilities or packing houses).

Handwashing facility means a facility providing either a basin, container, or outlet with an adequate supply of potable water, soap and single-use towels. Potable water means water that meets the standards for drinking purposes of the state or local authority having jurisdiction or water that meets the quality standards prescribed by the U.S. Environmental Protection Agency's National Interim Primary Drinking Water Regulations, published in 40 CFR part 141. Toilet facility means a fixed or portable facility designed for the purpose of adequate collection and containment of the products of both defecation and urination which is supplied with toilet paper adequate to employee needs. Toilet fa-

---

[1]29 CFR 1928.110 (2002).

cility includes biological, chemical, flush and combustion toilets and sanitary privies.

(c) Requirements. Agricultural employers shall provide the following for employees engaged in hand-labor operations in the field, without cost to the employee:

(1) Potable drinking water. (i) Potable water shall be provided and placed in locations readily accessible to all employees.

(ii) The water shall be suitably cool and in sufficient amounts, taking into account the air temperature, humidity and the nature of the work performed, to meet the needs of all employees.

(iii) The water shall be dispensed in single-use drinking cups or by fountains. The use of common drinking cups or dippers is prohibited.

(2) Toilet and handwashing facilities. (i) One toilet facility and one handwashing facility shall be provided for each twenty (20) employees or fraction thereof, except as stated in paragraph (c)(2)(v) of this section.

(ii) Toilet facilities shall be adequately ventilated, appropriately screened, have self-closing doors that can be closed and latched from the inside and shall be constructed to insure privacy.

(iii) Toilet and handwashing facilities shall be accessibly located and in close proximity to each other. The facilities shall be located within a one-quarter-mile walk of each hand laborer's place of work in the field.

(iv) Where due to terrain it is not feasible to locate facilities as required above, the facilities shall be located at the point of closest vehicular access.

(v) Toilet and handwashing facilities are not required for employees who perform field work for a period of three (3) hours or less (including transportation time to and from the field) during the day.

(3) Maintenance. Potable drinking water and toilet and handwashing facilities shall be maintained in accordance with appropriate public health sanitation practices, including the following:

(i) Drinking water containers shall be constructed of materials that maintain water quality, shall be refilled daily or more often as necessary, shall be kept covered and shall be regularly cleaned.

(ii) Toilet facilities shall be operational and maintained in clean and sanitary condition.

(iii) Handwashing facilities shall be refilled with potable water as necessary to ensure an adequate supply and shall be maintained in a clean and sanitary condition; and

(iv) Disposal of wastes from facilities shall not cause unsanitary conditions.

(4) Reasonable use. The employer shall notify each employee of the location of the sanitation facilities and water and shall allow each employee reasonable opportunities during the workday to use them. The employer also shall inform each employee of the importance of each of the following good hygiene practices to minimize exposure to the hazards in the field of heat, communicable diseases, retention of urine and agrichemical residues:

(i) Use the water and facilities provided for drinking, handwashing and elimination;

(ii) Drink water frequently and especially on hot days;

(iii) Urinate as frequently as necessary;

(iv) Wash hands both before and after using the toilet; and

(v) Wash hands before eating and smoking.

(d) Dates--(1) Effective Date. This standard shall take effect on May 30, 1987.

(2) Startup Dates. Employers must comply with the requirements of paragraphs:

(i) Paragraph (c)(1), to provide potable drinking water, by May 30, 1987;

(ii) Paragraph (c)(2), to provide handwashing and toilet facilities, by July 30, 1987;

(iii) Paragraph (c)(3), to provide maintenance for toilet and handwashing facilities, by July 30, 1987; and

(iv) Paragraph (c)(4), to assure reasonable use, by July 30, 1987.

Appendix IV

OSHA's Construction Sanitation Standard[1]

(a) Potable water. (1) An adequate supply of potable water shall be provided in all places of employment.

(2) Portable containers used to dispense drinking water shall be capable of being tightly closed, and equipped with a tap. Water shall not be dipped from containers.

(3) Any container used to distribute drinking water shall be clearly marked as to the nature of its contents and not used for any other purpose.

(4) The common drinking cup is prohibited.

(5) Where single service cups (to be used but once) are supplied, both a sanitary container for the unused cups and a receptacle for disposing of the used cups shall be provided.

(6) Potable water means water which meets the quality standards prescribed in the U.S. Public Health Service Drinking Water Standards, published in 42 CFR part 72, or water which is approved for drinking purposes by the State or local authority having jurisdiction.

(b) Nonpotable water. (1) Outlets for nonpotable water, such as water for industrial or firefighting purposes only, shall be identified by signs meeting the requirements of subpart G of this part, to indicate clearly that the water is unsafe and is not to be used for drinking, washing, or cooking purposes.

(2) There shall be no cross-connection, open or potential, between a system furnishing potable water and a system furnishing nonpotable water.

(c) Toilets at construction jobsites. (1) Toilets shall be provided for employees according to the following table:

Table D-1

Number of employees	Minimum number of facilities
20 or less	1.
20 or more	1 toilet seat and 1 urinal per 40 workers.
200 or more	1 toilet seat and 1 urinal per 50 workers.

[1]29 CFR 1926.51 (2002).

(2) Under temporary field conditions, provisions shall be made to assure not less than one toilet facility is available.

(3) Job sites, not provided with a sanitary sewer, shall be provided with one of the following toilet facilities unless prohibited by local codes:

(i) Privies (where their use will not contaminate ground or surface water);

(ii) Chemical toilets;

(iii) Recirculating toilets;

(iv) Combustion toilets.

(4) The requirements of this paragraph (c) for sanitation facilities shall not apply to mobile crews having transportation readily available to nearby toilet facilities.

(d) Food handling. (1) All employees' food service facilities and operations shall meet the applicable laws, ordinances, and regulations of the jurisdictions in which they are located.

(2) All employee food service facilities and operations shall be carried out in accordance with sound hygienic principles. In all places of employment where all or part of the food service is provided, the food dispensed shall be wholesome, free from spoilage, and shall be processed, prepared, handled, and stored in such a manner as to be protected against contamination.

(e) Temporary sleeping quarters. When temporary sleeping quarters are provided, they shall be heated, ventilated, and lighted.

(f) Washing facilities. (1) The employer shall provide adequate washing facilities for employees engaged in the application of paints, coating, herbicides, or insecticides, or in other operations where contaminants may be harmful to the employees. Such facilities shall be in near proximity to the worksite and shall be so equipped as to enable employees to remove such substances.

(2) General. Washing facilities shall be maintained in a sanitary condition.

(3) Lavatories. (i) Lavatories shall be made available in all places of employment. The requirements of this subdivision do not apply to mobile crews or to normally unattended work locations if employees working at these locations have transportation readily available to nearby washing facilities which meet the other requirements of this paragraph.

(ii) Each lavatory shall be provided with hot and cold running water, or tepid running water.

(iii) Hand soap or similar cleansing agents shall be provided.

(iv) Individual hand towels or sections thereof, of cloth or paper, warm air blowers or clean individual sections of continuous cloth toweling, convenient to the lavatories, shall be provided.

(4) Showers. (i) Whenever showers are required by a particular standard, the showers shall be provided in accordance with paragraphs (f)(4) (ii) through (v) of this section.

(ii) One shower shall be provided for each 10 employees of each sex, or numerical fraction thereof, who are required to shower during the same shift.

(iii) Body soap or other appropriate cleansing agents convenient to the showers shall be provided as specified in paragraph (f)(3)(iii) of this section.

(iv) Showers shall be provided with hot and cold water feeding a common discharge line.

(v) Employees who use showers shall be provided with individual clean towels.

(g) Eating and drinking areas. No employee shall be allowed to consume food or beverages in a toilet room nor in any area exposed to a toxic material.

(h) Vermin control. Every enclosed workplace shall be so constructed, equipped, and maintained, so far as reasonably practicable, as to prevent the entrance or harborage of rodents, insects, and other vermin. A continuing and effective extermination program shall be instituted where their presence is detected.

(i) Change rooms. Whenever employees are required by a particular standard to wear protective clothing because of the possibility of contamination with toxic materials, change rooms equipped with storage facilities for street clothes and separate storage facilities for the protective clothing shall be provided.

Appendix V

Data Methodology and Sources: OSHA Inspection Reports and Citations

---

Part V's empirical discussion of OSHA's enforcement of employers' obligation to "make toilet facilities available so that employees can use them when they need to do so," is based in part on OSHA's inspection reports, which are public records. They are available on the OSHA website and also on Lexis-Nexis.[1] The OSHA website reports include some data (for example, the number of employees exposed to the violation and the number of instances of exposure) not available on the Lexis database, but the usefulness of the former is radically undermined by the fact that it is impossible to search for individual employers by standard number; consequently, it is impossible to search for the citations issued under section 1910.141(c)(1)(i).[2] Since such standard searches are possible on Lexis, it is necessary to identify all the employers to which OSHA issued citations and then go back to the OSHA website (which is searchable by employer name). An example of an inspection report from each of these databases is reproduced at end of Appendix V. These OSHA inspector reports, however, do not reveal in what way the employer violated the standard.

Since most citations issued by OSHA under section 1910.141(c)(1)(i) after April 6, 1998 have been triggered by a failure to provide the required number of toilets and only a very few by a denial of access, it was necessary to make Freedom of Information Act requests to the Federal OSHA area offices and state public or open records act requests to the state-plan programs that issued the citations for a copy of the Citation and Notification of Penalty, which includes an alleged violation description (AVD) and becomes a public record as soon as the employer

---

[1]A version is also available on Westlaw, but it is less complete.

[2]The searches on the OSHA website that are possible by standard number are confined to the immediately preceding fiscal year and generate only summary data on the number of citations and inspections and the total dollar amount of penalties disaggregated by SIC-code and establishment size; searches can also not be performed by subsections—that is, sect. 1910.141 can be searched, but not sect. 1910.141(c)(1)(i).

has received it. The AVD is usually quite brief (often one sentence) and states the specific way in which the employer violated the standard.

The process of collecting all the citations was delayed and made much more difficult by the incompetence of several Federal OSHA officials in the National Office. After several employees lost the author's Freedom of Information Act request and another stated that he was in the process of forwarding the requests to the individual OSHA area offices, Richard Fairfax, the Director of the Directorate of Enforcement Programs, informed the author: "Your Freedom of Information Act (FOIA) request dated September 9, was referred to this office for a direct response. You requested information for files in cases pertaining to citations and settlement agreements based on the April 6, 1998 interpretation of 20 [sic; should be 29] CFR 1910.141(c)(1)(i), which requires employers to make toilet facilities available so that employees can use them when they need to use them. We have enclosed the information that you requested. The data on the computer report contains the pertinent information for the period January 1, 2000 until September 22, 2002, the date of the data search."[3] In fact, however, these data were merely public access data available on OSHA's website (which the author had already collected on his own) and lacked the one crucial piece of information that exists only in OSHA's paper files and that constituted all that the author wanted—namely, the aforementioned alleged violation description, which explains what the employer did or did not do that triggered the violation. The data that OSHA sent did not distinguish between cases in which employers were cited for not having provided the proper number of toilets and those in which the employer denied workers access to toilets. They were thus totally superfluous, irrelevant, and useless.

This nonresponsiveness somehow dawned on someone at OSHA, prompting Fairfax to write again a week later: "Please be aware that the National Office of the Occupational Safety and Health Administration (OSHA) does not maintain records that are responsive to your request. We previously provided you with the citation history for the requested standard, but we do not have the abstracts in which you are interested. To facilitate your request, we are providing a computer data report of the cited cases, with the issuing area offices identified in the left hand column, along with a list of the addresses of all the OSHA area offices."[4] In fact, Fairfax himself, continuing a policy established in the Memorandum itself, had informed OSHA offices in 2000 that "copies of any citations issued based on the April 6, 1998 interpretation should still be sent to DCP [the Directorate of Compliance Programs was the former title of Fairfax's office]. This topic continues to generate interest from the public. Early this year we had a Free-

---

[3]Letter from Richard Fairfax to Marc Linder (Oct. 21, 2002).

[4]Letter from Richard Fairfax to Marc Linder (Oct. 29, 2002).

dom of Information Act (FOIA) request for copies of citations issued. Therefore, please continue to send copies of any citations issued pursuant to the 1998 interpretation to the National Office."[5] The author then spent six weeks gathering the AVD pages for these citations from the 43 (of a total of 85) Federal OSHA area offices that had issued them.

The sources and methodology for compiling the information in tables 1, 2, and 3[6] were as follows. The author searched the Lexis-Nexis OSHAIR file with the search term "19100141 c01 i," which is the form in which this standard number is entered[7] and a date limitation of "after April 5, 1998." This search was conducted in the early fall of 2002; because of the dilatory and uncooperative responses of some OSHA agencies to requests for information, inducing them to repeat the searches later to update for new citations was precluded; therefore the end date for the searches is also fall 2002. The search proved to be somewhat overinclusive because the date limitation identified cases that included any actions recorded on dates after April 5, 1998, whereas in fact the information sought was cases that had been opened after that date on the assumption that OSHA's new interpretation was not applied retroactively. In order to segregate out the cases that had been opened after that date, the author examined the contents of each individual report. The initial search was also underinclusive because three states renumbered the Federal OSHA standards to fit their own regulatory codes: California adopted section 3364(b), Michigan R4201(3)(a)(i), and Washington 296-24-12007(1)(a) (which was recodified as 296-800-23020 in 2001). After all these adjustments, a total of 84 Federal OSHA cases (for 2000, 2001, and 2002) and 329 state OSHA cases (from April 6, 1998 through the fall of 2002) were identified, in which employers were cited for violating 29 CFR 1910.141(c)(1)(i) (or its state equivalent). No search was conducted for Federal OSHA cases for the period from April 6, 1998 through December 31, 1999, because the OSHA National Office had already identified such cases pursuant to FOIA requests by the author in 1998 and another lawyer in 2000.[8]

The author also furnished the 21 state-plan OSHA agencies plus New York (whose program covers only public employees) with a list of the employers and inspection numbers identified on Lexis-Nexis and the agencies then conducted their own searches of their own internal databases, which also permit searches by standard number; with a few quirky exceptions, the lists matched. With respect to those five states (Hawaii, New Mexico, Utah, Vermont, and Wyoming) which had not issued a single citation for any violation of section 1910.141(c)(1)(i) for

---

[5]See above Appendix II.

[6]See above ch. 11.

[7]One case from Indiana was erroneously entered as "19100141 c01i."

[8]See above ch. 11.

any reason since April 5, 1998, the author confirmed this finding and its plausibility with the relevant officials, thus ending that part of the search. For the other seventeen states (with the exception of California) the author then requested the alleged violation description page from the Citation and Notification of Penalty for all citations of the standard. The extent of cooperation extended by the state OSHA agencies varied widely, depending in part on the officials' own interest in the question. In Oregon, for example, Chris Ottoson, Health Analyst, Enforcement Policy Section, Oregon OSHA, reviewed all the inspections and "found that none involved denial of access. ... In one unusual case, a female employee was killed while using the portable toilet; a truck offloading bark dust hit a concrete barrier serving as a retaining wall for the bark dust (the toilet was located between a service building and the retaining wall), causing the concrete wall to topple onto the toilet, which crushed the employee. To summarize, denial of access wasn't an issue in these inspections."[9] But, on request, Ottoson also later sent the violation descriptions. In Minnesota, where only one citation was issued, the OSHA official read the AVD to the author on the telephone indicating that the citation was for the lack of a latch.[10] In Arizona, the official "researched those employers who have been cited under standard number 1910.141(c) beginning on 4/6/98 to present time. Of the 6 employers who were cited, only 1 employer was actually cited for preventing employees['] unrestricted use of the toilet facilities."[11] She then faxed that one narrative report. In Utah, an uncooperative official, after much prodding, reviewed the two citations and stated that neither was for denial of access, adding that there has never been such a citation in Utah.[12] He did not send the narratives. In Tennessee, Mike Maenza, Manager for Standards and Procedures, reviewed the files and stated that none was for denial of access, and later sent the narratives on request.[13] Officials in all the other states sent all the narratives.

The aforementioned exception was Cal/OSHA, which issued far more citations (181 excluding two that were spuriously classified as sect. 3364, but had in fact been issued under sect. 3664 dealing with forklifts) under this standard than all other states combined. Although the author was eager to study all the citations and narratives reports, officials in many Cal/OSHA offices were not eager to search for any of them. As an initial compromise, the author identified eight

---

[9]Email from Chris Ottoson to Marc Linder (Sept. 25, 2002).

[10]Telephone interview with Kelly Taylor (Oct. 1, 2002).

[11] Fax from Dianne Marks, Industrial Commission of Arizona, Div. of OSH, to Marc Linder (Sept. 25, 2002).

[12]Fax from William W. Adams, Jr., UOSH Technical Adviser, to Marc Linder (Sept. 20, 2002); telephone interview with William Adams (Oct. 9, 2002).

[13]Telephone interview with Mike Maenza (Sept. 23, 2002).

cited employers, the nature of whose businesses appeared to offer the greatest probability of restriction of access. Over many weeks the staff located them, determining that none was for access.[14] Dissatisfied with this process and result, the author then requested copies of the citations from 20 individual Cal/OSHA offices. After several months of cajoling, the author finally received all of the citations (including two that lacked an alleged violation description)—except 20 that had been lost or purged. Ultimately it turned out that none of the citations had been issued for restricting access because in fact Cal/OSHA refuses to enforce the Federal OSHA interpretive Memorandum.[15]

---

[14]Fax from Robert Hayes, Cal/OSHA, to Marc Linder (Sept. 23, 2002); telephone interview with Robert Hayes, Cal/OSHA (Oct. 11, 18, and 29, 2002).

[15]See above ch. 15.

[Lexis-Nexis]

OCCUPATIONAL SAFETY & HEALTH ADMINISTRATION (OSHA)
OSHA Inspection Reports

CONVERGYS CORPORATION
4501 ROY J. SMITH DR
KILLEEN, TEXAS 76543

County: BELL
Reporting Office ID: 0625400
Activity Number: 0304389711
Unique Establishment Identifier: N122 000019420
SIC: 7389 (BUSINESS SERVICES, NEC)
Owner Type: PRIVATE
Unionized: NO

******** INSPECTION INFORMATION ********

Inspection Type: COMPLAINT
Category: HEALTH
Scope: PARTIAL
Inspector: HEALTH OFFICER
Open Inspection Date: 04/12/2002
Closed Inspection Date: 04/12/2002
Closed Activity Date: 05/09/2002
Advance Notice: NO
Employee Rep Present: YES
Employee(s) Interviewed: YES
Inspection Hours: PREP 10 HRS.; TRAVEL 25 HRS.; ON-SITE 40 HRS.; RESEARCH 20 HRS.; CONFERENCES 10 HRS.
Total Inspection Time: 105 HRS.

******** CITATION INFORMATION ********

ID Number	Citation Std/Subsection	Type	Issuance Date
01001	19100141 C01 I	SERIOUS	04/17/2002

ID Number	Contested	Abate Date	Abatement Complete
01001	NO	04/25/2002	ABATEMENT, PPE, REPORT COMPLETED

******** PENALTY INFORMATION ********

ID Number	Initial Penalty	Current Penalty	Failure to Abate Penalty	FTA Issue Date
O1001	$ 1,125.00			

Establishment Search Inspection Detail-- Public View

Definitions

Inspection: 304389711 - Convergys Corporation

Inspection Information - Office: Austin	
Nr: 304389711 Report ID: 0625400 Open: 04/12/2002	
Convergys Corporation 4501 Roy J. Smith Dr Killeen , TX 76543 SIC: 7389/Business Services, Nec Mailing: 4501 Roy J. Smith Dr. , Killeen , TX 76543	Union Status: NonUnion
Inspection Type: Complaint Scope: Partial Ownership: Private Safety/Health: Health	Advance Notice: N Close Conference: 04/12/2002 Close Case: 05/09/2002

Related Activity: Type	ID	Date	Safety Health
Complaint	203901780	04/10/2002	Yes

	Violation Summary					
	Serious	Willful	Repeat	Other	Unclass	Total
Initial Violations	1					1
Current Violations	1					1
Initial Penalty	1125.00					1125.00
Current Penalty						
FTA Amount						

Violation Items

ID	Type	Standard	Issuance	Abate	Curr$	Init$	Fta$	Contest	LastEvent
1 01001	Serious	19100141 C01 I	04/17/2002	04/25/2002	0.00	1125.00	0.00		I-Informal Settlement

Violation - Convergys Corporation

Standard Cited: 19100141 C01 I *Sanitation*

Violation Information		
Nr: 304389711 Citation: 01001 Issuance: 04/17/2002 Report ID: 0625400		
Viol Type: Serious	Nr Instances: 2	Contest Date:
Abatement Date: 04/25/2002 X	Nr Exposed: 2	Final Order:
Initial Penalty: 1125.00	REC:	Emphasis:
Current Penalty:	Gravity: 01	Haz Category:

Penalty and Failure to Abate Event History

Type		Event	Date	Penalty	Abatement	Type	FTA Insp
PEN	Z	Issued	04/17/2002	1125.00	04/25/2002	Serious	
PEN	I	Informal Settlement	05/09/2002		04/25/2002	Serious	

Index